Heat and Mass Transfer
Series Editors: D. Mewes and F. Mayinger

Ching Shen

Rarefied Gas Dynamics

Fundamentals, Simulations and Micro Flows

With 53 Figures

Series Editors

Prof. Dr.-Ing. Dieter Mewes
Universität Hannover
Institut für Verfahrenstechnik
Callinstr. 36
30167 Hannover, Germany

Prof. em. Dr.-Ing. E.h. Franz Mayinger
Technische Universität München
Lehrstuhl für Thermodynamik
Boltzmannstr. 15
85748 Garching, Germany

Author
Prof. Ching Shen
Chinese Academy of Sciences
Institute of Mechanics
Zhongguancun Rd. 15
100080 Beijing
People's Republic of China

Library of Congress Control Number: 2004116858

ISBN 3-540-23926-X Springer Berlin Heidelberg New York

This work is subject to copyright. All rights are reserved, whether the whole or part of the material is concerned, specifically the rights of translation, reprinting, reuse of illustrations, recitation, broadcasting, reproduction on microfilm or in other ways, and storage in data banks. Duplication of this publication or parts thereof is permitted only under the provisions of the German Copyright Law of September 9, 1965, in its current version, and permission for use must always be obtained from Springer. Violations are liable to prosecution act under German Copyright Law.

Springer is a part of Springer Science + Business Media

springeronline.com

© Springer-Verlag Berlin Heidelberg 2005
Printed in Germany

The use of general descriptive names, registered names, trademarks, etc. in this publication does not imply, even in the absence of a specific statement, that such names are exempt from the relevant protective laws and regulations and therefore free for general use.

Typesetting: camera ready copy supplied by author
Cover-Design: deblik, Berlin
Production: medionet AG, Berlin

Printed on acid-free paper   62/3141 Rw   5 4 3 2 1 0

PREFACE

Aerodynamics is a science engaged in the investigation of the motion of air and other gases and their interaction with bodies, and is one of the most important bases of the aeronautic and astronautic techniques. The continuous improvement of the configurations of the airplanes and the space vehicles and the constant enhancement of their performances are closely related with the development of the aerodynamics. In the design of new flying vehicles the aerodynamics will play more and more important role.

The undertakings of aeronautics and astronautics in our country have gained achievements of world interest, the aerodynamics community has made outstanding contributions for the development of these undertakings and the science of aerodynamics. To promote further the development of the aerodynamics, meet the challenge in the new century, summary the experience, cultivate the professional personnel and to serve better the cause of aeronautics and astronautics and the national economy, the present Series of Modern Aerodynamics is organized and published.

The Series of Modern Aerodynamics consists of about 20 monographs divided into theoretical and experimental parts. The theoretical part includes: Theory of transonic aerodynamics, Theory of inviscid hypersonic aerodynamics, Rarefied gas dynamics, Computational fluid dynamics-fundamentals and applications of finite difference methods, Spectral methods of computational fluid dynamics, Finite element methods of fluid mechanics, Heat transfer of hypersonic gas and ablation thermal protection, Dynamics of multiphase turbulent reacting fluid flows, High temperature nonequilibrium air flows, Turbulence, Vortex stability, Wing engineering and industrial aerodynamics, Aerodynamics for airplane design, and others. The experimental part includes: Wind tunnel testing, Wind tunnel balance, Interference and correction on wind tunnel testing, Impulsive wind tunnels, Modern flow visualization, and others. The editors and writers of this series uphold the following principles of writing. Firstly, it serves as a bridge from the fundamental aerodynamics to the frontiers of the modern aerodynamics. Secondly, it is a series of special monographs each devoted to a specialized topic. Thirdly, the series pays attention not only to the already existing achievements but also to the modern developments. Fourthly, each monograph introduces systematically the knowledge and development of the special topic. Fifthly, the series forms a complete whole, the monographs combined together cover various fields of modern aerodynamics. To organize and promote the writing of the series an editorial board with academician Zhuang Fenggan as its president was formed in charge of working out the plane of writing, selecting writers, examining and approving the manuscripts, recommending to the Judging Panel of the Publication Foundation of National Defense Technical Books to apply for the financial support. The Chinese Aerodynamics Research and Development Center supported the work of the editorial board in the aspects of personnel and expenditure. Each monograph of this series after applying and receiving the support from the Publication Foundation of National Defense Technical Books was chosen and edited by the National Defense Industry Press.

This English version of *Rarefied Gas Dynamics* published by Springer-Verlag, Berlin/Heidelberg, is a translation from its Chinese edition revised and updated by the author. In particular, the 7.7 section is rewritten as a new Chapter 8 'Microscale Slow Gas Flows, Information Preservation Method' to give more comprehensive account of the subject and reflect some advances obtained after the publication of the Chinese edition.

The flight, maneuver and braking of aerospace vehicles at high altitudes demands from the gas dynamics the answer on questions of force and heat action of the low density gas flow. When the density is lowered to a level that the mean free path of the gas molecules is not a small magnitude in comparison with the characteristic length of the flow, the ordinary methods of the continuum gas dynamics are no longer suitable, the methods of discrete molecular gas dynamics, or, of rarefied gas dynamics are required. Meanwhile the condition of high speed flight leads to the necessity of consideration of the physical processes taking place inside the molecules such as the excitation of the internal energies, the chemical reactions and the excitation and transition of electronic levels. This leads to the development and expansion of the gas dynamics towards micro scale scopes in two aspects, i.e. the research of the gas flows with the discrete molecular effect taken into account by involving the methods of rarefied gas dynamics, and the consideration of the internal structure of the molecules. The application areas of the rarefied gas dynamics besides aeronautics and astronautics include some frontier realms of the advanced technical development such as: plasma material processing in vacuum, micro-electronic etching, micro-electronic mechanic systems and chemical industry. This makes the research on low speed rarefied gas dynamics very important. In the world literature the research on gas dynamics on the molecular level and on the internal physical processes inside molecules in the gas flows is very active. Relatively the domestic works on these aspects is fewer. From the viewpoint of the prospects of the subject and the demands of the development of science and technology research on gas dynamics on the micro-scale is a direction needs to be strengthened and further developed in our country, and it is appropriate to advocate and promote in the scientific research layout and education arrangement.

This book elucidates the methods of molecular gas dynamics or rarefied gas dynamics which treat the problems of gas flows when the discrete molecular effects of the gas prevail under the circumstances of low density, the emphases being stressed on the basis of the methods, the direct simulation Monte Carlo method applied to the simulation of non-equilibrium effects and the frontier subjects related to low speed microscale rarefied gas flows. As the basis of the discipline two chapters on molecular structure and basic kinetic theory are introduced. The first chapter devotes a minimum space in brief and summarized description of the molecular energy state structure and energy distribution as the necessary basis for the investigation of the non-equilibrium in high enthalpy rarefied gas flows. The second chapter discusses the basis of the kinetic theory focusing on binary collisions, Boltzmann equation and the equilibrium state of the gas, including the phenomenological molecular models: the VHS model of G. A. Bird and also the VSS model, the GHS model and the GSS model. The third chapter discusses various realistic models of gas surface interactions, including the reciprocity principle reflecting the detailed equilibrium, the CLL model based on this principle and the application in direct simulation in the cases including incomplete energy accommodation and internal energy exchange. The fourth chapter deals with the free molecular flows. Chapter 5 discusses the continuum equations and slip boundary conditions applied for slip flow regime, including Burnett equation the usability of which has been proved to be penetrated to more rarefied range. This chapter also includes the discussion of some simple problems solved by the Navier-Stokes equation plus slip boundary conditions and the problem of thermophoresis. Chapter 6 introduces with fair comprehensiveness and generality various analytical and numerical methods developed in transition regime. Chapter 7 introduces the direct simulation Monte Carlo (DSMC) method with emphasis stressed on the specific issues encountered in dealing with non-equilibrium rarefied gas dynamics, including the work of the author and his colleague in treating the excitation and relaxation of the internal energy, the chemical reactions and the general coding of simulation of the flow field around complex configurations, i.e. the generalized acceptance-rejection method in sampling from the distribution with singularities, the sterically dependent chemical reaction model and the derivation of the Ahrrenius-Kooij form of reaction coefficient and the new version of the position element algorithm of the general code of the DSMC simulation etc. Chapter 8 is dedicated to the simulation of low velocity micro scale rarefied gas flows which becomes an actual and urgent task encountered by the rarefied gas dynamics in the 21st century with the rapid development of the micro-electro-mechanical system (MEMS). Some methods of solution of the rarefied flow problems is examined from the point of view of utilization in simulating the flows in MEMS. The information preservation (IP) method is introduced with a general description, some validation of the method and a program demonstrating the method. The resolving of the boundary condition regulation problem in MEMS by using the conservative scheme and the super relaxation method is illustrated on the example of flow in long micro channels. The thin film air bearing problem with authentic length of the write/read head of the Winchester hard disc drive is solved, and the use of the degenerated Reynolds equation is suggested to solve the microchannel flow and to serve as a criterion of the merit of strict kinetic theory for testing various methods intending to treat the rarefied gas flows in MEMS.

In choosing the contents of the book the author proceeds out of the consideration to elucidate various essential basics of the subject and is influenced by his own interests, so the quotation of literature is far from complete, hoping the forgiveness and understanding of all the scientists having made contributions to the development of the subject.

This book provides a solid basis for engaging in the studying of the molecular gas dynamics for the senior students and graduates in the aerospace and the mechanical engineering departments of universities and colleges, giving them an overall acquaintance of the modern development of rarefied gas dynamics in various regimes and leading them to reach the frontier topics of the non-equilibrium rarefied gas dynamics and the low speed microscale gas dynamics. It will be also of benefit to the scientific and technical workers engaged in the aerospace high altitude aerodynamic force and heating design and in the research on gas flow in the MEMS when treating practical gas dynamics problems.

The author would appreciate Prof. Hu Zhenhua and Prof. Fan Jing for many helpful discussions concerning the contents of some sections of the book. The author thanks Mr. Tian Dongbo, Mr. Jiang Jianzheng, Mr. Xie Chong, and Dr. Liu Hongli who helped in making the drawings and layout of the book.

Zhongguancun, Beijing, China   C. Shen
September, 2004

NOMENCLATURE

a	sound speed, $=\sqrt{\gamma kT/m}$; radius of a sphere particle
a_d	dynamic factor
b	distance of the closest approach, miss-distance
c	the magnitude of the molecule velocity
$\mathbf{c}$	molecule velocity
c'	molecule thermal velocity
c_0	molecule average velocity, average velocity
c_r	relative velocity
$\bar{c}$	average thermal speed, $=\sqrt{8kT/\pi m}$
c_m	most probable molecular thermal speed, $=\sqrt{2kT/m}$
$\bar{c}_r$	mean value of the relative speed
c'_s	mean square root thermal speed, $=\sqrt{3kT/m}$
c_p	specific heat at constant pressure
c_v	specific heat at constant volume
C	flow conductance, $=Q/\Delta p$
C_m	velocity slip coefficient
C_s	thermal creep coefficient
C_t	temperature jump coefficient
d	molecular diameter
d_{HS}	molecular diameter of the hard sphere model
d_{VHS}	molecular diameter of the VHS model
d_{VSS}	molecular diameter of the VSS model
D^T	thermal diffusion coefficient

e		specific energy
E_a		activation energy
f		velocity distribution function
f_K		another definition of velocity distribution function, $= f/n$
$F^{(1)}$		single particle distribution function, $= f/N$
F_i		incident flux
F_{sp}		specularly reflected flux
F_w		diffusely reflected flux; impact stress originated from collision
F		external force
g		equilibrium state Maxwellian distribution
g_j		degeneracy
G		dimensionless temperature gradient, $= |\nabla T| a / T_0$
h		Planck constant, $= 6.6260755 \times 10^{-34} Js$; height of the channel; height of the write/read head over the drive platter
h_o		height of the write/read head over the drive platter at the outlet
H		Boltzmann H function; dimensionless height $= h/h_o$
I		moment of inertia
J		rotational quantum number
k		Boltzmann constant, $= 1.380658 \times 10^{-23} JK^{-1}$; the ratio of the conductivity of gas to that of the particle, $= K_g / K_p$,
k_e		force constant
k_f		forward reaction rate constant
f_r		reverse reaction rate constant
K		conductivity
Kn		Knudsen number, $= \lambda / L$
l		unit normal vector
L		characteristic length of the flow field; length of the write/read head in the hard disc drive

NOMENCLATURE

m	molecular mass
m_r	reduced mass
Ma	Mach number, $= U/a$
M	molecular weight
mol	gram molecular weight
n	number density; vibrational quantum number
n_0	Loschmidt number, $= 2.68666 \times 10^{25} m^{-3}$
N	total number of degrees of freedom of the internal degree variable ξ, $= (5-3\gamma)/(\gamma-1)+1$
N	Avogadro number, $= 6.022137 \times 10^{23} mol^{-1}$
p	pressure
$psig$	pound per square inch, gauge, $\approx 6.895 kPa$ above atm
$\mathbf{P}$	pressure tensor
Pr	Prandtl number, $= c_p \mu / K$
P	dimensionless pressure, p/p_e
q_i	heat flux vector
q	$= q_i$
Q	function of molecular velocity; partition function; flow rate
Q_C	flow rate of the Couette flow
Q_P	flow rate of the Poiseuille flow
R	gas constant, $= R/M = k/m$
$ranf$	random fraction between 0 and 1
Re	Reynolds number, $= \rho U L / \mu$
R	universal gas constant, $= 8.314511 Jmol^{-1}K^{-1}$
S	entropy; speed ratio, $= U/c_m = U/\sqrt{2RT}$
$S(c_r)$	$= c_r \sigma_T$
T	temperature
U	velocity of oncoming flow

u_s		gas velocity near the surface, gas velocity at outer edge of Knudsen layer
V		potential energy of system
W		dimensionless variable, $=b/r$
X		$=x/L$
Z		relaxation collision number
Z_v		vibrational relaxation collision number
Z_r		rotational relaxation collision number
α		thermal accommodation coefficient; the power exponent of the cosine of the deflection angle in VSS scattering model
B		beta function
β		reciprocal of the most probable thermal speed, $=(2RT)^{-1/2}$
γ		ratio of specific heats
Γ		gamma function
δ		spacing between molecules
δ_{ij}		Kronecker tensor
ε		molecular energy
ε_d		molecular dissociation energy
ε_e		molecular electronic energy
ε_r		molecular rotational energy
ε_t		molecular translational energy
ε_v		molecular vibrational energy
ξ		negative power exponent in the dependence of σ_T on ε_t
ζ		number of degrees of freedom; slip coefficient
η		power exponent in the inverse power law model
Θ		characteristic temperature
κ		constant in the force power law
λ		molecular mean free path

λ_{ei}	mean free path in collision of the scattered molecules with the oncoming molecules
Λ	bearing number, $= 6\mu UL/p_o h_o^2$
μ	viscosity, for hard sphere model $\mu \approx 1/2(\rho \bar{c} \lambda)$
ν	collision frequency
ρ	density
σ	tangential momentum accommodation coefficient
σ'	normal momentum accommodation coefficient
σ_D	diffusion collision cross section
σ_M	momentum collision cross section
σ_R	reaction collision cross section
σ_T	total collision cross section
σ_μ	viscosity collision cross section
T	viscous stress tensor, $\equiv \tau_{ij}$
τ	mean collision time, $= \lambda/\bar{c}$; temperature
τ_{ij}	shear stress tensor
ϕ_v	vibrational energy exchange probability
χ	deflection angle in collision
ψ	wave function
ω	power exponent of temperature in the viscosity law; super relaxation factor
Ω	number of ways of distributions of particles on different levels

SUPERSCRIPTS AND SUBSCRIPTS

e	emitted, scattered; electronic
g	gas
i	incident; internal

int		internal
l		left
p		particle
r		rotational; right
t		translational
v		vibrational
∞		oncoming flow value
$*$		values after collision
$\bar{a}$		average of a
$\lfloor a \rfloor$		cut off value of a

ENGLISH ABBREVIATIONS

AFE vehicle	aero-assisted flight experiment vehicle
BGK equation	Bhatnagar-Gross-Krook equation
BKW equation	Boltzmann-Krook-Welender equation
CALTECH	California Institute of Technology
CFD	computational fluid dynamics
CLL model	Cercignani-Lampis-Lord model
DSMC method	direct simulation Monte Carlo method
EDM	electric discharge machining
FORTRAN language	formula translation language
GHS model	generalized hard sphere model
GSS model	generalized soft sphere model
HS model	hard sphere model
IP method	information preservation method
LB method	Larsen-Borgnakke method
LBM	lattice Boltzmann method

LIGA	Lithographie Galvanoformung Abformung (German), Lithographic electroforming
MD method	molecular dynamics method
MEMS	micro-electro-mechanical system
MIT	Massachusetts Institute of Technology
NEMS	nano-electro-mechanical system
NTC method	no time counter method
RSF method	randomly sampled frequency method
STP	standard temperature and pressure
SSTO vehicle	single stage to orbit vehicle
TC method	time counter method
UCLA	University of California, Los Angeles
VHS model	variable hard sphere model
VSS model	variable soft sphere model

CONTENTS

Preface v

Nomenclature xi

0 Introduction 1
    0.1   The Conception of Rarefied Gas Dynamics 1
    0.2   The Molecular Model of Gases 3
    0.3   Mean Free Path of Molecules 4
    0.4   Division of Flow Regimes 5
    0.5   Nonequilibrium Phenomena and Rarefied Gas Dynamics 9
    0.6   Similarity Criteria 13
    References 18

1 Molecular Structure and Energy States 21
    1.1   Diatomic Molecules 21
    1.2   Energy Distribution of Molecules 30
           1.2.1   Boltzmann's Relation 32
           1.2.2   Calculation of The Number Ω of Microscopic States 34
           1.2.3   Boltzmann Distribution 37
    1.3   Internal Energy Distribution Functions 43
    References 50

2 Some Basic Concepts of Kinetic Theory 51
    2.1   The Velocity Distribution Function 51
    2.2   Macroscopic Properties 53
    2.3   Binary Elastic Collisions of Molecules 61
    2.4   Collision Cross-sections and Molecule Models 69
           2.4.1   Hard Sphere Model 72

	2.4.2	The Inverse Power Law Model 74
	2.4.3	Maxwell Model 76
	2.4.4	Variable Hard Sphere (VHS) Model 77
	2.4.5	Variable Soft Sphere (VSS) Model 80
	2.4.6	Generalized Hard Sphere (GHS) Model 85
	2.4.7	Generalized Soft Sphere (GSS) Model 87
2.5	The Eight Velocity Gas Model 88	
2.6	Boltzmann Equation 92	
2.7	Collision Integral and The Total Number of Collisions 98	
2.8	Evaluation of Collision Integrals 100	
2.9	The Maxwell Transport Equation – The Moment Equation 104	
2.10	Maxwell Distribution 106	
2.11	Equilibrium State of Gases 112	
	2.11.1	Some Peculiar Speeds of Gas 112
	2.11.2	Molecular Collision Frequency and The Mean Free Path 115
	2.11.3	The Mean Value of Collision Quantities 120
	2.11.4	The Reference Diameter of The VSS Model and The VHS Model 123
2.12	Gas Mixture 124	
	2.12.1	The Macroscopic Properties 124
	2.12.2	The Boltzmann Equations 126
	2.12.3	Number of Collisions, Collision Frequency and Mean Free Path 126
	2.12.4	Collision Frequency of a Molecule of species A with Molecules of Species B in Gas Mixture of VSS (or VHS) Molecules 127
References 128		

3 Interaction of Molecules with Solid Surface   131

3.1	Introduction 131
3.2	Specular and Diffuse Reflection 132
3.3	The Reciprocity Principle 140
3.4	The CLL Gas Surface Interaction Model 142

References 158

4 Free Molecular Flow — 159

- 4.1 The Number Flux and The Momentum Flux of Molecules in Gases 160
- 4.2 The Aerodynamic Forces Acted on Bodies 163
- 4.3 Heat Transfer to Surface Element 170
- 4.4 Free Molecular Effusion and Thermal Transpiration 174
- 4.5 Couette Flow and Heat Transfer between Plane Plates 177
- 4.6 The general solutions, unsteady flow 182
- References 190

5 Continuum Models — 191

- 5.1 Introduction 191
- 5.2 Basic Equations 192
  - 5.2.1 Equations of Mass, Momentum and Energy Conservation .. 192
  - 5.2.2 Chapman-Enskog Expansion 193
  - 5.2.3 Euler Equation 194
  - 5.2.4 Navier-Stokes Equations 194
  - 5.2.5 Burnett Equations 196
  - 5.2.6 Grad's Thirteen Moment Equations 201
  - 5.2.7 The Asymptotic Theory for Small Knudsen Numbers 203
- 5.3 Slip Boundary Conditions 204
  - 5.3.1 The Simple Derivation 204
  - 5.3.2 The Conservation of Momentum and Energy Fluxes in The Knudsen Layer 206
  - 5.3.3 The Derivation of The Slip Velocity Formula 207
  - 5.3.4 The Derivation of The Temperature Jump Expression 209
  - 5.3.5 The Extension to Cases of Multi-component Gases and Non-equilibrium Flows 212
- 5.4 The Solution of Some Simple Problems 212
  - 5.4.1 Couette Flow 213
  - 5.4.2 The Poiseuille Flow 215
  - 5.4.3 The Rayleigh Problem 218
- 5.5 Thermal Creep and Thermophoresis 220
- 5.6 Second Order Slip-jump Conditions 227

	References	228
6	**Transitional Regime**	**231**
	6.1 General Overview	231
	6.2 Linearized Boltzmann Equation	233
	6.3 The Moment Method	239
	6.4 Model Equations	247
	6.5 The Finite Difference Method	255
	6.6 Discrete Ordinate Method	257
	6.7 Integral Methods	263
	6.8 Direct Simulation Methods	264
	References	269
7	**Direct Simulation Monte-Carlo (DSMC) Method**	**275**
	7.1 Introduction	275
	7.2 Sampling of Collisions	278
	7.3 Example of Solution of Problem by The DSMC Method	281
	7.4 The Excitation and Relaxation of The Internal Energies	288
	7.4.1 Introduction of Phenomenological Models	288
	7.4.2 Implementation of Larsen-Borgnakke Model	289
	7.4.3 Cases of Distributions with Singularities, Generalized Acceptance-rejection Method	293
	7.4.4 Larsen-Borgnakke Method for Discrete Energy Levels	295
	7.4.5 Relaxation Collision Number and Vibrational Exchange Probability	297
	7.5 Simulation of Chemical Reactions	299
	7.5.1 Chemical Reaction Rate Coefficient	299
	7.5.2 Phenomenological Chemical Reaction Model of Bird	300
	7.5.3 A Sterically Dependent Chemical Reaction Model	302
	7.6 Computation of Complicated Flow Fields	310
	References	313
8	**Microscale Slow Gas Flows, Information Preservation Method**	**317**
	8.1 Introduction	317

8.2	Methods for Solving The Rarefied Gas Flows in MEMS	321
8.3	Information Preservation (IP) Method	326
	8.3.1 The Description of The Method	326
	8.3.2 The Validation of The Method	329
	8.3.3 Program Demonstrating The Method	332
8.4	Unidirectional Flows	333
8.5	The Microchannel Flow Problem	338
8.6	Thin Film Air Bearing Problem	348
8.7	Use of Degenerated Reynolds Equation in Channel Flow	355
8.8	Some Actual Problems and Concluding Remarks	360
	References	363

Appendix I Gas Properties 367
  References 368

Appendix II Some Integrals 369
  II.1 The gamma Function and Error Function 369
  II.2 Some Definite Integrals 370
  II.3 The beta Function 373
  References 374

Appendix III Sampling from a Prescribed Distribution 375
  III.1 Inversion of Cumulative Distribution Function 375
  III.2 Acceptance-rejection Method 378
  III.3 Generalized Acceptance-rejection Method 378
  References 381

Appendix IV Program of The Couette Flow 383

Subject Index 399

0 INTRODUCTION

0.1 THE CONCEPTION OF RAREFIED GAS DYNAMICS

Under usual circumstances the continuum model is adopted for treating gas flows. The properties of gases observed by the ordinary measuring apparatus are continuous and smooth, so the continuum hypothesis is natural. In practice or in the laboratory, the characteristic scale of flow under usual conditions has at least the size of $1 cm$, and the change of physical or dynamic properties of gas in the range of $10^{-3} cm$ would be very small, the sensing volume of $10^{-9} cm^3$ would give the measurement of the local properties of the gas. From the fact that the Loschmidt number n_0 (the number of gas molecules in cm^3 under standard state) is $2.687 \times 10^{19} cm^{-3}$ it can be seen, that under standard state in this very small volume $10^{-9} cm^3$ there are about 3×10^{10} molecules. This number is big enough to ensure that the average property of the gas is not influenced by the concrete number of molecules in this volume. But under the special condition of extremely low density (or when the size of the body is extremely tiny), the discrete particle effect becomes remarkable, and we are obliged to give up the continuum hypothesis and to adopt the discrete model, i. e., the method of molecular gas dynamics or the method of rarefied gas dynamics.

Strictly speaking, the failure of the continuum model happens when the shear stress and the heat flux in the continuum equations of mass, momentum and energy conservations can no longer be characterized by the macroscopic magnitudes of lower order (velocity, temperature), that is, when the expressions of the transport coefficients are no longer valid. This happens when the *scale length of the gradients of the macroscopic quantities* becomes so small that it is comparable with the *molecular mean free path* of the gas. The scale length of the gradients of the macroscopic magnitudes is

$$L = \rho/(d\rho/dx) \,. \tag{0.1}$$

To characterize the degree of rarefaction the *Knudsen number* is introduced and is defined as the ratio of the molecular mean free path λ (the average distance traveled by a molecule between two collisions) and the characteristic length of the flow L

$$Kn = \lambda/L \,. \tag{0.2}$$

Generally, when $Kn > 0.1$, the Navier-Stokes equation and the continuum model is no longer valid. When L in Eq. (0.2) is taken as a global size of the flow, Kn is not able to characterize the details of the flow, if L is taken as the scale length of the local flow gradient (from Eq. (0.1)), then Kn is a local Knudsen number, and can characterize the degree of non-continuum of the local flow field. When Kn is less than 0.1 but is larger than 0.01, some discrete effects such as the velocity slip and temperature jump appear, although the continuum model is still valid. Such effects also belong to the category of rarefied gas dynamics effects.

The density of the high altitude atmosphere decreases with descent of the height, the molecular mean free path increases from about $0.07 \times 10^{-6} m$ at the sea level to about $1mm$ at $70km$, and about $1cm$ at $85km$, and the rarefied gas dynamics effects become important. In the problem of force action and the heating of missiles, spacecrafts, satellites, space shuttles and space stations, rarefied gas dynamics plays an important role. The modern rarefied gas dynamics is initiated from the studying of the force and heating action exerted on bodies flying at high altitude [1, 2]. In vacuum systems, with the enhancement of the vacuum degree, the mean free path becomes comparable with the characteristic length of the flow, and the molecular gas dynamics method is to be applied. When considering problems like such of the shock wave structure, the macroscopic magnitudes experience consistently drastic change in several mean free paths, so the molecular description is to be invoked even under usual density conditions. Generally, under usual density conditions, when the size of the subject of our investigation is small enough to be comparable with the mean free path, the gas discrete effect reveals itself, as in the case of the study of aerosol particles (of the size of micrometer or less). In the last ten-odd years, the micro-electro-mechanical-systems (MEMS) develops rapidly, and will play promoting role in the fields of micro electronics, aeronautics and astronautics, optics, biology, medicine and other realms. Owing to the tiny size of MEMS, when gas is used as the working media, the study of their properties requires the involvement of the rarefied gas effects.

0.2 THE MOLECULAR MODEL OF GASES

The hypothesis that all matter (including gas) is composed of tiny particles separated by vacuum traces the origin from Greek materialistic philosopher Democritus (460-370 B.C.). He supposes that these particles are of different sizes and shapes and are in persistent rapid motion, and attributes the differences in the properties of matter to the differences in the kinds of particles and the differences in their motion. Owing to the absence of any convincing evidence this closest to truth hypothesis could be considered only as one of the many philosophic conjectures about the structure of matter and had been rejected by the authority at that time Aristotle (384-322 B.C.). The real quantitative argumentation of the kinetic theory of gases can be thought initiated by Daniel Bernoulli (1700-1782). In 1738 Bernoulli explained Boyle's law, i.e. the pressure and density of a gas are in direct proportion at constant temperature. From the fifties to eighties of the 19th century the work of Clausius (1822-1888), Maxwell (1831-1879) and Boltzmann (1844-1906) accomplished the founding and the overall development of the kinetic theory of gases. In 1857 Clausius introduced the concept of mean free path (the average value of the distance traveled by a molecule between two successive collisions) [3]. In 1859 Maxwell obtained the expressions of the transport coefficients (viscosity μ, conductivity K and diffusivity D) using the concepts of mean free path and the velocity distribution function f introduced by himself, and the equilibrium state velocity distribution function, i.e. the Maxwell distribution $f^{(o)}$ [4]. In 1872 Boltzmann proposed and proved the H-theorem, improved the proof of the Maxwell distribution, and put forward the integral-differential equation for the velocity distribution function f, i.e. the Boltzmann equation [5]. At the beginning of 20th century the experiments of Perrin el al. (1908) concerning the Brownian movement allowed people to really observe (indirectly) the unceasing motion of the molecules hypothesized by the kinetic theory of gases, playing a key role in convincing the correctness of this theory. Chapman (1889-1970) and Enskog (1887-1947) in 1916 and 1917 respectively derived the exact expressions of the transport coefficients through entirely different mathematical ways [6, 7]. Especially their derivation for gas mixture resulted in the discovery of a new transport coefficient, i. e. the thermal diffusivity D^T, and its expression. The phenomenon of thermal diffusion is widely used in the separation of uranium and other isotopes, this is considered as an important achievement of the kinetic theory of gases.

0.3 MEAN FREE PATH OF MOLECULES

As mentioned in the previous section the *molecular mean free path* is an important fundamental concept of the kinetic theory of gases. According to the definition, the molecular mean free path is the average distance traveled by a molecule between two successive collisions. For general molecular model the sphere of influence of the molecular force is very big, the happening of a collision is judged by stipulating that a certain value the deflection angle must be surpassed. The elucidation of the concept of mean free path is easy when using the hard sphere model of molecule. Suppose the molecules are hard spheres of diameter d, and first calculate the number of collisions happened in unit time, i. e., the so called collision frequency v. Consider the collision of one test molecule with other field molecules. A collision happens when the center of a field molecule is located inside the sphere of radius d with the center of the test molecule as its center. Obviously, multiplying the volume swept by this sphere (with collision section $\sigma_T = \pi d^2$) in unit time by the number density n yields the requested v. To obtain the correct averaged value, suppose the test molecule moves with the mean value of the relative speed $\overline{c_r}$. Then the volume swept by the sphere of radius d in unit time is $\overline{c_r}\pi d^2$, and the collision frequency v is

$$v = \overline{c_r}\sigma_T n = \overline{c_r}\pi d^2 n. \tag{0.3}$$

For general molecular model, the collision section σ_T is dependent on c_r and the averaging extends over both c_r and σ_T, and

$$v = c_r \overline{\sigma_T} n,$$

see Eq. (2.207) to be derived in Chapter 2. As the distance traveled by a molecule in unit time is the mean thermal speed $\overline{c}$, obviously the mean free path is

$$\lambda = \frac{\overline{c}}{v} = \frac{\overline{c}}{\overline{c_r}} \frac{1}{\pi d^2 n}. \tag{0.4}$$

As $\overline{c_r} = \sqrt{2}\,\overline{c}$, see Eq. (2.214) in Chapter 2, the expression of the mean free path λ is obtained

$$\lambda = \frac{1}{\sqrt{2}\pi d^2 n}, \tag{0.5}$$

or

$$\lambda = \frac{m}{\sqrt{2}\pi d^2 \rho}. \tag{0.6}$$

Note, that from Eq. (0.6) one sees that $\lambda\rho = constant$, i. e. the mean free path and the density of the gas are in reverse proportion.

0.4 DIVISION OF FLOW REGIMES

Qian Xuesen (H. S. Tsien) [1] first divided the rarefied gas flows into three realms according to the degree of rarefaction, i. e., the *slip flow regime*, the *transitional regime* and the *free molecular regime*. According to the range of the properly chosen Knudsen number Kn, the three regimes are

$$0.01 < Kn < 0.1 \quad \text{Slip flow regime}, \tag{0.7}$$

$$0.1 < Kn < 10 \quad \text{Transitional regime}, \tag{0.8}$$

$$Kn > 10 \quad \text{Free molecular regime}. \tag{0.9}$$

In the *slip flow regime* in the gas flow appear some phenomena somewhat different from those in the ordinary flows, which manifest themselves mainly at the vicinities of the boundaries, i. e. the velocity slip and the temperature jump phenomena. In this regime, the ordinary gas dynamics equations are still valid, or equations of order higher than Navier-Stokes can be obtained by using high order terms of the Chapma-Enskog expansion, but it is necessary to introduce some modifications into the boundary conditions. At the verge of large Knudsen number where the gas is very rarefied, the collisions of the gas molecules with the surface of the body prevail. And when λ is much larger than L, the molecules reflected from the surface of the body collide with other molecules only after flying across a large distance, the velocity distribution function of the oncoming molecules is not influenced by the presence of the body, i.e., is the known equilibrium Maxwell distribution, and the momentum and energy action of the oncoming flow on the body can be easily calculated. This regime is called the *free molecular flow*. Of course, to obtain a complete solution certain assumption must be made about the law of reflection of molecules on the surface. In the *transitional regime* between the slip flow regime and the free molecule regime, the collisions of the molecules with the surface of the body and mutual collisions of the molecules in the oncoming flow have the equal importance. The analysis becomes difficult, and the molecular gas dynamics methods must be adopted.

When discussing the division of the flow regimes, it is to be noted that the flow parameters in the flow are extremely non-uniform. When the characteristic length is taken as the scale length of the gradients of the macroscopic quantities $L = \rho/(d\rho/dx)$ (see Eq. (0.1)), Kn number can characterize the local properties of the flow. According to Eq. (0.7), to make the Navier-Stokes equation valid, Kn must be less than 0.1. But the upper limit of the local Knudsen number for the validity of the continuum hypothesis might be enhanced to as large as 0.2.

For problems of hypersonic flow around bodies the demand for the flow to satisfy the free molecular condition is more severe than the condition Eq. (0.9). In Eq. (0.9) λ is the mean free path of the molecules in the oncoming flow. But in the context of hypersonic flow around body, the mean free path λ_{ei} of the reflected molecules in the collision with the oncoming molecules is the essentially relevant mean free path. This λ_{ei} is much less than λ. Let us estimate the magnitude of it. As the oncoming flow has the hypersonic speed, the relative speed of the molecules of the oncoming flow and the body is much larger than the thermal speed of the molecules. At the same time the speed of the molecules scattered from the surface is also small in comparison with the speed of oncoming flow. And the relative speed between the scattered molecules and the oncoming molecules can approximately be taken as the speed U of the oncoming flow. If d is the diameter of the molecule, then the number of collisions of a scattered molecule with the oncoming molecules in unit time is

$$\pi d^2 U n_i . \tag{0.10}$$

where n_i is the number density of the oncoming molecules. The frequency is independent of the coordinate system adopted. But the mean free path would vary with the coordinate system used. In the coordinate system connected with the body, the average speed of the scattered molecules is $\overline{v_e} = \sqrt{8RT_e/\pi}$ (see Eq. (2.200)), where R is the gas constant, T_e is the temperature of the body. In this coordinate system the mean free path of the scattered molecule in collision with the oncoming molecules is

$$\lambda_{ei} = \frac{\overline{v_e}}{\pi d^2 U n_i} . \tag{0.11}$$

Comparison between Eq. (0.5) and Eq. (0.11) yields

$$\lambda_{ei} = \sqrt{2} \frac{\overline{v_e}}{U} \lambda . \tag{0.12}$$

The basic condition for free molecular flow requires

$$\frac{\lambda_{ei}}{L} > 10 ,$$

or

$$\frac{\lambda}{L} > 10 \frac{U}{\sqrt{2} \overline{v_e}} . \tag{0.13}$$

For cold wall assumption $U \gg \overline{v_e}$, Eq. (0.13) is much severer than Eq. (0.9), i.e., more degree of rarefaction is required to ensure the occurrence of the condition of free molecule flow. When the temperature of the body is extremely low ($T_e \approx 0$), even though when the gas is very rarefied, Eq. (0.13) could not be satisfied, that is, free molecular flow would not appear. This is the so called *cold wall paradox of free molecular flow*.

In the problem of hypersonic flow around blunt body, the mean free path behind the shock wave λ_s and the shock detachment distance Δ to the body are the appropriate mean free path and the characteristic length of the problem, so the basic requirement for the continuum is

$$\lambda_s / \Delta \ll 1.$$

The hypersonic flow theory states that the estimate of the ratio of Δ to the radius of the nose R_b is $\Delta / R_b \sim \rho_\infty / \rho_s$, where ρ_∞ and ρ_s are the density of the oncoming flow and the density behind the shock respectively, so this basic requirement is equivalent to (*see* (0.27))

$$\lambda_s / \Delta \sim \lambda / R_b \sim Ma/\mathrm{Re} \ll 1. \tag{0.14}$$

For unsteady flows, the judgment of the division of regimes is accomplished through comparison of the average collision time $\tau = \lambda / \bar{c}$ and the time of motion. Take the unsteady motion of the gas caused by the instantaneous movement of a flat plate (the *Rayleigh Problem*) as an example. A flat plate of infinite length acquires instantaneously velocity U and moves along its own plane, thus causes the gas above it to move. The problem is unsteady and contains two independent parameters: time t counted from the instant of the start of the motion and the vertical coordinate y. The ratio of the average time τ between two collisions and time t is an important parameter, it can be expressed through Mach number $Ma = U/a$ and Reynolds number $\mathrm{Re} = \rho U^2 t / \mu$

$$\frac{Ma^2}{\mathrm{Re}} = \frac{U^2}{a^2} \frac{\mu}{\rho U^2 t} \approx \frac{\frac{1}{2}\bar{c}\lambda}{a^2 t} \sim \frac{\tau}{t}. \tag{0.15}$$

Here the relation $\mu \approx (1/2)\rho\bar{c}\lambda$ (see Eq. (2.222)) has been used. According to the value of this ratio the motion can be divided into free molecular flow regime, transitional regime and continuum regime.

1. Free molecular flow regime ($\tau/t \gg 1$). Just after the beginning of the motion, t does not reach the average time of collision, the molecules

reflected from the flat plate have not enough time to collide with other molecules. No matter how large is the gas density, at the initial stage there always appears the free molecular flow regime.

2. Transitional regime ($\tau/t = O(1)$). When the number of mutual collisions between molecules and the number of collisions of molecules with the plate surface are almost the same, the molecular dynamics method different from free molecular method must be used.

3. Continuum regime ($\tau/t \ll 1$). The number of collisions is big enough, the ordinary gas dynamics equations can be used.

Take another example of dispersion of acoustic wave, the parameter characterizing the degree of rarefaction is $p/\omega\mu$, where ω is the frequency of the acoustic wave. This parameter in fact is the ratio of the collision frequency to the acoustic frequency, i. e. the ratio of the length of the acoustic wave to the mean free path

$$\frac{p}{\omega\mu} \sim \frac{p}{\omega\overline{\rho c \lambda}} \sim \frac{1/\tau}{\omega} \sim \frac{\lambda_a}{\lambda}. \tag{0.16}$$

0.5 NONEQULIBRIUM PHENOMENA AND RAREFIED GAS DYNAMICS

The speed of the spacecraft flying over the high altitude atmosphere is approximately between $7km/s$ and $10km/s$. At the stagnation point such high kinetic energy is to be converted into the internal energy. Using the energy conservation equation of ideal gas, the *stagnation temperature* T_0 is derived as

$$T_0 = T + \frac{U^2}{2c_p} = T\left(1 + \frac{U^2}{2c_p T}\right) = T\left(1 + \frac{\gamma-1}{2}Ma^2\right). \tag{0.17}$$

Take the specific heat $\gamma = 1.4$, the static temperature of oncoming flow $T = 300K$, it follows that $T_0 = 24700K$, when $U = 7km/s$, and $T_0 = 50100K$, when $U = 10km/s$. But the stagnation point temperature is by far not so high, this is because the energy goes to excite the internal degrees of the gas molecules and is absorbed by the endothermic reaction such as the dissociation process in the gas.

Owing to the quantum nature of the molecular structure, we proceed from the Schrodinger equation. For the simplest rigid rotator model and harmonic oscillator model representing the rotation and vibration of the molecule, Schrodinger equation gives eigenfunctions with degeneracies, yielding the probabilities of the configuration of molecules (see section 1.1). For gas dynamics of most importance is the fact that the energy eigenvalues corresponding to the eigenfunctions, for rigid rotator and harmonic oscillator, can attain only discrete energies:

Rigid rotator

$$\varepsilon_{r,J} = \frac{h^2 J(J+1)}{8\pi^2 \mu r^2} = \frac{h^2 J(J+1)}{8\pi^2 I}, J = 0,1,2,\cdots. \quad (0.18)$$

Harmonic oscillator

$$\varepsilon_{v,n} = \left(n + \frac{1}{2}\right)hv, n = 0,1,2,\cdots. \quad (0.19)$$

Corresponding characteristic temperature of rotation Θ_r and characteristic temperature of vibration Θ_v (the typical value of energy divided by k) are

$$\Theta_r = h^2/(8\pi^2 Ik), \quad (0.20)$$

$$\Theta_v = hv/k. \quad (0.21)$$

In the above h is the Planck constant; k is the Boltzmann constant; I is the moment of inertia of molecule; v is the frequency of vibration, J and n are the quantum number of rotation and the quantum number of vibration, respectively, that can take only integer numbers. For N_2, NO and O_2, the characteristic temperatures of rotation have the values in the range of only several degrees: $\Theta_r = 2.88K$, $2.44K$, $2.07K$; and the characteristic temperatures of vibration are: $\Theta_v = 3371K$, $2719K$ and $2256K$ (see appendix I).

Just because the nitrogen and oxygen molecules have such low characteristic temperatures of rotation, the rotational energy is fully excited under room temperature, the rotational energy presents two degrees of freedom $e_r = RT$,

$c_{v,r} = R$, $c_v = (5/2)R$, $c_p = (7/2)R$, $\gamma = 1.4$. Here e_r is the rotational energy, R is the gas constant, $c_{v,r}$ is the rotational constant volume specific heat, c_v is the constant volume specific heat, c_p is the constant pressure specific heat. And as the characteristic temperatures of vibration have relatively high values, the vibrational energy are being excited only at relatively high temperatures, a certain portion of gas energy at high temperature goes to excite the vibrational energy.

The uncertainty principle (Heisenberg principle) tells us that the product of the uncertainty $|\Delta r|$ in coordinate and the uncertainty $|\Delta mc|$ in momentum of a particle is of the size of h

$$|\Delta r| \cdot |\Delta mc| \sim h. \tag{0.22}$$

In the problems of molecular gas dynamics, the typical scale to ascertain the position of a molecule is the *mean molecular spacing* $\delta \sim n^{-1/3}$, and the typical value of the average momentum is $m\sqrt{\overline{c^2}} = m(3kT/m)^{1/2} = (3mkT)^{1/2}$, see Eq. (2.201). If their product is much larger than h, the quantum effects are not important, thus yielding the condition for negligible quantum effects

$$(3mkT)^{1/2} / (n^{1/3} h) \gg 1 \tag{0.23}$$

It is seen by substituting the known data that under standard state this condition is satisfied. Under situations of high temperature and low density which interest us, it is all the more satisfied. This makes clear that we need not to use the quantum mechanics method to describe the behavior of air molecules in treating the aerodynamic problems. However, the above discussion of the rigid rotator and harmonic oscillator shows, that the results of quantum mechanics, i.e., the magnitudes of the characteristic temperatures of rotation and vibration obtained therefrom, do influence the behavior of gas in totality at standard and high temperatures.

The endothermic dissociation reaction occurred in the air explains further why the stagnation temperature in a hypersonic flow does not attain the value expected from Eq. (0.17). The dissociation energies ε_d and respective characteristic dissociation temperatures Θ_d (dissociation energies divided by k, see Table I.1 in Appendix I) of O_2, NO and N_2 are

$$\varepsilon_d = 5.12eV, \quad 6.5eV, \quad 9.76eV,$$

$$\Theta_d = 59500K, \quad 75500K, \quad 113500K.$$

When consider the dissociation in the stagnation region, there are two reasons for the fact that though the temperature seems to be much lower than Θ_d, there are still apparent dissociations taking place. The first is the high probability of collisions of particles with energies much higher than kT. The second season is that the recombination reaction is an exothermic reaction and the participation of a third body is needed to carry away the dissociation heat, and as the probability of the ternary collision is extremely small, mostly only dissociation reaction occurs. And the higher characteristic temperature of dissociation (and hence the dissociation energy) makes the decreasing temperature effect of dissociation more remarkable.

When temperature attains even higher values, ionization reaction occurs and becomes important. The ionization potentials and characteristic temperatures of O_2, NO and N_2 are

$$\varepsilon_i = 12.3eV, \quad 9.34eV, \quad 15.7eV,$$

$$\Theta_i = 142000K, \quad 108000K, \quad 181000K.$$

The ionization manifests itself, when the temperature attains the value of about $10,000K$.

From the discussion of the variation of the temperature in the stagnation region of the hypersonic flow it is clear that the physico-chemical changes such as the internal energy excitation, dissociation and ionization occurred in the air drastically influence the magnitude of the temperature and the behavior of air in the stagnation region. These phenomena commonly exist in the high altitude flight. The excitation of the internal states and chemical reactions are all rate processes, to accomplish a process certain time is needed and is measured as relaxation time. As all the processes are accomplished in the collisions of molecules, the relaxation time is proportional to the average collision time τ, the distance across which the relaxation occurs is proportional to the mean free path. The relaxation time of rotation is usually about 5~10 mean collision times, and the vibrational relaxation at normal temperature is usually two or three order of magnitude slower than the translational and rotational relaxation. When the relaxation time is large in comparison with the characteristic time of flow and the relaxation distance (the distance traveled by the molecule during relaxation time) is large in comparison with the characteristic length of the flow, the internal energy and chemical reaction are out of equilibrium, i.e., there appears the internal mode and chemical non-equilibrium. And this is the case where rarefied gas dynamics effects prevail. The non-equilibrium phenomena always become more severe when the flow condition changes towards rarefaction, and the internal energy and chemical processes are usually in the state of non-equilibrium. The maneuver and braking of aerospace vehicles at high altitudes demands the gas dynamics to answer the question of force and heat action of the low density gas flow. The high speed condition leads to the necessity to consider the physico-chemical processes, sometimes including the quantum effects, such as the internal energy mode excitation, chemical reaction, the excitation and transition of electronic energy levels. The low density condition leads to the necessity to take into account that the gas is a discrete system composed of individual molecules and to discard the continuum hypothesis. This leads in two aspects to the development of gas dynamics towards microscale. Rarefied gas dynamics enriches our knowledge of the microscale level in these two aspects. Taking into account the discrete structure of gas and the internal mode change, chemical reaction and ionization and transition of quantum levels is the premise of the gas dynamics research not only in the realm of aerospace but also in the new technology realms to play remarkable role in the 21st century. These realms include plasma material processing, micro-electronic etching, MEMS, laser techniques, chemical industry, combustion and others

0.6 SIMILARITY CRITERIA

The Knudsen number introduced above

$$Kn = \lambda / L,$$

characterizes the degree of rarefaction in the flow. *Knudsen number* can be related to the well known *Mach number* $Ma = U/a$ and *Reynolds number* $Re = \rho UL/\mu$ [1]. According to the calculation of kinetic theory of gases [8], the viscosity μ is related to the density ρ, the thermal mean speed $\bar{c}$ and the mean free path through (see Eq. (2.222))

$$\mu \approx \frac{1}{2}\rho \bar{c} \lambda. \qquad (0.24)$$

To be more accurate, 1/2 is to be replaced by 0.491. According to the kinetic theory of gases $\bar{c} = \sqrt{8kT/\pi m}$, and the sonic speed is $a = \sqrt{\gamma kT/m}$, here T is the absolute temperature of the gas, γ is the specific heat, k is the Boltzmann constant, m is the molecule mass, so we obtain

$$a = \sqrt{\frac{\pi \gamma}{8}} \bar{c}. \qquad (0.25)$$

From Eq. (0.24) and Eq. (0.25) one obtains

$$\lambda = 1.26\sqrt{\gamma}\mu / a\rho. \qquad (0.26)$$

It is easy to obtain the following fundamental relationship

$$Kn = \frac{\lambda}{L} = 1.26\sqrt{\gamma} Ma/Re. \qquad (0.27)$$

The characteristic length L in Re is the same as that in Kn. Thus, the similarity of two flows in rarefaction requires that the ratio of Ma to Re remains the same.

For not too high a speed the similarity of two flows requires the equality of Mach number and Reynolds number, and thus, according to Eq. (0.27), the equality of Knudsen number. Owing to Eq. (0.27) the similarity of flows requires that any two of the three parameters Ma, Re and Kn are the same. Owing to the difficulty in simulation of the slow micro-scale rarefied flow, Hadjiconstantinou and Garsia [9] suggested to simulate such flow by keeping the

Re number unchanged but allowing a larger Ma number in a scope where the compressibility can be neglected. Such a way of doing simulation can not be proved to be lawful. For with an allowance of larger Ma, the Knudsen number would also be enhanced (see Eq. (0.27)), two flows will be different unless the flow can be proved to be independent of Kn for Kn less than this enhanced value.

As it can be seen in section 0.4 of the division of flow regimes, the criterion Kn for the division must have various modifications relative to different problems. For example, the basic requirement of the free molecular flow around bodies in hypersonic gas streams is $\lambda/L > 10U/\sqrt{2v_e}$, or Eq. (0.13), but not Eq. (0.9), the relevant Kn number in Rayleigh problem is τ/t or M^2/Re (see Eq. (0.15)), etc.

For similarity of two flows, it is not always lawful to assure the same Knudsen number by changing the length scale and the gas density (thus the mean free path). When the characteristic scale of the problem considered is extremely small, the real number of molecules in the characteristic volume may be so small, that the problem of fluctuation of the macroscopic magnitudes arises. This happens when considering the gas flows in MEMS. For example, when the characteristic length is $0.1 \ \mu m$, the cube with the sides of this length contains only 3×10^4 molecules under atmospheric pressure, the fluctuation would be of the order 1%, as the variance of the macroscopic magnitude is in reverse proportion to the squire root $\sqrt{N}$ of the number of particles. This fluctuation is a real one in this problem of small size. The mean free path is $\lambda \sim 0.07 \mu m$ under the pressure of one atmosphere, so for this problem $Kn = 0.7$. If we proceeded to simulate the Kn number under the pressure of 10^{-3} atmosphere, we would find that $\lambda \sim 70 \mu m$, and the flow scale becomes accordingly $100 \mu m$, the cube with the sides of this length would contain 3×10^{13} molecules, the statistic variance reduces to $2 \times 10^{-5} \%$, and the fluctuation which is real disappears. For NEMS (nano-electro-mechanical system) it is even more unlawful to keep Kn the same by changing the length of the flow, the equality of the number of real molecules will be the essential requirement, the fluctuations of the flow quantities are real.

For flows of very high speed in the astronautic realm, it is not sufficient to have the similarity in Kn number and Ma number. The similarity criterion for high speed or high enthalpy reaction flows is the *binary scaling law*, i.e., two flows are similar, when they have the same static temperature and the same

$$U_\infty, \rho_\infty L . \tag{0.28}$$

This was first put forward by Birkhoff in the second edition of his famous book Hydrodynamics [10] and was called there *binary collision modeling law*, which is in fact more precise than the term binary scaling law, but the latter has gained the wider spread. This similarity criterion requires the equality of the speed of oncoming flow, allowing only the change of distance scale and density in reverse proportion. For flows where only binary collisions are essential, this conclusion can be drawn from the kinetic theory of gases. The equality of U_∞ is an extremely severe requirement for experiment, but is necessarily to be satisfied. This practically is the demand for the equality of the total enthalpy to assure that the chemical reactions of the flows have the same supply of energies. The equality of $\rho_\infty L$ ensures the equality of the ratio of the characteristic distance of the relaxation process to the characteristic length of the body (flow field) (see Eq. (0.26), where λ is understood as relaxation distance). Binary scaling law can be derived from the continuum equations (*see* [11]). Zhang Hanxin [12] obtained from the Navier–Stokes equations with chemical reactions, that when only binary collisions are important, the simulation parameters are

1) $Ma_\infty, U_\infty, \rho_\infty L, Ma_\infty = 10\sim13$. (0.29)

2) $\widetilde{V}'_\infty, U_\infty, \rho_\infty L, Ma_\infty = 13\sim15$. (0.30)

where $\widetilde{V}'_\infty = \sqrt{C} Ma_\infty / \sqrt{Re_\infty}$. Shen has shown in [11], that there is not the need to distinguish the similarity in Ma_∞ and $\widetilde{V}'_\infty$. For if we have Eq. (0.29), the similarity of Ma_∞ and U_∞ leads to the similarity of the static temperature. And when there is similarity of $\rho_\infty L, U_\infty$ and equality of static temperature, the Reynolds number $Re_\infty = \rho_\infty L U_\infty / \mu$ is also the same, so the similarity of $\widetilde{V}'_\infty$ is also satisfied. That is, Eqs. (0.29) and (0.30) are summed up as the equality of the static temperature and the similarity of

$$U_\infty, \rho_\infty L , \tag{0.31}$$

i.e., the binary collision modeling law or binary scaling law follows.

It is important to determine the applicability of the binary scaling law under different conditions. Ellington [13] determined its scope of application at relatively low speed. Shen et al. [14] calculated the electron densities around the blunt nose at the altitude of $70km$ and higher using the thin viscous layer approximation with chemical reactions and showed on numerical example the applicability of the binary scale law in the continuum regime. The calculation of Moss [15] by the direct simulation Monte- Carlo method [16] showed that at certain altitude it is applicable to the discrete molecular method.

Owing to the important position of the direct simulation Monte-Carlo (DSMC) method in rarefied gas dynamics, it is appropriate to discuss the similarity law of simulation. The DSMC method uses simulated molecules of number much fewer than the number of real molecules to represent the behavior of real molecules. In early works the product of the number density and the cross section of the molecules was kept constant to ensure the similarity of Kn number in the simulated flow and the real flow. This led to the much larger seeming size of the simulated molecules in comparison with the real molecules. But this had no essential consequences, it was used only to ensure the correct collision frequency under small number of simulated molecules. The current way of keeping similarity is to determine the number of real molecules one simulated molecule represents, and take into account that a simulated molecule represents a large number of real molecules when counting the number of collisions and calculating the macroscopic magnitudes. Appropriate choice of collision partners and implement of certain number of collisions during one time step in a cell guarantees the simulation of real flow. At least in the case of one-dimensional and two-dimensional flows, small number of simulated molecules represents the certain real situation in the real flow. In the one-dimensional flow, when we take very narrow gap between the two cross sections along the direction of variation of the flow field, the number of the real molecules is also very small, and the equality of the simulated molecules and the real molecules can be reached. In the two-dimensional flow, when we take very thin slice parallel with the flow plane as the representative flow field, the same situation happens. Then in the real flow field, as the number of molecules is extremely small, the fluctuations of the macroscopic magnitudes are real. From the theory of probability it is known, that the standard deviation from its mathematical expectation of the measured value of a physical magnitude in a volume containing N particles is proportional to $1/\sqrt{N}$. This is the reason why the number of simulated molecules and the sample size should be taken sufficiently large. Owing to the limit in computer storage, under the condition of certain number of simulated molecules, two types of averaging processes, i.e., the time average and the ensemble average are adopted to reduce the variance. *Time averaging* is an averaging obtained by summing the molecular properties over many time steps and is used in the steady problems to obtain the macroscopic characteristics of steady flows. *Ensemble* averaging is here referred to as an averaging process taken over a large number of instantaneous averages in the same space element at the same time instant in an unsteady problem to increase the sample size. Note, sometimes it is preferable to use large number of simulated molecules and avoid the ensemble averaging in unsteady flow problems to reveal the process in real time and study the physical instability. Too large a difference between the number of simulated and real molecules represents in some problems essential difficulty. For example, in the ionized gas, to use the usual proportion of the numbers of simulated and real molecules to simulate the re-entry ionized flow field, the electric field caused by the fluctuation of the charged particles is much larger than the real electric field in the flow. It is a problem under investigation to simulate the electric field of the ionized gas around a re-entry body by the DSMC method.

REFERENCES

1. Tsien H S (1946) Superaerodynamics, mechanics of rarefied gases. J Aero Sci 13: 653-664
2. Tsien H S (1948) Wind-tunnel testing problems in superaerodynamics. J Aero Sci 15: 573-580
3. Clausius S (1858) Annln Phys 105:239
4. Maxwell J C (1860) Phil Mag 19:22
5. Boltxmann L (1872) Wien Sitz 66: 275
6. Chapman S (1916) Phil Trans R Society 216, 279 (Dec. 1916) Proc. R. Soc. A 98, 1
7. Enskog D. (1917) Inaugural Dissertation, Uppsala
8. Chapman S, Cowling T G (1970) The mathematical Theory of Non-uniform Gases. Cambridge Univ. Press

9. Hadjiconstantionou NG Garcia AL (2001), Statistical error in particle simulations of low Mach number flows, Proceedings of the lst MIT Conference on computational Fluid and Solid Mechanics Elsevier
10. Birkhoff G (1960) Hydrodynamics, 2nd ed., revised and enlarged. Princeton Univ. Press
11. Shen C (1991) Non-equilibrium effects of real fluid. In: Modern advances in fluid mechanics. Science press, Beijing, pp. 108-114 (in Chinese)
12. Zhang H X (1990) The similarity law for real gas flow. Acta Aerodynamica Sinica 8: 1-8 (in Chinese)
13. Ellington (1967) Binary scaling limits for hypersonic flight. AIAA J, 5: 1705-1706
14. Shen C, Hu ZH, Wang XY (1970) The ionized flow field around sphere-cone at altitude of 59-90km. Report of Institute of Mech-7006 (in Chinese)
15. Moss J N et al. (1989), Nonequilibrium effects for hypersonic transitional flows. AIAA paper 87-0404
16. Bird G A (1994) Molecular Gas Dynamics and the Direct Simulation of Gas Flows. Clarendon Press. Oxford

1 MOLECULAR STRUCTURE AND ENERGY STATES

1.1 DIATOMIC MOLECULES

Air is a gas mixture composed of 76.3 % (mole fraction) of N_2 and 23.7% of O_2. At relatively high temperature the internal energy states change, air dissociates into N and O, and NO is produced. Under even higher temperature ionization occurs, and ions and electrons are produced accordingly (with radiation occurring). To study the internal energy exchange, dissociation and ionization that are going on in the air, some basic knowledge of quantum mechanics, statistical mechanics and chemical kinetics is needed, it is not possible to introduce them comprehensively. As a start and guidance, only the outline of the basic knowledge of the internal energies is given here. This is an attempt to build the understanding of the internal energies, internal degrees of freedom and the distribution of internal energies on the basis of the same profoundness as the understanding of the translational energy in the next chapter. For the classic exposition of the molecular structure see the book of G. Herzberg [1].

The starting point of description of the structure and the internal energy states of an atom-molecule system is the *quantum mechanics* or wave mechanics. For a particle of mass m, the *wave function* ψ depends on the coordinates x, y, and z. Schrodinger postulates that ψ satisfies the following equation [1]

$$\frac{\partial^2 \psi}{\partial x^2} + \frac{\partial^2 \psi}{\partial y^2} + \frac{\partial^2 \psi}{\partial z^2} + \frac{8\pi^2 m}{h^2}(E-V)\psi = 0, \qquad (1.1)$$

where V is the potential energy; E is the energy of the particle; h is the *Planck constant* = 6.626×10^{-34} Js. Equation (1.1) is called *Schrodinger equation*, the solution of it is naturally to be single valued, finite, continuous and to have zero value at the infinity. This is possible only for certain values of the energy E, and such values E are called the *eigenvalues*, i. e. the energy values of stable state. Such values are usually discrete, that is, the energies are quantized. The corresponding solutions ψ of the equation (or $\Psi = \psi e^{-2\pi i E/h)t}$) are called the *eigenfunctions*. According to *Bohr's hypothesis*, $\Psi\Psi^* dr = |\Psi|^2 dr = |\psi|^2 dr$ gives the probability of finding the particle in the volume element dr at the position given by the coordinates arguments of the function ψ.

The nucleons of the diatomic molecule repel each other due to the presence of the protons, the interaction of the outer shell electrons keeps the atoms remain in one molecule. The Schrodinger equation governing the diatomic molecule can be written as

$$\frac{1}{m}\sum_i\left(\frac{\partial^2\psi}{\partial x_i^2}+\frac{\partial^2\psi}{\partial y_i^2}+\frac{\partial^2\psi}{\partial z_i^2}\right)+$$
$$\sum_k\frac{1}{M_k}\left(\frac{\partial^2\psi}{\partial x_k^2}+\frac{\partial^2\psi}{\partial y_k^2}+\frac{\partial^2\psi}{\partial z_k^2}\right)+\frac{8\pi^2}{h^2}(E-V)\psi=0, \tag{1.2}$$

where m is the mass of the electron; M_k is the mass of the kth nucleus; x_i, y_i, z_i are the coordinates of the electrons; x_k, y_k, z_k are the coordinates of the nuclei; E is the energy of the system; V is the potential energy of the system, i. e., the sum of the potential V_e of the electrons (as the function of x_i, y_i, z_i) and the Coulomb potential V_n of the nuclei, i. e.

$$V = V_e + V_n. \tag{1.3}$$

One may try to use the method of separation of variables to solve Eq. (1.2). Write the solution in the form

$$\psi = \psi_e(\cdots,x_i,y_i,z_i,x_k,y_k,z_k,\cdots)\psi_{vr}(\cdots,x_k,y_k,z_k,\cdots), \tag{1.4}$$

where ψ_{vr} depends only on the coordinates of the nuclei. Let ψ_e and ψ_{vr} satisfy the following equations respectively

$$\sum_i\left[\frac{\partial^2\psi_e}{\partial x_i^2}+\frac{\partial^2\psi_e}{\partial y_i^2}+\frac{\partial^2\psi_e}{\partial z_i^2}\right]+\frac{8\pi^2 m}{h^2}(E^{el}-V_e)\psi_e = 0, \tag{1.5}$$

$$\sum_k\frac{1}{M_k}\left[\frac{\partial^2\psi_{vr}}{\partial x_k^2}+\frac{\partial^2\psi_{vr}}{\partial y_k^2}+\frac{\partial^2\psi_{vr}}{\partial z_k^2}\right]+\frac{8\pi^2}{h^2}(E-E^{el}-V_n)\psi_{vr} = 0. \tag{1.6}$$

Eq.(1.5) is the Schrodinger equation describing the motion of the electrons in the field V_e when the nuclei are fixed and motionless. V_e depends on the distance between the nuclei. The solution of Eq. (1.5), i.e., the eigenfunctions ψ_e and the eigenvalues E^{el}, also depends on the distance between the nuclei. Eq. (1.6) is the Schrodinger equation describing the motion of the nuclei. In comparison with Eq. (1.1), one can see that the sum $E^{el}+V_n$ plays the role of the potential of the force exerted by the nucleus, i. e., the sum of the electron energy E^{el} and the Coulomb potential V_n

$$E^{el}+V_n \qquad (1.7)$$

plays the role of the potential governing the vibration of the nuclei. The curve of the variation of the value of Eq. (1.7) with the distance between nuclei is usually called the *potential curve of the molecule*. It is seen that the condition for the satisfaction of Eq. (1.2), when the ψ_e and ψ_{vr} in Eq. (1.4) satisfy Eqs. (1.5) and (1.6), is that

$$\sum_k \frac{1}{M_k} \left\{ 2\left(\frac{\partial \psi_e}{\partial x_k} \frac{\partial \psi_{vr}}{\partial x_k} + \frac{\partial \psi_e}{\partial y_k} \frac{\partial \psi_{vr}}{\partial y_k} + \frac{\partial \psi_e}{\partial z_k} \frac{\partial \psi_{vr}}{\partial z_k} \right) + \psi_{vr} \left[\frac{\partial^2 \psi_e}{\partial x_k^2} + \frac{\partial^2 \psi_e}{\partial y_k^2} + \frac{\partial^2 \psi_e}{\partial z_k^2} \right] \right\} \qquad (1.8)$$

is small in comparison with Eq. (1.5) and Eq. (1.6). This condition usually is satisfied.

Eq. (1.6) is the Schrodinger equation describing the behavior of the motion of the nuclei under the action of the potential $E^{el}+V_n$. Two nuclei of the diatomic molecule can accomplish motions of two modes, i.e., the rotation around the axis passing through the mass center of two nuclei and perpendicular to the line joining the nuclei (internuclear axis) and the vibration along the internuclear axis. Correspondingly, under the first approximation the eigenfunction ψ_{vr} can be expressed as

$$\psi_{vr} = \frac{1}{r} \psi_v \psi_r, \qquad (1.9)$$

where ψ_v is the *eigenfunction of the linear harmonic oscillator*, dependent only on $(r-r_e)$, r being the instantaneous distance between nuclei, r_e being its value at equilibrium; ψ_r is the *eigenfunction of the rigid rotator*, when the internuclear distance retains the value at equilibrium, and ψ_r only depends on the orientation of the molecule in space, $\psi_r = \psi_r(\vartheta,\varphi)$, where ϑ,φ being the zenith angle and the azimuth angle. In the following the harmonic oscillator and the rigid rotator will be discussed successively.

Harmonic oscillator is referred to as a mass point acted upon by a force F proportional to the distance x from the equilibrium location and directed to the equilibrium location

$$F = -kx, \tag{1.10}$$

where k is the force constant, or, equivalently, the potential V can be expressed as

$$V = \frac{1}{2}kx^2. \tag{1.11}$$

That is, the harmonic oscillator model uses parabola to approximate the potential energy curve. The harmonic oscillation of a diatomic molecule with the masses M_1 and M_2 of two atoms can be reduced to the motion deviating from the equilibrium location of a single mass point with the mass μ

$$\mu = \frac{M_1 M_2}{M_1 + M_2} \tag{1.12}$$

called as the *reduced mass*. The classic mechanics gives

$$\mu \frac{d^2 x}{dt^2} = -kx. \tag{1.13}$$

The vibrational frequency is obtained as

$$v = \frac{1}{2\pi}\sqrt{\frac{k}{\mu}}. \tag{1.14}$$

The wave mechanics description of the harmonic oscillation of the two nuclei of a diatomic molecule is given by the one-dimensional Schrodinger equation

$$\frac{d^2 \psi_v}{dx^2} + \frac{8\pi^2 \mu}{h^2}\left(E - \frac{1}{2}kx^2\right)\psi_v = 0. \tag{1.15}$$

Making use of Eq. (1.14) and introducing the notations $\lambda = 8\pi^2 \mu E / h^2$, $\alpha = 4\pi^2 \mu v / h$, and $\xi = \sqrt{\alpha} x$, the above equation can be written as

$$\frac{d^2\psi_v}{d\xi^2} + (\lambda/\alpha - \xi^2)\psi_v = 0. \tag{1.15}'$$

The solution of this equation can be written in the following form

$$\psi_v = Ce^{\xi^2/2}\frac{d^n}{d\xi^n}(e^{-\xi^2}), n = 0,1,2,\cdots.$$

The second differentiation gives

$$\frac{d^2\psi_v}{d\xi^2} = Ce^{\xi^2/2}$$

$$\left\{(\xi^2+1)\frac{d^n}{d\xi^n}(e^{-\xi^2}) + 2\xi\frac{d^{n+1}}{d\xi^{n+1}}(e^{-\xi^2}) - 2\frac{d^{n+1}}{d\xi^{n+1}}(\xi e^{-\xi^2})\right\}.$$

Applying to the last term the *Leibnitz formula* for the n th order derivative of the production of functions

$$(uv)^{(n)} = \sum_{i=0}^{n} C_n^i u^{(i)} v^{(n-i)}$$

and substituting ψ_v and $d^2\psi_v/d\xi^2$ into Eq. (1.15), one sees that when

$$\lambda = \left(n+\frac{1}{2}\right)2\alpha,$$

the equation is satisfied. Thus the solutions of Eq. (1.15) that are single-valued, finite, continuous and vanishing at ∞ are possible only when the values E of the *energy of the oscillator* take the following discrete eigenvalues

$$\varepsilon_{v,n} = \left(n+\frac{1}{2}\right)h\nu, \quad n = 0,1,2,3,\cdots \tag{1.16}$$

i. e., the quantized (discrete) energy levels of the harmonic oscillator have equidistance spacing, n is the *vibrational quantum number*, giving the ordinal number of the energy levels.

The *rigid rotator* or the dumbbell model is the simplest model to represent the rotation of a diatomic molecule: two mass points of masses M_1 and M_2 are fastened at the ends of a weightless rigid rod (the length r being kept unchanged). The *moment of inertia* of the system is

$$I = M_1 r_1^2 + M_2 r_2^2,$$

where

$$r_1 = \frac{M_2}{M_1 + M_2} r, \quad r_2 = \frac{M_1}{M_1 + M_2} r,$$

or when expressed by μ (Eq. (1.12))

$$I = \mu r^2. \tag{1.17}$$

There is a relation between the energy $E = (1/2)I\omega^2$ and *angular momentum* $P = I\omega$ for the rotator in classical mechanics

$$E = \frac{P^2}{2I}. \tag{1.18}$$

As the rotator is rigid, so $V = 0$, and the Schrodinger equation has the form

$$\frac{\partial^2 \psi_r}{\partial x^2} + \frac{\partial^2 \psi_r}{\partial y^2} + \frac{\partial^2 \psi_r}{\partial z^2} + \frac{8\pi^2 \mu}{h^2} E \psi_r = 0 \tag{1.19}$$

Quantum mechanics states, that the solutions ψ_r of Eq. (1.19) that are single-valued, finite, continuous and vanishing at ∞ are possible only when the values E of the *energy of the rigid rotator* take the following discrete eigenvalues

$$\varepsilon_{r,J} = \frac{h^2}{8\pi^2 I} J(J+1), \quad J = 0,1,2,\cdots \tag{1.20}$$

i. e., the energy levels of the rotator increase in a quadric law (the spacing of the energy levels increases as an arithmetic progression), J is the *rotational quantum number*. The corresponding eigenfunctions are the surface harmonic functions

$$\psi_r = N_r P_J^{|M|}(\cos\theta) e^{iM\varphi}, \tag{1.21}$$

where N_r is the normalizing constant, $P_J^M(\cos\theta)$ is the associated Legendre function

$$P_J^M(\xi) = \frac{1}{2^J J!} (1-\xi^2)^{\frac{M}{2}} \frac{d^{J+M}}{d\xi^{J+M}} (\xi^2 - 1)^J.$$

M is the *second quantum number* and can take $(2J+1)$ different values

$$M = J, (J-1), (J-2), \cdots, -(J-1), -J, \tag{1.22}$$

i. e., corresponding to one eigenvalue of the rotation (see Eq. (1.20)) there are $(2J+1)$ eigenfunctions (see Eq. (1.21)). The probability of appearance of the particle in the orientation θ, φ is

$$\psi_r \psi_r^* = N_r^2 \left[P_J^{|M|}(\cos\theta) \right]^2, \tag{1.23}$$

which is independent of φ. That is, the probability distribution is axisymmetric, the θ dependent distributions are presented by the solid lines in Fig. 1.1. The situation that a particle with the same energy can have different states is called *degenerate state*, the number of linearly independent eigenfunctions corresponding to one eigenvalue is called *degeneracy*. Thus, the rigid rotator having a rotational quantum number J is in the degenerate state with degeneracy of $(2J+1)$

$$g_J = 2J+1. \tag{1.24}$$

This corresponds to the *space quantization* of the angular momentum vector. From Eq. (1.18) it is seen, that when E takes $\varepsilon_{r,J}$ (see Eq. (1.20)), the angular momentum is

$$P = \frac{h}{2\pi}\sqrt{J(J+1)} \approx J\frac{h}{2\pi},$$

i.e., $(h/2\pi)$ appears as the unit of angular momenta, and the angular momentum is approximately J times of $(h/2\pi)$. If M represents the angular momentum vector of value P, then according to the quantum mechanics, the projection of M

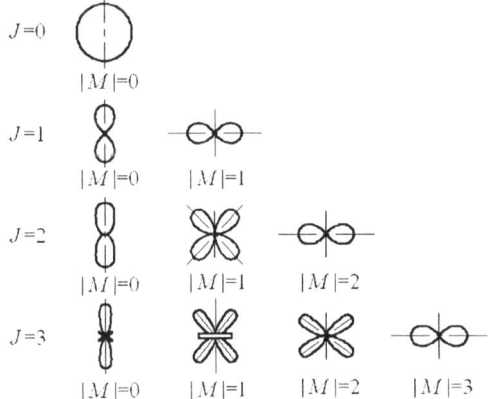

Fig. 1.1 The probability distributions of the rigid rotator for the rotational quantum numbers $J = 0, 1, 2, 3$

on any direction (for example the electric or magnetic field) can only be the integer times of $(h/2\pi)$, that is $M_J \cdot (h/2\pi)$, where

$$M_J = J, (J-1), (J-2), \cdots, -J. \tag{1.25}$$

That is, M can only form certain angles with the given direction. Fig1.2 shows the space quantization of the angular momentum when $J = 1, 2, 3$.

In the above discussion we have divided the molecular energy into three modes: electronic, vibrational and rotational, leaving the electronic energy not defined exactly. Generally, the *electronic energy* is defined as the minimum of $E^{el} + V_n$, i. e. the minimum of the potential energy function of a given steady electronic state, and is denoted by ε_e. In the first approximation the *vibratioinal energy* ε_v and the *rotational energy* ε_r are given by the Eqs. (1.16) and (1.20). Now let us estimate the relative order of magnitude of the three modes of energies. Use d to denote the linear scale of the molecule. The electronic energy ε_e is associated with the motion of the electron with mass m in the molecular scope d, the order of magnitude of the electron velocity v, according to the Heisenberg uncertainty principle (see Eq. (0.22)), can be estimated as

$$v \sim \frac{h}{md}.$$

The order of magnitude of ε_e is mv^2, that is

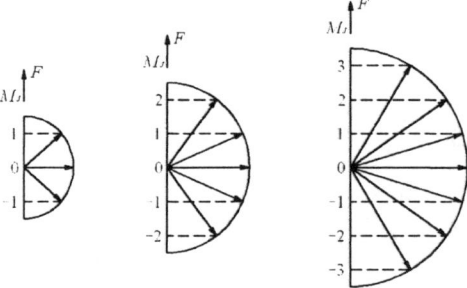

Fig. 1.2 Space quantization of the angular momentum vector of value $h\sqrt{J(J+1)}/2\pi$ in the field F

$$\varepsilon_e \sim \frac{h^2}{md^2}. \tag{1.26}$$

The order of magnitude of the vibrational energy ε_v, according to Eq. (1.16) and Eq. (1.14), is $h(k/M)^{1/2}$, here M is the mass of the nucleus. The order of the force constant k can be obtained in this way: the change of the internuclear distance by a value of d would lead to the distortion of the form of the electronic wave, and the change of the potential energy is $\sim kd^2$, the order of it should be correspond to the electronic energy ε_e. So we have

$$k \sim \varepsilon_e / d^2,$$

from where the estimate of ε_v is obtained

$$\varepsilon_v \sim h(k/M)^{1/2} \sim \frac{h\sqrt{\varepsilon_e}}{\sqrt{M}d} \sim \sqrt{\frac{m}{M}}\varepsilon_e. \tag{1.27}$$

The order of magnitude of the rotational energy ε_r according to Eq. (1.20) and Eq. (1.17) is

$$\varepsilon_r \sim \frac{h^2}{Md^2} \sim \frac{m}{M}\varepsilon_e. \tag{1.28}$$

As the mass M of the nucleus is four orders of magnitude larger than the mass m of the electron, $M/m \sim 10^4$, according to Eq. (1.27) and Eq. (1.28), one can ascertain, that ε_v is two orders of magnitude smaller than ε_e, and ε_r is another two orders of magnitudes smaller than ε_v.

Fig 1.3 is the diagram of two electronic energy levels with their vibrational and rotational energy levels. A and B are the electronic energy levels; $n_A, n_B = 0,1,2,3,\cdots$, are several first vibrational energy levels, $J_A, J_B = 0,1,2,3,\cdots$ are several first rotational energy levels. Any energy level in the figure represents the energy of the molecule of energy ε, and corresponds to certain electronic energy level, certain vibrational quantum number and certain rotational quantum number. ε can be presented as the sum of the electronic energy, the vibrational energy and the rotational energy

$$\varepsilon = \varepsilon_e + \varepsilon_v + \varepsilon_r. \tag{1.29}$$

30   1 MOLECULAR STRUCTURE AND ENERGY STATES

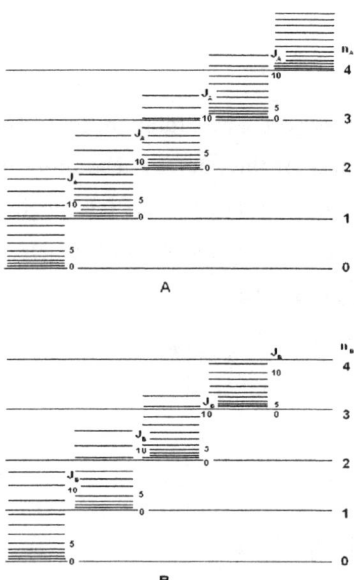

Fig. 1.3 Two electronic levels A and B with vibrational energy levels $n_A, n_B = 0,1,2,3,\cdots$ and rotational energy levels $J_A, J_B = 0,1,2,3,\cdots$

This certainly is an approximation, for there are mutual interactions between three energy modes, and the vibrational mode can adopt the anharmonic oscillator model with more realistic potential energy curve (to modify the parabola adopted by the harmonic oscillator model), the rotational mode also can be improved by the non-rigid rotator model. The vibrating rotator model (or the rotating oscillator model) taking into account simultaneous rotation and vibration gives results in better agreement with the observed fine structures of the spectra.

1.2 ENERGY DISTRIBUTION OF MOLECULES

In this section the methods of *statistical mechanics* are introduced to calculate the equilibrium energy distribution of molecules. The statistical mechanics gets cognition of the macroscopic properties of matter basing on the staring point that matter is composed of tiny particles without penetrating into the mutual interaction be-

1.2 ENERGY DISTRIBUTION OF MOLECULES

tween the particles, so the results of statistical mechanics are applicable not only to gas molecules.

Consider a system composed of N particle with a total energy of E in the volume V, each particle can have only discrete energies $\varepsilon_1, \varepsilon_2, \varepsilon_3, \cdots, \varepsilon_j, \cdots$, the degeneracies of them are $g_1, g_2, g_3, \cdots, g_j, \cdots$. At a specific instant the numbers of molecules occupying corresponding energy levels are $N_1, N_2, N_3, \cdots, N_j, \cdots$

$$N = \sum_j N_j, \tag{1.30}$$

$$E = \sum_j N_j \varepsilon_j. \tag{1.31}$$

The problem we are concerned is: what is the most probable state of the distribution of the energy over various energy levels? Or, in another formulation, what portion in the whole number N of particles does the number N_i of molecules possessing energy level ε_i constitute? To answer this question, the methods of statistical mechanics are to be used, a bridging relation connecting statistical mechanics with thermodynamics – the Boltzmann relation – is needed, and an even more fundamental question must be answered: i.e., what in the number Ω of different possible ways to distribute the N particles over certain allowable energy levels $\varepsilon_1, \varepsilon_2, \varepsilon_3, \cdots, \varepsilon_j, \cdots$, provided that they obey Eq. (1.30) and Eq. (1.31). If any possible distribution of the particles over the allowable energy levels in consistence with the macroscopic state $(N, V$ and $E)$ of the system is defined as a *microscopic state* of the system, then the latter question is reduced to: for a given macroscopic state what is the *number Ω of the microscopic states*? In the following we start the discussion of statistical mechanics from the Boltzmann relation, then we get the total number of microstates from the Bose-Einstein statistics and the Fermi-Dirac statistics, and give the answer to our question, i. e., how are the energies distributed over the energy levels. About the contents of this section readers are referred to reference [2]. It is noted that the statistical mechanics is not concerned with the nature of the particles and the interaction between them, so the results are not limited to gas molecules. The exposition of this section is not an overall discussion of the statistical mechanics, the main concern here is only the problem posed in the beginning of the section.

1.2.1 BOLTZMANN'S RELATION

Boltzmann's relation is an important relation connecting statistical mechanics and thermodynamics, it presents *entropy S* of thermodynamics through the quantity Ω of statistical mechanics

$$S = k \ln \Omega, \tag{1.32}$$

where Ω is the number of ways in which the N particles are distributed over the energy levels $\varepsilon_1, \varepsilon_2, \varepsilon_3, \cdots, \varepsilon_j, \cdots$, as defined above, or, it is the number of possible microstates of the system, and obviously it represents the "randomness" of the system. Boltzmann's relation is a fundamental law of physics, its formation from hypothesis into theory is determined by the verification in the practice and the success of its theoretical system. Here the elucidation but not the proof of this relation is given. To show the increase of the entropy with the randomness quantitatively the example of perfect gas is used, but this is done only with the purpose of illustration. The validity of the relation is not limited to the case of perfect gas. The definition of entropy S is given by the following differential relationship

$$dS = \frac{dQ}{T}, \tag{1.33}$$

where dQ is the heat received by the system, and according to the first law of thermodynamics

$$dQ = dE + pdV, \tag{1.34}$$

where E is the internal energy of the system; p is pressure. Substituting Eq. (1.34) into Eq. (1.33) yields

$$dS = \frac{1}{T}(dE + pdV). \tag{1.35}$$

1.2 ENERGY DISTRIBUTION OF MOLECULES

On the example of perfect gas after integration (and by utilizing the well known thermodynamics relations) one obtains

$$S = \frac{R}{\gamma-1}\ln\frac{p}{\rho^\gamma} + \text{const}, \qquad (1.36)$$

or

$$S = \frac{\gamma R}{\gamma-1}\ln T - R\ln p + \text{const}, \qquad (1.37)$$

where R is the gas constant; γ is the specific heat ratio.

Now we cite two examples when in a closed system (the number N of particles remains the same) the entropy increases and, correspondingly, the degree of the randomness or the disorder of the system increases. First consider an example of perfect gas, the temperature remains constant and the volume increases to V_2, say to two times of the original volume V_1, and accordingly the pressure becomes from p_1 to $p_2 = (1/2)p_1$. Then the entropy of the system increases: due to the isothermal condition, from Eq. (1.37) the entropy increase is obtained

$$S_2 - S_1 = R\ln(p_1/p_2) = (\ln 2)R.$$

As the volume is increased, it is possible to find the gas particles in larger volume, the degree of uncertainty of the particles increases, correspondingly the degree of disorder or randomness is also increased. Now consider the reversible heating of a system of perfect gas having constant volume. This time due to the constancy of the volume, we have $p_2/p_1 = T_2/T_1$, from Eq. (1.37) it is seen, that with the increase of the temperature, the entropy increases. The constancy in the volume keeps the degree of uncertainty of the gas molecules in position constant, but the increase in the temperature means the increase of the speed of the thermal motion, i. e., the increase of the degree of uncertainty of the velocities of the molecules in the system, so the degree of the disorder also increases. Thus, if the number Ω of the microstates is used to measure the degree of the disorder or the randomness of the system, one can write

$$S = f(\Omega), \qquad (1.38)$$

here f is an undetermined function of Ω.

If combine two systems with numbers N_1, N_2 of particles, entropies S_1, S_2, and numbers Ω_1, Ω_2 of microstates into one combined system, the entropy of the new system is $S_1 + S_2$, and the number of microsystems is $\Omega_1\Omega_2$. This specifies the condition that f should satisfy

$$f(\Omega_1\Omega_2) = f(\Omega_1) + f(\Omega_2). \tag{1.39}$$

Differentiating Eq. (1.39) first with respect to Ω_1 and then to Ω_2 yields

$$\Omega_1\Omega_2 f''(\Omega_1\Omega_2) + f'(\Omega_1\Omega_2) = 0,$$

or, by changing the argument,

$$\Omega f''(\Omega) + f'(\Omega) = 0.$$

The general solution of it is

$$f(\Omega) = A \ln\Omega + B. \tag{1.40}$$

By substituting Eq. (1.40) into Eq. (1.39) one can prove that A is an absolute constant independent of the number N of the particles of the system, denote it by k, i. e., the Boltzmann constant. The constant B can be taken as $B = 0$, which means that the entropy of a completely ordered system ($\Omega = 1$) is defined to be 0. So one obtains

$$S = k \ln\Omega,$$

i. e., the Boltzmann relation (1.32) follows.

1.2.2 CALCULATION OF THE NUMBER Ω OF MICROSCOPIC STATES

The task of calculation of the number Ω of microstates is decomposed into three steps. The first step is simple and concrete, i. e., to find, how many ways are there to distribute N_j particles over the energy states g_j of energy ε_j? Or, pointing at the more general cases, to find, how many ways are there to distribute N_j particles over the energy range of small energy extents ($\delta\varepsilon_j$) near ε_j (the average energy of this range) including a number g_j of energy states? The second step is, relative to a certain combination of N_j to sum these known numbers of ways of distribution for different energy ranges ε_j, to obtain the total number of all possible microstates for a certain combination of N_j in accordance with the macroscopic state. The third step is to sum up the results of the second task for all possible combinations of N_j, to obtain the total number of microscopic states of the system.

For the accomplishment of the first step, according to the difference in the quantum properties of the particles, different statistics, i. e., *Bose-Einstein statistics* or *Fermi-Dirac statistics,* must be employed. Bose-Einstein statistics applies to *bosons,* i. e., atoms or molecules composed of *even* number of elementary particles (electrons, neutrons, electrons). For bosons, any degenerated state of energy level ε_j can be occupied by any number of particles. Fermi-Dirac statistics applies to *fermions,* i. e., atoms or molecules composed of *odd* number of elementary particles. Fermions obey the *Pauli exclusion principle*, i. e., one quantum energy state can be occupied only by one particle.

The first task (to find, how many ways are there to distribute N_j undistinguishable particles over specified energy states g_j) is mathematically equivalent to the question of, how many ways are there to distribute N_j undistinguishable particles into g_j marked containers? Or, if use the range between two neighboring vertical bars aligned in one raw to represent a container (the space to the left of the first bar and the space to the right of the last bar also represent two containers), the task is reduced to: how many ways of arrangement are there to line up N_j particles and $(g_j - 1)$ bars?

In Bose-Einstein statistics there is not restriction to the number of particles between the neighboring bars. At first assume artificially that the N_j particles and $(g_j - 1)$ bars can be distinguished from each other (having individual properties), to line them up there are

$$(N_j + g_j - 1)!$$

different ways of arrangement (permutation of $(N_j + g_j - 1)$ elements). But the assumption that the particles are distinguished from each other is incorrect, it increased artificially the number of ways of arrangement by $N_j!$ too many times, similarly, the distinguishing of (g_j-1) bars also increased artificially the number of ways of arrangement by $(g_j-1)!$ too many times. So

$$\frac{(N_j+g_j-1)!}{N_j!(g_j-1)!} \tag{1.41}$$

is the number of microscopic states when distributing N_j undistinguishable particles into g_j different energy states of the energy group ε_j with no restriction to the number of particles in every energy state (Bose-Einstein statistics).

In the Fermi-Dirac statistics, at most only one particle can be put into each container, so $N_j \leq g_j$. At first assume artificially that the N_j particles can be distinguished from each other, the number of ways to put them into g_j containers (at most one particle into one container) is obviously

$$g_j(g_j-1)\cdots(g_j-N_j+1),$$

for the first particle has the choice of g_j containers, the second one has the choice of (g_j-1) containers, .., the N_jth particle has the choice of (g_j-N_j+1) containers. The above formula can be rewritten as

$$g_j!/(g_j-N_j)!.$$

But in reality the N_j particles are undistinguishable, so the number of arrangement has been increased by $N_j!$ times too many. So

$$g_j!/\left[(g_j-N_j)!N_j!\right] \tag{1.42}$$

is the number of microscopic states (Fermi-Dirac statistics), when distributing N_j undistinguishable particles into g_j different energy states of the energy group, when in every energy state there can be at most only one particle (Pauli principle).

Now we proceed to the second task. For the whole macroscopic system, the numbers of ways of arrangement in various energy levels (or energy ranges with intervals $(\delta\varepsilon_i)_j$ having ε_j as average values) are independent of each other. For

1.2 ENERGY DISTRIBUTION OF MOLECULES

certain combination of N_j satisfying Eq. (1.30), the number of ways of the energy arrangement of the macroscopic system is the continued multiplication of the numbers of microstates of various energy levels. From Eq. (1.41) and Eq. (1.42), one obtains the total number of microstates under certain combination of N_j, for Bose-Einstein statistics and Fermi-Dirac statistics respectively

$$W_{BE}(N_j) = \prod_j \frac{(N_j + g_j - 1)!}{N_j!(g_j - 1)!}, \qquad (1.43)$$

$$W_{FD}(N_j) = \prod_j \frac{g_j!}{(g_j - N_j)! N_j!}. \qquad (1.44)$$

The third task is to sum over all possible macroscopic states, that is, to sum up over all combinations of N_j to obtain the total number of the microstates of the system

$$\Omega = \sum_{N_j \text{ satisfy Eq. (1.30) and Eq. (1.31)}} W(N_j). \qquad (1.45)$$

It seemed very difficult to carry the calculation of the sum of Eq. (1.45). Fortunately, closed result can be obtained when $N \to \infty$, and the N for systems we are treating are as a rule very large. Under such circumstances in this sum only the maximum term W_{max} has the most important contribution to Ω. To be concrete, a proof can be given (see at the end of this section) to show, that when in Ω there are N terms important and can be presented as NW_{max} and other terms can be neglected, the following formula is valid

$$\ln \Omega / \ln W_{max}\Big|_{N \to \infty} \to 1. \qquad (1.46)$$

So, when utilizing Boltzmann's relation $\ln W_{max}$ can be used instead of $\ln \Omega$.

1.2.3 BOLTZMANN DISTRIBUTION

The number W of microstates counted in the last subsection gains the maximum value W_{max} under some specific combination of N_j. This state is defined as the most probable state. Now consider the condition under which $\ln W$ gains the maximum. From Eq. (1.43) and Eq. (1.44) one has

$$\ln W_{BE} = \sum_j \left[\ln(N_j + g_j - 1)! - \ln(g_j - 1)! - \ln N_j! \right], \qquad (1.47)$$

$$\ln W_{FD} = \sum_j \left[\ln g_j! - \ln(g_j - N_j)! - \ln N_j! \right]. \qquad (1.48)$$

With the help of *Stirling formula*, when $x \to \infty$

$$\ln x! \approx x \ln x - x, \qquad (1.49)$$

Eq. (1.47) and Eq. (1.48) can be written

$$\ln W = \sum_j \left[N_j \ln\left(\frac{g_j}{N_j} \pm 1\right) \pm g_j \ln\left(1 \pm \frac{N_j}{g_j}\right) \right]. \qquad (1.50)$$

The condition under which $\ln W$ gains maximum is

$$d\ln W = \sum \frac{\partial \ln W}{\partial N_j} dN_j = 0.$$

After using Eq. (1.50) this condition becomes

$$\sum_j \ln\left(\frac{g_j}{N_j} \pm 1\right) dN_j = 0. \qquad (1.51)$$

From Eq. (1.30) and Eq. (1.31) two physical conditions that must be satisfied are obtained

$$\sum dN_j = 0, \qquad (1.52)$$

$$\sum \varepsilon_j dN_j = 0. \qquad (1.53)$$

Utilizing the Lagrange's method of undetermined coefficients, multiplying Eq. (1.52) and Eq. (1.53) by α and β respectively and combining with Eq. (1.51) yields

$$\sum_j \left[\ln\left(\frac{g_j}{N_j} \pm 1\right) - \alpha - \beta\varepsilon_j \right] dN_j = 0. \qquad (1.54)$$

1.2 ENERGY DISTRIBUTION OF MOLECULES

Let the coefficient of dN_j to be 0, the relation which N_j must satisfy is obtained

$$\ln\left(\frac{g_j}{N_j} \pm 1\right) - \alpha - \beta \varepsilon_j = 0.$$

From this relation the specified value of N_j that makes $\ln W$ the maximum $\ln W_{max}$, or the value of N_j in the most probable macroscopic state, is

$$\frac{N_j}{g_j} = \frac{1}{e^{\alpha + \beta \varepsilon_j} \mp 1}. \tag{1.55}$$

From the discussion of the concrete energy levels of diatomic molecules in section 1.1 it is seen, that the spacing between the energy levels is extremely small, the particles are sparsely distributed among the allowable energy states, that is

$$g_j \gg N_j \tag{1.56}$$

Under such circumstances the factor 1 in the denominator of Eq. (1.55) can be omitted, and identical result is obtained for both Bose-Einstein and Fermi-Dirac statistics

$$\frac{N_j^*}{g_j} = \frac{1}{e^{\alpha + \beta \varepsilon_j}}. \tag{1.57}$$

This limiting result is called the *Boltzmann limit* and can be obtained utilizing the *Boltzmann statistics*, or the classical statistics. In the Introduction it is shown, that if the following equation (Eq. (0.23))

$$(3mkT)^{1/2} / (n^{1/3}h) \gg 1 \tag{1.58}$$

is satisfied, the quantum effects are not remarkable, and the classical methods can be used to treat the motion of particles. It has also been shown there, that Eq. (1.58) usually is satisfied in situations that are going to be discussed by us. This shows, that in the problems concerned by us the classical statistics is correct, that is, the limiting Eq. (1.57) is valid, and hence the condition Eq. (1.56) is valid.

Substituting Eq. (1.57) into the closure condition Eq. (1.30) and the energy condition Eq. (1.31) leads to the equations for determining α and β

$$\sum \frac{g_j}{e^{\alpha+\beta\varepsilon_j}} = N, \quad \sum \frac{g_j\varepsilon_j}{e^{\alpha+\beta\varepsilon_j}} = E. \tag{1.59}$$

From the first equation of Eq. (1.59) α is easily determined

$$e^{-\alpha} = \frac{N}{\sum_j g_j e^{-\beta\varepsilon_j}}. \tag{1.60}$$

Substituting it into Eq. (1.57), one obtains the most probable distribution of particles over various energy levels

$$N_j^* = N\frac{g_j e^{-\beta\varepsilon_j}}{\sum_j g_j e^{-\beta\varepsilon_j}}. \tag{1.61}$$

Substitution of Eq. (1.60) into the second equation of Eqs. (1.59) yields

$$N\frac{\sum \varepsilon_j g_j e^{-\beta\varepsilon_j}}{\sum_j g_j e^{-\beta\varepsilon_j}} = E. \tag{1.62}$$

To determine β we need to apply Boltzmann's relation and in the calculation use $\ln W_{max}$ instead of $\ln \Omega$ (*see* Eq. (1.46)). For this reason we write $\ln W$ (see Eq. (1.50)) under condition $g_j \gg N_j$ (Eq. (1.56)) as

$$\ln W = \sum \left(N_j \ln \frac{g_j}{N_j} + N_j \right) = \sum N_j \left(\ln \frac{g_j}{N_j} + 1 \right). \tag{1.63}$$

Substituting the value of N_j^* into the above equation, one obtains $\ln W_{max}$

$$\ln W_{max} = \sum N_j \left[\ln \frac{\left[\left(\sum g_j e^{-\beta\varepsilon_j}\right) \cdot e^{\beta\varepsilon_j}\right]}{N} + 1 \right] =$$

$$\sum N_j \left[\ln \frac{\sum_j g_j e^{-\beta\varepsilon_j}}{N} + \beta\varepsilon_j + 1 \right] =$$

$$N \left[\ln \frac{\sum g_j e^{-\beta\varepsilon_j}}{N} + 1 \right] + \beta E.$$

1.2 ENERGY DISTRIBUTION OF MOLECULES

Owing to Eq. (1.46), one can replace $\ln\Omega$ by the above expression. Thus Boltzmann's relation yields

$$S = k\left[N\left[\ln\frac{\sum g_j e^{-\beta\varepsilon_j}}{N}+1\right]+\beta E\right]. \qquad (1.64)$$

From Eq. (1.35) we have

$$\left(\frac{\partial S}{\partial E}\right)_{V,N} = \frac{1}{T}. \qquad (1.65)$$

Now differentiating Eq. (1.64), noting that β is a function of E (see Eq. (1.62)), gives

$$\left(\frac{\partial S}{\partial E}\right)_{V,N} = k\cdot\left\{\beta+\frac{\partial}{\partial\beta}\left[N\left[\ln\frac{\sum g_j e^{-\beta\varepsilon_j}}{N}+1\right]+\beta E\right]\frac{\partial\beta}{\partial E}\right\}.$$

With the help of Eq. (1.62) it is easily seen, that the partial derivative with respect to β in the above expression is actually 0, i. e.

$$\frac{\partial}{\partial\beta}\left[N\left[\ln\frac{\sum g_j e^{-\beta\varepsilon_j}}{N}+1\right]+\beta E\right] = N\left[\frac{-\sum \varepsilon_j g_j e^{-\beta\varepsilon_j}}{\sum g_j e^{-\beta\varepsilon_j}}\right]+E = -N\frac{E}{N}+E = 0.$$

Thus we have

$$\left(\frac{\partial S}{\partial E}\right)_{V,N} = k\beta \qquad (1.66)$$

Comparison of Eq.(1.64) and Eq. (1.65) gives

$$\beta = \frac{1}{kT}. \qquad (1.67)$$

Consequently, the most probable distribution of energy over different energy levels is (see Eq. (1.61))

$$N_j^* = N\frac{g_j e^{-\varepsilon_j/kT}}{\sum_i g_j e^{-\varepsilon_j/kT}} \qquad (1.68)$$

This is the renowned *Boltzmann distribution*. The sum in the denominator is called the *partition function*

$$Q = \sum_j g_j e^{-\varepsilon_j/kT} . \tag{1.69}$$

This result is obtained in the understanding that ε_j is an average value of the energy range of interval $(\delta \varepsilon_i)_j$ including a number g_j of energy states. If now consider the case in which the energy values ε_i of different energy states in the energy ranges are all different, the partition function can be expressed as

$$Q = \sum_i e^{-\varepsilon_i/kT} , \tag{1.70}$$

where $\sum_i$ means summation over all energy states. After introducing Q the Boltzmann distribution can be written as

$$N_j^* = N g_j e^{-\varepsilon_j/kT} / Q \tag{1.68'}$$

Note that after the determination of β and the introduction of partition function Q the energy E of the system can be expressed as

$$E = \frac{N}{Q} \sum \varepsilon_j g_j e^{-\varepsilon_j/kT} . \tag{1.71}$$

Owing to the limitation in space the following fact is not expounded and proved mathematically here, i. e., the number W of microstates as a function of N_j is very steep in the vicinity of N_j^*. It is namely the values of W corresponding to those N_j that are very near to N_j^* make the main contribution to Ω (see Eq. (1.45)). That is, when there is even though only very small deviation of particle distribution N_j from the most probable distribution N_j^*, the number W of the microstates would decrease drastically. Besides, we are standing on the foundation of the *ergodic principle* that constitutes the basic starting point of statistical mechanics, that is, all possible microscopic states have the same probability. Starting from these two points we inevitably come to the conclusion: the system defined in the beginning of this section, composed of N particles and having total energy E, most time is necessarily in the macroscopic state satisfying Boltzmann distribution or in the macroscopic state that deviates extremely little from it. That is, Boltzmann distribution Eq. (1.68) is the most probable distribution of energies.

Now we proceed to prove, that when in the total number Ω (expressed by Eq. (1.45)) of microstates of the system there are N terms important and can be presented as NW_{max}, while other terms can be neglected, the formula Eq. (1.46) is valid. According to this condition, we have

$$\frac{\ln \Omega}{\ln W_{max}} = 1 + \frac{\ln N}{\ln W_{max}},$$

where $\ln W_{max}$ is expressed by Eq. (1.64), then on the basis of Eq. (1.69) it can be written as

$$\ln W_{max} = N(\ln Q - \ln N + 1) + \beta E.$$

Because $E = \sum N_j E_j$ should be larger than the energy when only the translational energy is fully excited, $E > (3/2) NkT$. As Q and β are independent of N, so we have Eq. (1.46).

1.3 INTERNAL ENERGY DISTRIBUTION FUNCTIONS

In section 1.1 it is shown, that when the interaction between the various energy modes are neglected, the internal energy of molecules can be presented as the sum of electronic energy ε_e, vibrational energy ε_v and rotational energy ε_r, if add the kinetic energy ε_{tr} of the motion of molecules, the energy of a molecule can be presented as

$$\varepsilon = \varepsilon_e + \varepsilon_v + \varepsilon_r + \varepsilon_{tr}. \tag{1.72}$$

Correspondingly, it is seen from the form of the expression Eq. (1.70) of the partition function Q, that the complete partition function has the factorization property. i. e., it can be expressed as the product of the partition functions of the different energy modes

$$Q = Q_e Q_v Q_r Q_{tr} \tag{1.73}$$

44   1 MOLECULAR STRUCTURE AND ENERGY STATES

Although the electronic energy of a molecule is relatively large in the order of magnitude, but under not too high temperature the electronic state usually is not excited, for the spacing between the lowest energy level and the second lowest energy level is much larger than kT. For N_2, O_2, No, N and O, when the basic state energy ε_0 is taken as 0, the partition functions of the electronic state (see Eq. (1.69)) can be written

$$\begin{cases} N_2: & Q_e = 1 & + O(e^{-10000/T}), \\ O_2: & Q_e = 3 + 2e^{-1139/T} & + O(e^{-19000/T}), \\ NO: & Q_e = 2 + 2e^{-174/T} & + O(e^{-65000/T}), \\ N: & Q_e = 4 & + O(e^{-28000/T}), \\ O: & Q_e = 5 + 3e^{-228/T} + e^{-326/T} & + O(e^{-23000/T}), \end{cases} \quad (1.74)$$

where the last terms of the right hand sides represent the orders of magnitude of the neglected terms. These are the results obtained from the spectra data [3] of the transition between different energy levels of the 5 components. For N_2 and N, the electronic partition functions Q_e are approximately constants. This is true in some degree also for O_2 when the temperature is not too high (even though the second term has been retained in the expression of Eq. (1.74)). For O and NO, under relatively high temperature ($T \approx 2000K \sim 3000K$) of practical production of O and NO, the exponential functions $e^{-\Theta/T}$ in the second and the third terms of the third and fifth expressions of Eq. (1.74) are approximately 1, so Q_e are also approximately constants. At higher temperature, for O_2, the error caused by neglect of $e^{-\Theta/T}$ has little consequences owing to the smallness of the concentration of O_2 under such temperatures. Thus, the electronic partition functions of the above 5 components can be approximately taken as constants. Due to the fact that the value of energy (and also the specific energy and specific heat) is related to the derivative of Q (see Eq. (1.75) below) the contribution of the electronic state to the energy and specific heat usually can be neglected. Under higher temperature, electronic state may transit to higher energy level, and can radiate when transiting from higher energy level to lower energy level, the contribution of the electronic state is to be taken into account.

1.3 INTERNAL ENERGY DISTRIBUTION FUNCTIONS 45

For the kinetic energy of the molecular motion, the methods of the kinetic theory can be used to get detailed information by analyzing the velocity distribution function, as it will be done in the next chapter.

Thus, the internal energy discussed in this chapter is referred mainly to the vibrational energy ε_v and rotational energy ε_r. Now let us try to get the expressions for total vibrational energy E_v and total rotational energy E_r, and correspondingly, the vibrational energy $e_v = E_v/(Nm)$ per unit mass (the *specific vibrational energy*) and the rotational energy $e_r = E_r/(Nm)$ per unit mass (the *specific rotational energy*) from the expressions Eq. (1.16) of ε_v and Eq. (1.20) of ε_r for individual molecules. And finally we give the expression of the internal energy distribution function. In the present section m denotes the mass of the molecule.

To get the total energy, Eq. (1.71) is used. At first note, that differentiation of Eq. (1.69) with respect to T yields

$$\frac{\partial Q}{\partial T} = \frac{1}{kT^2} \sum \varepsilon_j g_j e^{-e_j/kT} .$$

Thus, substituting the above expression into Eq. (1.71) one obtains the following expression for the total energy

$$E = \frac{N}{Q} kT^2 \frac{\partial Q}{\partial T} = NkT^2 \frac{\partial \ln Q}{\partial T} . \tag{1.75}$$

Correspondingly, the energy per unit mass of the gas (the specific energy) can be written

$$e = \frac{E}{Nm} = \frac{k}{m} T^2 \frac{\partial \ln Q}{\partial T} = RT^2 \frac{\partial \ln Q}{\partial T} , \tag{1.76}$$

where $R = k/m$ is the gas constant per unit mass.

Note, e is calculated from the total energy E of the system, so it is an energy in the average sense, i. e., the specific energy is the mean energy in unit mass.

For the vibrational energy of molecule, we have Eq. (1.16)

$$\varepsilon_{v,n} = \left(n + \frac{1}{2}\right) h\nu , \quad n = 1,2,3,\cdots .$$

The zero point energy of a molecule (when $n=0$) is $\varepsilon_{v,0} = (1/2)h\nu$, but this value of $\varepsilon_{v,0}$ is nonessential, and is neglected here, for only the change of the state of the system is essential. Thus the allowable vibrational energy state can be written as

$$\varepsilon_{v,n} = nh\nu, \quad n = 0,1,2,3,\cdots \quad (1.77)$$

From Eq. (1.69) and Eq. (1.77) the vibrational partition function is known as

$$Q_v = \sum_{n=0}^{\infty} e^{-nh\nu/kT}.$$

If introduce the characteristic temperature Θ_v

$$\Theta_v = h\nu/k, \quad (1.78)$$

and put $x = e^{-h\nu/kT} \equiv e^{-\Theta_v/T}$, as $x < 1$, so Q_v can be written

$$Q_v = \sum_{n=0}^{\infty} x^n = 1 + x + x^2 + \cdots = \frac{1}{1-x} = \frac{1}{1-e^{-\Theta_v/T}}. \quad (1.79)$$

According to Eq. (1.76), the *specific vibrational energy* is obtained as

$$e_v = \frac{R\Theta_v}{e^{\Theta_v/T} - 1}. \quad (1.80)$$

For the rotational energy from Eqs. (1.69), Eq. (1.20) and Eq. (1.24), the partition function can be written

$$Q_r = \sum_{j=0}^{\infty}(2J+1)\exp\left[\frac{-J(J+1)h^2}{8\pi^2 IkT}\right]. \quad (1.81)$$

If introduce the rotational characteristic temperature Θ_r

$$\Theta_r = \frac{h^2}{8\pi^2 Ik} \quad (1.82)$$

and replace the summation over J by the integration with respect to J (that is lawful for very small Θ_r/T, and it is the case, as Θ_r is very small for N_2, NO and O_2, see table I.1 in the Appendix I)

$$Q_r = \int_0^{\infty}(2J+1)e^{-J(J+1)\Theta_r/T}dJ = \int_0^{\infty} e^{-(\Theta_r/T)x}dx,$$

one obtains

1.3 INTERNAL ENERGY DISTRIBUTION FUNCTIONS 47

$$Q_r = T/\Theta_r . \quad (1.83)$$

This result is valid only for heteronuclear molecules, i. e. diatomic molecules with two different atoms. For the homonuclear molecules such as N_2 and O_2, the result must be divided by 2. But this does not influence the calculation of e_r and $c_{v,r}$, for only derivative of $\ln Q_r$ appears in the calculation. According to Eq. (1.76) the *specific rotational energy* is obtained

$$e_r = RT \equiv \frac{k}{m}T . \quad (1.84)$$

In the discussion of kinetic theory of gases in Chapter 2, the specific translational energy, or the internal energy associated with the *translational energy per unit mass*, is obtained (see Eq. (2.31))

$$e_{tr} = \frac{3}{2}RT = \frac{3}{2}\frac{k}{m}T .$$

As the *translational energy* of the molecule has *three degrees of freedom*, it is seen that there is an energy of $(1/2)kT/m$ shared by each degree of freedom. The comparison with Eq. (1.84) shows, that the *rotational energy* is fully excited and has *two degrees of freedom*. This is in agreement with the result of classical mechanics. As to the vibrational energy, circumstances are different. If suppose the *vibrational energy* has ζ_v degrees of freedom, then according to the principle of equipartition of energies the specific vibratioinal energy can be written

$$e_v = \frac{\zeta_v}{2}RT \equiv \frac{\zeta_v}{2}\frac{k}{m}T . \quad (1.85)$$

From the comparison with Eq. (1.80), it is seen that the expression of ζ_v is given by

$$\zeta_v = \frac{2\Theta_v/T}{e^{\Theta_v/T}-1} . \quad (1.86)$$

Only at very high temperature $\zeta_v \to 2$. According to the expressions of the translational energy, the vibrational energy and the translational energy, Eq. (2.31), Eq. (1.80) and Eq. (1.84), the corresponding specific heats $c_v = (\partial e/\partial T)_v$ are

$$C_{v,r} = \frac{3}{2}R,$$

$$C_{v,r} = R,$$

$$C_{v,v} = R\left(\frac{\Theta_v}{T}\right)^2 \frac{e^{\Theta_v/T}}{\left(e^{\Theta_v/T}-1\right)^2}. \tag{1.87}$$

Thus it can be seen, the contribution of the translational motion and the rotation of molecules to the specific heat is $(5/2)R$, and the vibrational specific heat changes with temperature. This conclusion arrived at through quantum mechanics is verified by the experiments

Now we require the general expression of the *distribution function of the internal energy of molecules*. If the number of molecules dN, out of a system composed of N molecules, whose energy of certain mode lies between ε_i and $\varepsilon_i + d\varepsilon_i$, can be written as

$$dN/N = f(\varepsilon_i)d\varepsilon_i, \tag{1.88}$$

then $f(\varepsilon_i)$ is called the distribution function of that energy mode.

For the *rotational energy* according to Eqs. (1.68), (1.20), (1.24), (1.82) and (1.83), we have

$$\frac{dN}{N} = \frac{(2J+1)e^{-\varepsilon_{r,J}/kT}dJ}{Q_r} = \frac{e^{-\varepsilon_{r,J}/kT}(2J+1)dJ}{T/\Theta_r} = \frac{e^{-\varepsilon_{r,J}/kT}d\varepsilon_v}{kT}, \tag{1.89}$$

or

$$f(\varepsilon_r) = \frac{e^{-\varepsilon_{r,J}/kT}}{kT}. \tag{1.90}$$

For the *translational energy*, we make use of the results of the kinetic theory of gases in Chapter 2. In section 2.11.1, the distribution function of the value of the velocity $\chi(c')dc'$ is derived, i. e., the number fraction of the molecules between c' and $c'+dc'$ is $\chi(c')dc'$ (see Eqs. (2.197) and (2.198))

$$dn/n = 4\pi\left(\frac{m}{2\pi kT}\right)^{3/2} c'^2 \exp\left(-\frac{c'^2}{2kT}\right)dc'.$$

Noting that the translational energy of a molecule is $\varepsilon_{tr} = mc'^2/2$, the above equation can be rewritten as

1.3 INTERNAL ENERGY DISTRIBUTION FUNCTIONS 49

$$dN/N = \frac{2}{\sqrt{\pi}} \frac{1}{(kT)^{3/2}} \varepsilon_{tr}^{1/2} e^{-\varepsilon_{tr}/kT} d\varepsilon_{tr}, \qquad (1.91)$$

or

$$f(\varepsilon_{tr}) = \frac{2}{\sqrt{\pi}} \frac{1}{(kT)^{3/2}} \varepsilon_{tr}^{1/2} e^{-\varepsilon_{tr}/kT}. \qquad (1.92)$$

In the light of the form of the distribution function $f(\varepsilon_r)$ of rotational energy with 2 degrees of freedom and the distribution function $f(\varepsilon_{tr})$ of translational energy with 3 degrees of freedom (Eq. (1.90) and Eq. (1.92)) one can assume, that generally the distribution function f of an energy mode ε with ζ degrees of freedom has the following form

$$f = C \frac{\varepsilon^{(\zeta/2-1)}}{(kT)^{\zeta/2}} e^{-\varepsilon/kT}. \qquad (1.93)$$

The value of constant C is determined from the normalization condition

$$C \int f d\varepsilon = C \int_0^\infty \frac{\varepsilon^{\frac{\zeta}{2}-1}}{(kT)^{\zeta/2}} e^{-\varepsilon/kT} d\varepsilon = C \int_0^\infty x^{\frac{\zeta}{2}-1} e^{-x} dx = 1,$$

or

$$C = 1/\Gamma(\zeta/2)$$

Thus, the *distribution function of energy* ε with ζ *degrees of freedom* is

$$f(\varepsilon) = \frac{1}{\Gamma(\zeta/2)} \frac{\varepsilon^{\frac{\zeta}{2}-1}}{(kT)^{\zeta/2}} e^{-\varepsilon/kT}. \qquad (1.94)$$

This is certainly true for vibrational energy ε_v with ζ_v degrees of freedom (see Eq. (1.86)). According to the *distribution function of vibrational energy* in the form of Eq. (1.94) (put $\varepsilon = \varepsilon_v$, $\zeta = \zeta_v$), the average value of the vibrational energy of molecules can be calculated as

$$e_v = \overline{\varepsilon_v} = \int_0^\infty \varepsilon_v f(\varepsilon_v) d\varepsilon_v =$$

$$\frac{kT}{\Gamma(\zeta_v/2)} \int_0^\infty \frac{\varepsilon^{\zeta/2}}{(kT)^{\zeta_v/2+1}} e^{-\varepsilon_v/kT} d\varepsilon_v =$$

$$\frac{kT}{\Gamma(\zeta_v/2)} \int_0^\infty x^{(\zeta_v/2)-1} e^{-x} dx =$$

$$\frac{kT}{\Gamma(\zeta_v/2)} \Gamma(\zeta_v/2 + 1) = \frac{\zeta_v}{2} kT. \tag{1.95}$$

This result is identical with the average vibrational energy (Eq. (1.85), where ζ_v is introduced according to Eq. (1.86)) obtained from the summation of molecular vibrational energy ε_v (Eq. (1.16)). This shows that the distribution function of energy in the form of Eq. (1.94) yields the correct average value of energy.

REFERENCES

1. Herzberg G (1953) Molecular Spectra and Molecular Structure. I Spectra of Diatomic Molecules, 2nd ed., Van Nostrand
2. Vincenti WG and Kruger GH (1965) Introduction to Physical Gas Dynamics, John Wiley and Sons, New York
3. Moore CE (1949, 1952, 1958) Atomic Energy Levels, National Bureau of Standards Circ. 467, vols, I, II, III

2 SOME BASIC CONCEPTS OF KINETIC THEORY

2.1 THE VELOCITY DISTRIBUTION FUNCTION

The concept of the *velocity distribution f* is similar with the number density in the physical space r, but it is the number density in the phase space c, r, and it has the probable sense, i.e., it is the *probabilistic number density in the phase space*.

For a system composed of N molecules, at any instant, the entire system is represented by a point in the phase space of $6N$ dimensions (formed by velocity c and position r). Consider a number of such systems, or ensemble of such systems. The probability of finding the system in the phase space element $dc_1 dc_2 \cdots dc_N dr_1 dr_2 \cdots dr_N$ near the phase space point $c_1, c_2, \cdots, c_N, r_1, r_2, \cdots, r_N$ is

$$F^{(N)}(c_1, c_2, \cdots, c_N, r_1, r_2, \cdots, r_N, t) dc_1 dc_2 \cdots dc_N dr_1 dr_2 \cdots dr_N. \tag{2.1}$$

This is the definition of the N particle distribution function $F^{(N)}$. The subscripts $1,2,3,\cdots,N$ denote the mark numbers of the molecules, N is the total number of molecules. The *reduced distribution function* of R molecules out of N molecules is defined as

$$F^{(R)}(c_1, c_2, \cdots, c_R, r_1, r_2, \cdots, r_R, t) = \int_{-\infty}^{\infty} \cdots \int_{-\infty}^{\infty} F^{(N)} dc_{R+1} \cdots dc_N dr_{R+1} \cdots dr_N. \tag{2.2}$$

That is, the probability of finding the molecules with mark numbers $1, 2, \cdots, R$ in the phase space element $dc_1 dc_2 \cdots dc_R dr_1 dr_2 \cdots dr_R$ near the phase space point $c_1, c_2, \cdots c_R, r_1, r_2 \cdots r_R$ irrespective of the positions of the other $(N-R)$ molecules is

$$F^{(R)} dc_1 dc_2 \cdots dc_R dr_1 dr_2 \cdots dr_R.$$

When $R = 1$, the *single particle distribution function* $F^{(1)}$ is obtained. That is, the probability of finding the molecule with mark number 1 in the phase space element $dc_1 dr_1$ near the phase space point c_1, r_1 irrespective of the positions of the other $(N-1)$ molecules is $F^{(1)} dc_1 dr_1$. Since the number of molecules in the entire phase space is N, and all the molecules are identical, the probable number of molecules in phase element $dc_i dr_i$ is $NF^{(1)} dc_i dr_i$. Kinetic theory of gases uses just the single particle distribution function to treat the motions of molecules.

The distribution function f used in this book is $NF^{(1)}$, i. e.

$$f(c, r, t) \equiv NF^{(1)}(c, r, t) \tag{2.3}$$

As f is the single particle distribution function multiplied by the number of molecules, $f dc dr$ is not any more a probability, but is a probable number of molecules in $dcdr$, so f is the probable density in the phase space. By definition of the *velocity distribution function* $f(c, r, t)$, the number of molecules in time t in the physical space element $dr \equiv dxdydz$, with velocity in the velocity space element $dv \equiv dudvdw$ (see Fig. 2.1) near velocity c is

$$f(c, r, t) dc dr . \tag{2.4}$$

Practically the number of molecules in the phase space $dcdr$ would fluctuate, $f dcdr$ is the number of molecules in the average meaning, the definition formula Eq. (2.4) contains the probabilistic sense and the distribution function is also called *probability function*. Integrating this formula over the entire velocity space, we obtain the number of molecules in dr, and the *number density* of the molecules in the physical space is obviously

$$n = \int f dc . \tag{2.5}$$

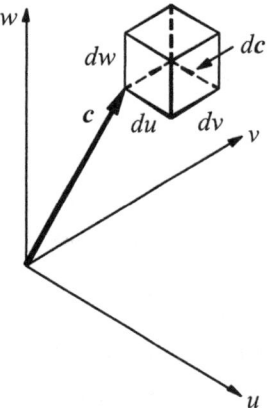

Fig. 2.1 Velocity element dc, $dc = dudvdw$, in the vicinity of c in the velocity space

The distribution function $f(c,r,t)$ introduced here is defined identically as in the book of Chapman and Cowling [1]. This definition starts from the point that $fdcdr$ is the (probable) number of molecules in $dcdr$, i. e., f is defined as the density in phase space. Some authors (e.g., Kennard [2],Vincenti and Kruger [3], and Bird [4]) use

$$f_K = f/n, \tag{2.6}$$

as the definition of the distribution function. We must note the difference of these two definitions. For example, instead of Eq.(2.5), for the distribution function f_K in Kennard or Bird, one would have

$$\int_{-\infty}^{\infty} f_K dc = 1.$$

2.2 MACROSCOPIC PROPERTIES

When the velocity distribution function f of the gas is known, all the macroscopic properties of the gas that are concerned in gas dynamics can be obtained by finding the moments of f. By finding moments of f we mean multiplying f by certain function of molecular velocity and taking integral over the entire velocity space.

According to the definition of f, Eq. (2.4) $fdcdr \equiv fdudvdwdxdydz$ is the number of molecules in time t in the physical space element $dr \equiv dxdydz$ near $r(x,y,z)$, with velocities in the velocity space element $dc \equiv dudvdw$ near velocity $c(u,v,w)$ (in the following such velocities will be referred to as velocities in c, dc, the molecules with such velocities will be referred to as molecules of class c). Integrating Eq. (2.4) over the whole velocity space, one obtains the number of molecules in dr. The *density* ρ of the gas in the physical space is obviously

$$\rho = m \int_{-\infty}^{\infty} \int_{-\infty}^{\infty} \int_{-\infty}^{\infty} fdudvdw \equiv m \int_{-\infty}^{\infty} fdc \equiv mn, \tag{2.7}$$

where m is the mass of the molecule.

For single-component gas c_0 is used to denote the mean molecular velocity $\bar{c}$, there are ndr molecules in the physical space element $dr = dxdydz$, their mass is $mndr$, each molecule carries a momentum of mc. The definition of c_0 is

$$\rho dr c_0 = \sum mc,$$

or

$$ndr c_0 = \sum c. \tag{2.7}'$$

$\sum c$ is a summation over ndr molecules. The number of molecules in dr with velocity ranging in c, dc is $fdcdr$. These molecules carry a velocity of $cfdcdr$, integrating over the whole velocity space one obtains the value $\sum c$

$$\sum c = dr \int cf dc.$$

Therefore from Eq. (2.7)′ we obtain

$$c_0 = \bar{c} = \frac{1}{n}\int cf dc. \tag{2.8}$$

The *mean molecular velocity* $c_0 \equiv \bar{c}$ is the *macroscopic stream velocity* of the gas, or the *mean velocity,* or the *mass velocity* of the gas.

Generally, if $Q(c)$ is an arbitrary function of the molecular velocity, its mean value $\bar{Q}$ in the gas is defined as

$$ndr\bar{Q} = \sum Q,$$

$\sum Q$ represents the summation over all ndr molecules in dr. The number of class c molecules in dr is $fdcdr$, their contribution to Q is $Q\ fdcdr$. Finding the moment over the whole space c, one obtains the total value of Q of all molecules in dr

$$\sum Q = dr \int Qf dc.$$

So for the mean value of arbitrary $Q(c)$ one has the following expression

$$\bar{Q} = \frac{1}{n}\int Qf dc. \tag{2.9}$$

Q may represent scalar, vector or tensor.

The velocity $c - c_0$ of a molecule relative to the macroscopic stream velocity is called the *thermal velocity*, or *peculiar velocity*, some times is also called the *random velocity* of the molecule and is denoted by c'

$$c' = c - c_0. \tag{2.10}$$

Obviously, the mean thermal velocity is zero

$$\overline{c'} = \overline{c} - c_0 = \overline{c} - \overline{c} = 0.$$

A class of macroscopic magnitudes that we are concerned is related to the transport of mass, momentum and energy in the flow caused by the molecular movements. All of them require the calculation of the flux of certain magnitude across some (elementary) area. Let us first obtain the expression of the flux across a differential area element dS (see Fig. 2.2) of the quantity $Q(c)$ associated with molecular velocity of each molecule, which may be the mass of the molecule, the momentum or the energy of the molecule. The unit normal vector of this area element is l. Consider molecules of class c (i. e. molecules with velocity near c in dc). Such molecules that cross the area dS in a small (differential) time interval dt are located, at the initial instant of dt, in a cylinder situated on the opposite to c side of dS with dS as the base and with a side length of cdt. If denote the volume of this cylinder $c \cdot ldtdS$ by dr, then the number of class c molecules that cross dS in dt is

$$fdcdr = fdcc \cdot ldtdS. \tag{2.11}$$

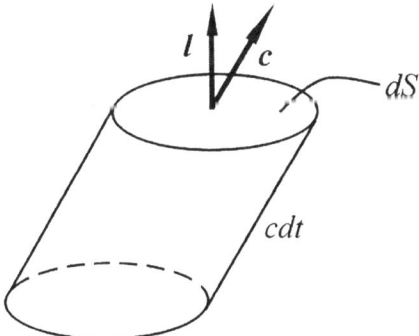

Fig. 2.2 The volume occupied by the molecules of class c in the initial instant of dt that traverse dS in time dt, l: the unit normal of the surface element

The flow quantity of Q carried by the molecules crossing dS in dt is therefore

$$Qfdcc \cdot l dt dS .\tag{2.12}$$

To find the entire flow quantity of Q carried by all classes of molecules crossing dS in dt, one should integrate Eq. (2.12) over the entire c space to obtain

$$dtdS\int Qfc\cdot l\, dc = dtdSn\overline{Qc\cdot l} .\tag{2.13}$$

When writing the last term, the expression Eq. (2.9) of the mean value of the quantity carried by the molecules has been used. Dividing Eq. (2.13) by $dSdt$, we obtain the *flux of* Q, i. e. the quantity of Q carried by the molecules crossing an area per unit area per unit time

$$\int Qfc\cdot l\, dc = n\overline{Qc\cdot l} .\tag{2.14}$$

Equation (2.14) is a total flux, it includes the fluxes of molecules crossing the area in the positive l direction and in the opposite l direction. Some times the flux crossing dS only in one direction is needed to be known. For example the flux of Q in the positive l direction is

$$\overline{n(Qc\cdot l)}_{c\cdot l>0} .\tag{2.15}$$

As l is an unit vector, Eq. (2.14) can be written as $n\overline{Qc}\cdot l$, and define $n\overline{Qc}$ as the *flux vector* of quantity Q

$$n\overline{Qc} .\tag{2.16}$$

The mass flux vector $(Q=m)$ is

$$nm\overline{c} = \rho c_0 .\tag{2.17}$$

Pointing at the cases of the stress and heat flux to be discussed in the following, the transports of momentum and energy of molecules concerned are the transports in the reference frame of the macroscopic motion, i.e., the transports by gas molecules moving with the mean velocity c_0 across an area dS also moving with velocity c_0. So we shall discuss the flux vector of the quantity Q in the coordinate system moving with the flow velocity c_0, i. e., we shall discuss $n\overline{Qc'}$.

2.2 MACROSCOPIC PROPERTIES　57

Let $Q = mc'$, we attain the expression of the momentum transport caused by the thermal motion. The expression so obtained is a tensor with nine components, and is called the *pressure tensor* $\boldsymbol{P}$

$$\boldsymbol{P} = \overline{nmc'c'} = \overline{\rho c'c'} = m\int c'c' f dc . \tag{2.18}$$

The physical meaning of the pressure tensor $\boldsymbol{P}$ will become more articulate after understanding clearly its individual components. Consider the transport of the momentum in the x direction across the surface dS with normal in the y direction in time dt. For the molecules of class c, this momentum is mu', the molecules of this class crossing dS in dt are located in the cylinder of volume $v'dtdS$, the number of the molecules is $fd\boldsymbol{c}\,dtdS$, the momentum in the x direction carried by them is $mfv'udcdtdS$. So the total flux of mu' is

$$\int mfu'v'dc = \overline{nmu'v'} = \overline{\rho u'v'} = P_{xy} . \tag{2.19}$$

Similarly, the components of the pressure tensor $\boldsymbol{P}$ can be written as

$$\boldsymbol{P} \equiv \begin{Bmatrix} P_{xx} & P_{xy} & P_{xz} \\ P_{yx} & P_{yy} & P_{yz} \\ P_{zx} & P_{zy} & P_{zz} \end{Bmatrix} = \begin{Bmatrix} \overline{\rho u'^2} & \overline{\rho u'v'} & \overline{\rho u'w'} \\ \overline{\rho v'u'} & \overline{\rho v'^2} & \overline{\rho v'w'} \\ \overline{\rho w'u'} & \overline{\rho v'w'} & \overline{\rho w'^2} \end{Bmatrix} . \tag{2.20}$$

Obviously, $\boldsymbol{P}$ is a symmetrical tensor, and can be expressed in the reduced form

$$P_{ij} = \overline{\rho c'_i c'_j} = m\int (c_i - c_{0i})(c_j - c_{0j}) f d\boldsymbol{c}, i,j = 1,2,3 , \tag{2.21}$$

where $c'_1 \equiv u'$, $c'_2 \equiv v'$, $c'_3 \equiv w'$, $c_{01} \equiv u_0$, $c_{02} \equiv v_0$, $c_{03} \equiv w_0$.

At an arbitrary point P inside the gas, the *pressure* (*pressure flux*) $\boldsymbol{p}_l$ acting on dS and pointing in the positive direction of the surface is the flux of molecular momentum mc' across this surface along the positive direction (per unit time, per unit area). To obtain this flux one should take $Q = mc'$ in Eq. (2.14), and let $c = c'$ (for we are considering the transport of momentum in the system connected with the flow velocity of the gas), so we have

$$\boldsymbol{p}_l = \overline{nc'_l mc'} = \overline{\rho c'_l c'} = \overline{\rho \boldsymbol{l} \cdot c'c'} = \boldsymbol{l} \cdot \overline{\rho c'c'} = \boldsymbol{l} \cdot \boldsymbol{P} . \tag{2.22}$$

The component of this pressure in the normal direction of this surface is

$$l \cdot p_l = l \cdot \overline{\rho c'_i c'} = \overline{\rho c'^2_l}. \tag{2.23}$$

That is, this pressure is always positive. Pondering the definition of the pressure flux, it shows that inside the gas the flux of the momentum crossing the surface in the positive direction is always positive, and the surface with the external normal l is subjected from the gas to a pressure (the reaction of the pressure flux) pointing in the opposite direction of l. This result shows, that the gas always exerts on a surface a normal stress of compressing but not stretching nature. Note that the direction of the stress exerted by the gas on the surface is opposite to the direction of the momentum flux of the gas, and only when the force subjected by the surface from the gas is pointed in the opposite direction of the external normal to the surface, the surface is said to be subjected to positive pressure. For the same reason, i. e., as the direction of the stress exerted by the gas on the surface is opposite to the direction of the momentum flux, we define the *viscous stress tensor* $T \equiv \tau_{ij}$ as the negative of the pressure tensor with the static pressure subtracted from the normal components

$$? \equiv \tau_{ij} = -(\overline{\rho c'_i c'_j} - \delta_{ij} p), \tag{2.24}$$

where δ_{ij} is the Kronecker tensor

$$\delta_{ij} = 1 (i = j), \delta_{ij} = 0 (i \neq j),$$

p is the *static pressure*, defined as the average of the three normal components of the pressure tensor

$$p = \frac{1}{3}\overline{\rho u'^2 + v'^2 + w'^2} = \frac{1}{3}\overline{\rho c'^2}. \tag{2.25}$$

To obtain the expression of the *heat flux* $q = q_i$, we start again from $n\overline{Qc'}$, this time one lets $Q = (1/2) mc'^2 + \varepsilon_{in}$, where ε_{in} is the internal (vibrational or rotational) energy related to one molecule, then

$$q = \frac{1}{2}\overline{\rho c'^2 c'} + n\overline{\varepsilon_{in} c'} = \frac{m}{2}\int c' c'^2 f dc + \int c' \varepsilon_{in} f dc, \tag{2.26}$$

or

$$q_i = \frac{m}{2}\int c'_i c'^2 fdc + \int c'_i \varepsilon_{in} fdc . \qquad (2.27)$$

Making use of the general expression of the mean value Eq. (2.9) and letting $Q = (1/2)mc'^2$, one obtains the expression of the (connected with the translational energy) internal energy per unit mass

$$e_{tr} = \frac{1}{2}\overline{c'^2} = \frac{1}{2n}\int c'^2 fdc . \qquad (2.28)$$

From the expression Eq. (2.25) for the pressure p one obtains

$$p = \frac{2}{3}\rho e_{tr} . \qquad (2.29)$$

The above discussions have been proceeded without any assumptions about the state of the gas, that is, they are applicable both for the equilibrium and non-equilibrium states. In the following we proceed to the discussion of the thermodynamic temperature that is essentially a gas property in equilibrium state. The ideal gas equation of state is

$$p = \rho RT = \rho\frac{k}{m}T = nkT . \qquad (2.30)$$

Here R is the gas constant per unit mass, $R = \mathcal{R}/M = k/m$, $\mathcal{R}$ is the universal gas constant, i.e., the gas constant for a mole of gas, $\mathcal{R} = 8.3145 Jmol^{-1}K^{-1}$, M is the molecular weight, k is the Boltzmann constant, $k = mR = \mathcal{R}/N$, i.e. the gas constant for one molecule, $k = 1.38066\times 10^{-23} JK^{-1}$, N is the Avogadro's number, $N = 6.0221\times 10^{23} mol^{-1}$. From Eq. (2.29)~Eq. (2.30) one can introduce the *translational kinetic temperature*

$$\frac{3}{2}\frac{k}{m}T_{tr} = e_{tr} = \frac{1}{2}\overline{c'^2} = \frac{1}{2}(\overline{u'^2} + \overline{v'^2} + \overline{w'^2}) . \qquad (2.31)$$

Sometimes it is convenient to write Eq. (2.31) (by using Eq. (2.10)) in a form expressed by the average velocity and the molecular velocity:

$$\frac{3}{2}kT_{t_r} = \frac{1}{2}m\overline{c'^2} = \frac{1}{2}m(\overline{c^2} - c_0^2) = \frac{1}{2}m(\overline{u^2} + \overline{v^2} + \overline{w^2} - c_0^2) . \qquad (2.32)$$

Note that such a definition of the temperature T_{tr} and the ideal gas equation of state are applicable for the case when gas is in the state of non-equilibrium. One can define separately the translational kinetic temperatures for three velocity components, for example

$$k(T_{tr})_x = \overline{mu'^2}. \tag{2.33}$$

The difference between $(T_{tr})_x$, $(T_{tr})_y$, $(T_{tr})_z$ and T_{tr} and the differences between themselves characterize the degree of non-equilibrium of the translational energy mode in the gas.

In the gas of monatomic molecules T_{tr} can be defined as the temperature of the gas. The diatomic and polyatomic molecules have rotational and vibrational energy modes and the corresponding internal degrees of freedom (see sections 1.1 and 1.3). According to the number ζ of degrees of freedom of some energy mode and in analogy with the definition formula Eq. (2.31) of the translational temperature T_{tr} (translational energy has 3 degrees of freedom), the temperature T_{int} of this internal energy mode can be defined as

$$\frac{\zeta}{2}\frac{k}{m}T_{int} = e_{int}. \tag{2.33}'$$

The energy equipartition principle implies the equality of the translational and internal temperature of the gas in equilibrium state. In the non-equilibrium gas, an overall kinetic temperature T_{ov} can be defined

$$T_{ov} = (3T_{tr} + \zeta T_{int})/(3+\zeta). \tag{2.34}$$

In the foregoing the derivation of the expressions of the macroscopic quantities through the moments of the velocity distribution function is given in detail. This is an essential step from the description on the level of velocity distribution function towards the description via macroscopic quantities. The results thus obtained are collected here

$$n = \int f dc, \tag{2.5}$$

$$\rho = mn \equiv m\int f dc, \tag{2.7}$$

$$c_0 \equiv \bar{c} = \frac{1}{n}\int c f dc, \tag{2.8}$$

$$P_{ij} = m\int (c_i - c_{0i})(c_j - c_{0j}) f d\mathbf{c} = \rho \overline{c'_i c'_j} = p\delta_{ij} - \tau_{ij} ,\qquad (2.21)$$

$$\tau_{ij} = -(\rho \overline{c'_i c'_j} - \delta_{ij} p) ,\qquad (2.24)$$

$$p = \frac{1}{3}\rho \overline{c'^2} ,\qquad (2.25)$$

$$q_i = \int c'_i (\frac{1}{2} m c'^2) f d\mathbf{c} + \int c'_i \varepsilon_{in} f d\mathbf{c} ,\qquad (2.27)$$

$$\frac{3}{2}\frac{k}{m}T = e_{tr} = \frac{1}{2n}\int c'^2 f d\mathbf{c} = \frac{1}{2}\overline{c'^2} = \frac{3}{2}\frac{p}{\rho} .\qquad (2.31)$$

It is noted that the above description and derivation have been aimed at single-component gas. The extension to the gas mixture is direct and simple (see section 2.12).

2.3 BINARY ELASTIC COLLISIONS OF MOLECULES

The gas properties change (or are in a state of dynamic balance) because of the encounters of the gas molecules with the solid surfaces or the collisions between the molecules. In the present section we are going to discuss the binary elastic collision which is the most important collision and is liable to be treated mathematically. Elastic collisions are collisions without interchange of translational mode and internal modes of energies, and the momentum and energy conservation equations are used in the following as the starting point of our mathematical derivations. To see the importance of binary collision we first analyze the condition under which the *ternary collisions* (three-body collisions) are rear events but the *binary collisions* prevail. A *three-body collision* happens when a molecule collides with a pair of molecules temporally encountered together in a binary collision. The lifetime of the collision pair of two colliding molecules can be vividly thought of the time of interlapping of the two molecules moving towards each other, the order of magnitude of it can be estimated as $\sigma^{1/2}/\overline{c_r}$, where σ is the cross section area of the molecule, $\overline{c_r}$ is the relative velocity of the two molecules. The collision frequency of the molecules is $v \approx n\overline{\sigma c_r}$ (see Eq. (0.3)). Obviously, the probability of the triple collision is the probability of the collision of a collision pair in its lifetime with another molecule, its value is $\left(\sigma^{1/2}/\overline{c_r}\right)\cdot\left(n\overline{\sigma c_r}\right) \approx \sigma^{3/2}n$ $\approx d^3 n = d^3/\delta^3$ where $\delta = n^{-1/3}$ is the average *molecular spacing* (each molecule in average occupies a volume of n^{-1}). That is, the condition under which the ternary collision is not important is

$$d^3/\delta^3 \ll 1. \tag{2.35}$$

It is seen, when the condition of the *dilute gas*

$$\delta \equiv n^{-1/3} \gg d, \tag{2.35'}$$

is satisfied, the condition that only the binary collision is important is also satisfied. Note, that the requirement of the dilute gas Eq. (2.35)' is more strict than the condition Eq. (2.35).

For binary elastic collision there is no energy exchange between the translational mode and the internal modes. In this case specifying the pre-collision velocities c_1 and c_2 of the two colliding molecules and using the momentum and energy conservation equations (four scalar equations) is not sufficient to determine the post-collision velocities c_1^* and c_2^* (six scalar unknowns). To determine entirely the outcome of the collision of two molecules two *impact parameters* and the intermolecular potential are to be known. The two impact parameters are the *distance b of the closest approach* of the undisturbed trajectories of the two molecules in centre of mass system of coordinates (or the *projected distance* between the pre-collision velocities, it is also called the *miss distance*) and the angle ε between the collision plane (in which the trajectories lie in this system of coordinates) and some reference plane. Here we obtain at first the binary collision laws without the specification of the impact parameters.

For two colliding molecules of mass m_1 and m_2 with pre-collision velocities c_1 and c_2, the momentum and energy conservation equations can be written as

$$m_1 c_1 + m_2 c_2 = m_1 c_1^* + m_2 c_2^* = (m_1 + m_2) c_m, \tag{2.36}$$

$$m_1 c_1^2 + m_2 c_2^2 = m_1 c_1^{*2} + m_2 c_2^{*2}, \tag{2.37}$$

where the velocity c_m of the centre of mass of the two molecules is introduced:

$$c_m = \frac{m_1 c_1 + m_2 c_2}{m_1 + m_2}.$$ (2.38)

Eq. (2.36) shows that this velocity does not change pre and post the collision. Introducing the pre-collision and the post-collision relative velocities c_r and c_r^*,

$$c_r = c_1 - c_2,$$
$$c_r^* = c_1^* - c_2^*,$$ (2.39)

one obtains from Eqs. (2.36), (2.38) and (2.39),

$$c_1 = c_m + \frac{m_2}{m_1 + m_2} c_r$$
$$c_2 = c_m - \frac{m_1}{m_1 + m_2} c_r,$$ (2.40)

$$c_1^* = c_m + \frac{m_2}{m_1 + m_2} c_r^*$$
$$c_2^* = c_m - \frac{m_1}{m_1 + m_2} c_r^*.$$ (2.41)

From Eq. (2.40) it is seen, that in the center of mass frame of reference the pre-collision velocities $c_1 - c_m$ and $c_2 - c_m$ are antiparallel and, if the molecules are centers of force, the force between them initially lies in the plane determined by these two velocities. The velocities of the colliding molecules remain in this plane and so does the force between them. Eq. (2.41) shows, that in the center of mass frame of reference the post-collision velocities are also antiparallel.

From Eqs. (2.40) and (2.41) one obtains

$$\begin{cases} m_1 c_1^2 + m_2 c_2^2 = (m_1 + m_2) c_m^2 + m_r c_r^2 \\ m_1 c_1^{*2} + m_2 c_2^{*2} = (m_1 + m_2) c_m^2 + m_r c_r^{*2}, \end{cases}$$ (2.42)

where the *reduced mass* m_r is introduced

$$m_r = \frac{m_1 m_2}{m_1 + m_2}.$$ (2.43)

From the energy conservation Eq. (2.37) and Eq. (2.42) one obtains that the value of the relative velocity pre to and after the collision remains the same

2 SOME BASIC CONCEPTS OF KINETIC THEORY

$$c_r^* = c_r. \tag{2.44}$$

From the conservation of the angular momentum

$$c_r b = c_r^* b^*,$$

where b^* is the projected distance of the post-collision velocities, and from Eq. (2.44) follows, that the projected distance b remains unchanged before and after the collision. Thus we can draw the trajectories of the molecules in the center of mass frame system as shown in Fig. 2.3. From Eq. (2.41) it is seen that it is sufficient to calculate the angle χ of deviation of the molecule to entirely determine the velocities after the collision.

Denoting by r_1 and r_2 the position vectors of the two molecules which are spherically symmetric point centers of force and by F the force between them, the equations of motion of the molecules can be written as

$$m_1 \ddot{r}_1 = F, m_2 \ddot{r}_2 = -F. \tag{2.45}$$

Introducing the vector of relative velocity $c_r = \dot{r}_1 - \dot{r}_2 \equiv \dot{r}$, one obtains

$$m_r \ddot{r} = F, \tag{2.46}$$

i.e., the motion of molecule m_1 relative to molecule m_2' is reduced to the motion of a molecule of mass m_r relative to a fixed center of force (see Fig.2.4).

If two molecules have c_1^* and c_2^* as their pre-collision velocities, and the projected distance in the mass center frame of reference is b (see Fig. 2.3), then owing to the energy and momentum conservation equations and the symmetry of the pre- and post-collision velocities, the post- collision velocities of them will be c_1 and c_2. This collision $\left(c_1^*, c_2^* \to c_1, c_2\right)$ is called the *inverse collision* of the *direct collision* $\left(c_1, c_2 \to c_1^*, c_2^*\right)$.

Now we calculate the *deflection angle* χ of the molecule in the assumption that the molecules are force centers. Making use of the polar coordinate system r, θ, the momentum and energy conservation equations of the particle of reduced mass m_r in the force field with center O can be written as

$$r^2 \dot{\theta} = bc_r, \tag{2.47}$$

2.3 BINARY ELASTIC COLLISIONS OF MOLECULES

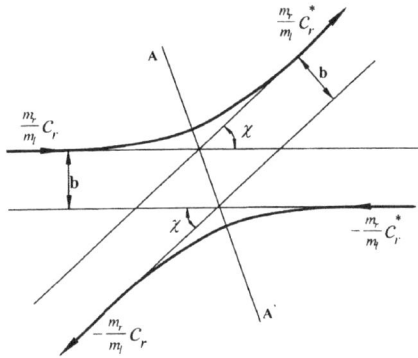

Fig. 2.3 Trajectories of molecules conducting a binary collision in the centre of mass framework (b : the miss-distance, χ : deflection angle)

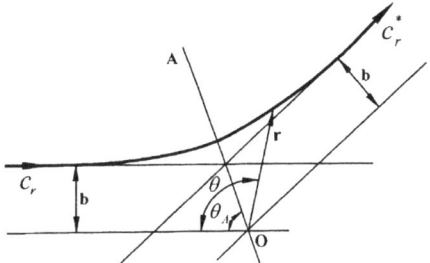

Fig. 2.4 Trajectory of a molecule with reduced mass m_r in the force field with fixed centre

$$\frac{1}{2}m_r(\dot{r}^2 + r^2\dot{\theta}^2) + \phi = \frac{1}{2}m_r c_r^2, \tag{2.48}$$

where bc_r and $(1/2)m_r c_r^2$ are the constant asymptotic values of the angular momentum and energy for $r \to \infty$; ϕ is the intermolecular potential and is related to the central force F by the following formula

$$F = -d\phi/dr. \tag{2.49}$$

Eq. (2.47) and Eq. (2.48) can be combined with the time variable eliminated to yield the orbit equation

$$\left(\frac{dr}{d\theta}\right)^2 = \frac{r^4}{b^2} - r^2 - \frac{\phi r^4}{\frac{1}{2}m_r c_r^2 b^2}. \tag{2.50}$$

If transfer from the variable r to the dimensionless variable W

$$W = b/r, \qquad (2.51)$$

then the orbit equation is transferred into

$$\left(\frac{dW}{d\theta}\right)^2 = 1 - W^2 - \frac{\phi}{\frac{1}{2}m_r c_r^2}. \qquad (2.52)$$

The solution can be written as

$$\theta = \int_0^W \left\{ 1 - W^2 - \frac{\phi}{\frac{1}{2}m_r c_r^2} \right\}^{-1/2} dW. \qquad (2.53)$$

At the intersection of the orbit with the apse line OA we have

$$dW/d\theta = 0, \text{ when } \theta = \theta_A,$$

so we have

$$\theta_A = \int_0^{W_1} \left\{ 1 - W^2 - \frac{\phi}{\frac{1}{2}m_r c_r^2} \right\}^{-1/2} dW, \qquad (2.54)$$

where W_1 is the positive root of the following equation

$$1 - W^2 - \frac{\phi}{\frac{1}{2}m_r c_r^2} = 0. \qquad (2.55)$$

And the *deflection angle* χ of the relative velocity is

$$\chi = \pi - 2\theta_A. \qquad (2.56)$$

Sometimes it is more convenient to use the supplementary angle δ of the deflection angle χ

$$\delta = \pi - \chi = 2\theta_A. \qquad (2.57)$$

Now we express the velocities c^*, c_1^* after the collision through the velocities c, c_1 before the collision and the angles χ (or δ) and ε. From Eq. (2.41) and Eq. (2.38) it is seen, that it is sufficient to find c_r^* to have the expression of c^*, c_1^*

(for the sake of convenience here c_1, c_2 are used to replace c, c_1 in the original formulae).

Let OX, OY, OZ be the three axes of the rectangular coordinate system. OG gives the opposite direction of c_r, OG^* gives the direction of c_r^*. The points X,Y,Z,G and G^* are located on the sphere with the center of O and the radius of c_r. The plane passing through c_r and OX, i.e. the plane GOX, is taken as the reference plane, ε is the angle between the collision plane GOG^* and the reference plane, i.e. the angle G^*GX see Fig. 2.5). Consider the spherical triangle G^*GX, the *cosine formula for the sides in the spherical trigonometry* reads

$$\cos G^*X = \cos GX \cos GG^* + \sin GX \sin GG^* \cos\varepsilon . \tag{2.58}$$

Obviously we have

$$\cos G^*X = u_r^*/c_r ,$$

$\cos GX = -u_r/c_r$, therefore $\sin GX = (v_r^2 + w_r^2)^{1/2}/c_r$,

$$\cos GG^* = \cos\delta , \quad \sin GG^* = \sin\delta .$$

From Eq. (2.58) one obtains

$$u_r^* = -u_r \cos\delta + (v_r^2 + w_r^2)^{1/2}\sin\delta \cos\varepsilon . \tag{2.59}$$

Similarly, for spherical triangle G^*GY we have

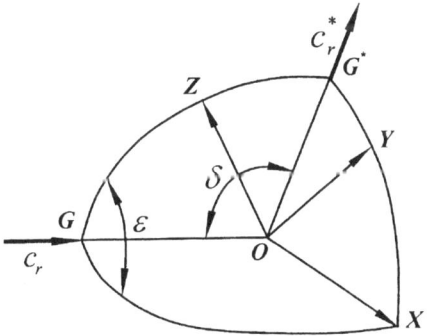

Fig. 2.5 The schematic of obtaining the post-collision relative velocity according to the spherical trigonometry (X,Y,Z,G,G^* are located on a sphere of radius c_r with O as the centre, in the figure only the part of spherical surface stretched by XGG^* is drawn, $OG = -c_r$, $OG^* = c_r^*$, GOG^* is the collision plane, GOX is the reference plane)

$$v_r^* = -v_r\cos\delta + (u_r^2 + w_r^2)^{1/2}\sin\delta\cos G\hat{G}Y, \qquad (2.60)$$

where $\cos G\hat{G}Y = \cos(\varepsilon - \xi_2)$, and ξ_2 denotes the angle XGY. To attain the value of $\cos\xi_2$, we consider the spherical triangle GXY and have

$$\cos XY = \cos GX \cos GY + \sin GX \sin GY \cos\xi_2,$$

i.e.,

$$0 = u_r v_r + (v_r^2 + w_r^2)^{1/2}(u_r^2 + w_r^2)^{1/2}\cos\xi_2,$$

therefore

$$\cos\xi_2 = \frac{-u_r v_r}{(v_r^2 + w_r^2)^{1/2}(u_r^2 + w_r^2)^{1/2}},$$

$$\sin\xi_2 = \frac{w_r c_r}{(v_r^2 + w_r^2)^{1/2}(u_r^2 + w_r^2)^{1/2}}.$$

Thus from Eq. (2.60) we have

$$v_r^* = -v_r\cos\delta + \sin\delta(w_r c_r \sin\varepsilon - u_r v_r \cos\varepsilon)/(v_r^2 + w_r^2)^{1/2}. \qquad (2.61)$$

Finally, for the spherical triangle $G\hat{G}Z$ we have

$$w_r^* = -w_r\cos\delta + (u_r^2 + v_r^2)^{1/2}\sin\delta\cos G\hat{G}Z, \qquad (2.62)$$

where $\cos G\hat{G}Z = \cos(\xi_3 - \varepsilon)$, and from the spherical triangle GZX one obtains

$$\cos\xi_3 = \frac{-u_r w_r}{(v_r^2 + w_r^2)^{1/2}(u_r^2 + v_r^2)^{1/2}},$$

$$\sin\xi_3 = \frac{v_r c_r}{(v_r^2 + w_r^2)^{1/2}(u_r^2 + v_r^2)^{1/2}}.$$

And Eq. (2.62) yields

$$w_r^* = -w_r\cos\delta + \sin\delta(v_r c_r \sin\varepsilon - u_r w_r \cos\varepsilon)/(v_r^2 + w_r^2)^{1/2}. \qquad (2.63)$$

Now we list together the components of the post-collision velocity c_r, and replace δ in Eqs. (2.59),(2.61) and (2.63) by χ (according Eq. (2.57))

$$u_r^* = u_r\cos\chi + \sin\chi\cos\varepsilon(v_r^2 + w_r^2)^{1/2}, \qquad (2.59)'$$

$$v_r^* = v_r \cos \chi + \sin \chi (c_r w_r \sin \varepsilon - u_r v_r \cos \varepsilon)/(v_r^2 + w_r^2)^{1/2}, \qquad (2.61)'$$

$$w_r^* = w_r \cos \chi + \sin \chi (c_r v_r \sin \varepsilon - u_r w_r \cos \varepsilon)/(v_r^2 + w_r^2)^{1/2}. \qquad (2.63)'$$

The expressions of c^* and c_1^* through c, c_1, χ and ε can be obtained from Eq. (2.41) and Eq. (2.38)

$$u^* = u + (u_1 - u)\cos^2 \frac{\chi}{2} + \frac{1}{2}[c_r^2 - (u_1 - u)^2]^{1/2} \sin \chi \cos \varepsilon, \qquad (2.59)^*$$

$$v^* = v + (v_1 - v)\cos^2 \frac{\chi}{2} + \frac{\sin \chi}{2}[c_r(w_1 - w)\sin \varepsilon - (u_1 - u)(v_1 - v)\cos \varepsilon]/[c_r^2 - (u_1 - u)^2]^{1/2}, \qquad (2.61)^*$$

$$w^* = w + (w_1 - w)\cos^2 \frac{\chi}{2} + \frac{\sin \chi}{2}[c_r(v_1 - v)\sin \varepsilon - (u_1 - u)(w_1 - w)\cos \varepsilon]/[c_r^2 - (u_1 - u)^2]^{1/2}, \qquad (2.63)^*$$

$$u_1^* = u_1 - (u_1 - u)\cos^2 \frac{\chi}{2} - \frac{1}{2}[c_r^2 - (u_1 - u)^2]^{1/2} \sin \chi \cos \varepsilon, \qquad (2.59)^{**}$$

$$v_1^* = v_1 - (v_1 - v)\cos^2 \frac{\chi}{2} - \frac{\sin \chi}{2}[c_r(w_1 - w)\sin \varepsilon - (u_1 - u)(v_1 - v)\cos \varepsilon]/[c_r^2 - (u_1 - u)^2]^{1/2}, \qquad (2.61)^{**}$$

$$w_1^* = w_1 - (w_1 - w)\cos^2 \frac{\chi}{2} - \frac{\sin \chi}{2}[c_r(v_1 - v)\sin \varepsilon - (u_1 - u)(w_1 - w)\cos \varepsilon]/[c_r^2 - (u_1 - u)^2]^{1/2}. \qquad (2.63)^{**}$$

2.4 COLLISION CROSS-SECTIONS AND MOLECULE MODELS

To determine entirely the binary collision, i.e., to determine the post-collision velocities, except the pre-collision velocities, the intermolecular potential and two impact parameters that stipulate the geometrical relations of the two colliding molecules must be given. One of these impact parameters is the miss distance b, that is, the distance of the closest approach of the trajectories unaffected yet by the intermolecular force of the two molecules in the center of mass reference frame. The smaller the miss distance b, the more manifest is the effect of the collision. The case of zero b is the case of head-on collision. With increase of b the deflection of the molecules resulted from collision decreases. When b exceeds certain extent, the molecules factually do not interact with each other. In the mass center frame of reference the plane in which lie the molecular trajectories is the plane of collision, fixing except b further the orientation of the collision plane will identify completely the collision. The angle ε between the collision plane and certain reference plane can be stipulated as the other impact parameter. Both b and ε are shown in Fig 2.6. Consider the collision of two molecules with relative velocity c_r, the relative velocity after crossing the differential area

$$bdbd\varepsilon \tag{2.64}$$

becomes c_r^*. The molecules crossing this area would scatter into the range of $d\chi$ and $d\varepsilon$ near c_r^*, or into the solid angle (see Fig.2.6)

$$d\Omega = \sin\chi d\chi d\varepsilon. \tag{2.65}$$

The *differential collision cross-section* σ is defined as the cross section corresponding to unit solid angle, that is,

$$\sigma d\Omega = bdbd\varepsilon, \tag{2.66}$$

so one has

$$\sigma = \frac{b}{\sin\chi}\left|\frac{db}{d\chi}\right|. \tag{2.67}$$

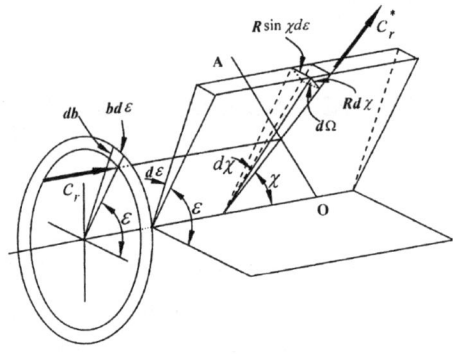

Fig. 2.6 Impact parameters b and ε, and the calculation of the collision cross section

2.4 COLLISION CROSS-SECTIONS AND MOLECULE MODELS

The *total collision section* σ_T is obtained from the differential one through integrating over the entire solid angle

$$\sigma_T = \int_0^{4\pi} \sigma d\Omega = 2\pi \int_0^{\pi} \sigma \sin\chi d\chi = 2\pi \int b db . \qquad (2.68)$$

In writing the last equality we had been making use of Eq. (2.66) and the integration over ε had been carried out. To find the collision cross-section b is of course not to be considered as simple geometrical coordinate and integrated. Only under certain model of the intermolecular force, that is, when the relation between the deflection angle χ and b is known, the definition of σ and σ_T through b has meaning.

In treating practical problems of molecular motion, not only the total cross-section Eq. (2.68) plays important role, but the cross-sections relevant to the transport properties of the gas also play important role. These are the so called *viscosity cross-section* σ_μ and *diffusion cross-section* σ_D. Their expressions through differential cross-section $\sigma d\Omega = 2\pi b db$ are

$$\sigma_\mu = \int_0^{4\pi} \sin^2\chi \sigma d\Omega = 2\pi \int_0^{\pi} \sigma \sin^3\chi d\chi = 2\pi \int \sin^2\chi b db , \qquad (2.69)$$

$$\sigma_D = \int_0^{4\pi} (1-\cos\chi) \sigma d\Omega = 2\pi \int_0^{\pi} \sigma(1-\cos\chi) \sin\chi d\chi = 2\pi \int (1-\cos\chi) b db . \qquad (2.70)$$

Viscosity cross-section σ_μ appears in the expression of the viscosity μ in the transport theory of Chapman-Enskog

$$\mu = \frac{(5/8)(\pi m k T)^{1/2}}{[m/(4kT)]^4 \int_0^{\infty} c_r^7 \sigma_\mu \exp[-mc_r^2/(4kT)] dc_r} , \qquad (2.71)$$

(see Eq. (7.2) of reference [3], p.404; for the expression of σ_μ see Eq. (8.8) of [3], p.358). Similarly, the diffusion collision cross-section σ_D appears in the expression of the diffusivity of binary gas mixture in the transport theory of Chapman-Enskog (for the expression of σ_D see Eq. (8.7) of reference [3], p.358). The latter cross-section is also called *momentum transfer cross-section*. From Fig. 2.3 we can see that the momentum change of the molecule in the direction of initial relative motion is proportional to $(1-\cos\chi)$, this is the origin of the latter name, and sometimes σ_D is designated also as σ_M.

2.4.1 HARD SPHERE MODEL

The simplest molecular model is the *hard sphere model*. When the distance of the closest approach of the two molecules is less than

$$d_{12} = \frac{1}{2}(d_1 + d_2),$$

the two molecules collide as two billiards. From Fig.2.7 it is seen

$$b = d_{12} \sin\theta_A = d_{12} \cos\left(\frac{1}{2}\chi\right),$$

$$|db/d\chi| = \frac{1}{2}d_{12} \sin\left(\frac{1}{2}\chi\right).$$

(2.72)

Equation (2.67) yields the differential cross-section

$$\sigma = d_{12}^2 / 4.$$

(2.73)

This expression shows that σ does not depend on the angle χ, that is, the scattered molecules after collision is uniformly distributed over all directions, or the direction of c_r^* has equal probability over any orientation. The total cross-section can be easily found from Eq. (2.68)

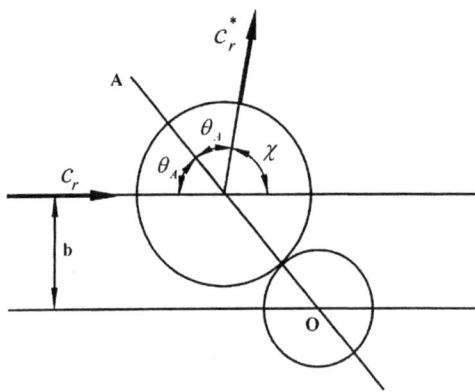

Fig. 2.7. Collision of hard sphere molecules

2.4 COLLISION CROSS-SECTIONS AND MOLECULE MODELS   73

$$\sigma_T = \pi d_{12}^2 .\tag{2.74}$$

This result can be obtained, of course, directly from the hard sphere model intuitively. The above result is valid for two identical molecules, in which case one has $d_{12} = d$.

What value one should take for d in practical applications? The practice shows, that it is essential to take such a value of d which ensures that the viscous coefficient of the gas has the real value.

Obviously, for the hard sphere model, in the second equality of Eq. (2.69), $\sigma = d_{12}/4$ can taken out of the integration and one has

$$\sigma_\mu = \frac{2}{3}\sigma_T .\tag{2.75}$$

Using the last equality of Eq. (2.69) with the help of Eq. (2.72), one can reach the same result.

Substituting Eq. (2.75) into Eq. (2.71), the integration is easily calculated (see Eq. (II.18) in the Appendix II), and the viscosity in the hard sphere model is

$$\mu = \frac{5}{16}(\pi m k T)^{1/2}/\sigma_T = \frac{5}{16}(mkT/\pi)^{1/2}/d^2 .\tag{2.76}$$

If in the vicinity of reference temperature T_{ref} consider the gas using the hard sphere model and the viscosity μ_{ref} under T_{ref} is known, then d can be taken as

$$d = \left[\frac{5}{16}\left(mkT_{ref}/\pi\right)^{1/2}/\mu_{ref}\right]^{1/2} .\tag{2.77}$$

And good result can be gained for slow motion when temperature variation is small.

The post-collision velocities of hard sphere molecules are easily calculated, as we have $c_r^* = c_r$ and the uniformity of all directions of c_r^* (see the discussion of the calculation of the post-collision velocities at the end of section 2.4.5). The shortcoming of the hard sphere model is that the dependence of μ on T is in power of 1/2 (see Eq. (2.76)), which is different from the power in the reality of approximately 0.75 for air. This is crucial for air at temperature of the order $1000K$ and higher.

2.4.2 THE INVERSE POWER LAW MODEL

The force and potential of the *inverse power law model* or the repulsion point center model is given by the following force expressions

$$F = \kappa / r^\eta .\tag{2.78}$$

Or in terms of potential

$$\phi = \kappa / [(\eta - 1) r^{\eta - 1}] .$$

The hard sphere model is a special case of this model with $\eta = \infty$.

To find the deflection angle one has Eq. (2.56) and Eq. (2.54). The ratio of the potential energy to the asymptotic value of kinetic energy can be written through W (see Eq. (2.51))

$$\frac{\phi}{\frac{1}{2} m_r c_r^2} = \frac{2\kappa}{(\eta - 1) r^{\eta - 1}} (m c_r^2)^{-1} = \frac{2}{\eta - 1} \left(\frac{W}{W_0} \right)^{\eta - 1} ,\tag{2.79}$$

where W_0 is the dimensionless miss distance

$$W_0 = b (m_r c_r^2 / \kappa)^{1/(\eta - 1)} .\tag{2.80}$$

Thus the deflection angle can be written

$$\chi = \pi - 2 \int_0^{W_1} \left[1 - W^2 - \frac{2}{\eta - 1} \left(\frac{W}{W_0} \right)^{\eta - 1} \right]^{-1/2} dW ,\tag{2.81}$$

where W_1 is the positive root of the following equation (see Eq. (2.55))

$$1 - W^2 - \frac{2}{\eta - 1} \left(\frac{W}{W_0} \right)^{\eta - 1} = 0 .$$

Note, for the inverse power law model with certain value of η, the collision is specified by the only impact parameter W_0 (dimensionless b). The differential cross-section of this model is obtained from Eq. (2.66)

$$\sigma d\Omega = W_0 \left[\frac{\kappa}{m_r c_r^2} \right]^{\frac{2}{\eta - 1}} dW_0 d\varepsilon .\tag{2.82}$$

2.4 COLLISION CROSS-SECTIONS AND MOLECULE MODELS 75

When η has finite value, the integral of the total cross-section diverges, as can be seen from Eq. (2.68). In such circumstances, it is usually necessary to make finite truncation either of miss distance b or deflection angle χ. From Eq. (2.81) it is seen, that χ is a function of W_0 (or b). The specification of the maximum value $W_{0,m}$ stipulates the truncation value of χ, and the total cross-section is

$$\sigma_T = \int_0^{2\pi} \int_0^{W_{0,m}} W_0 \left[\frac{K}{m_r c_r^2}\right]^{\frac{2}{\eta-1}} dW_0 d\varepsilon \tag{2.83}$$

$$= \pi W_{0,m}^2 \left(\frac{K}{m_r c_r^2}\right)^{\frac{2}{\eta-1}}. \tag{2.84}$$

Note, as the stipulation of $W_{0,m}$ is arbitrary, this total cross-section can not be used to define the collision frequency or the mean free path.

The integral in the expression Eq. (2.68) of the total cross-section for inverse power law molecule is divergent, but the viscosity and diffusion collision cross-sections are convergent. For these cross-sections σ_D and σ_μ (Eq. (2.70) and Eq. (2.69)), expressing the differential cross-section $\sigma d\Omega$ by Eq. (2.82), we have

$$\sigma_D = \int_0^{4\pi} (1 - \cos\chi) \sigma d\Omega = 2\pi \left(\frac{K}{m_r}\right)^{\frac{2}{\eta-1}} c_r^{-\frac{4}{\eta-1}} A_1(\eta), \tag{2.85}$$

$$\sigma_\mu = \int_0^{4\pi} \sin^2\chi \sigma d\Omega = 2\pi \left(\frac{K}{m_r}\right)^{\frac{2}{\eta-1}} c_r^{-\frac{4}{\eta-1}} A_2(\eta), \tag{2.86}$$

where

$$A_1(\eta) \equiv \int_0^\infty (1 - \cos\chi) W_0 dW_0, \tag{2.87}$$

$$A_2(\eta) \equiv \int_0^\infty \sin^2\chi W_0 dW_0. \tag{2.88}$$

The values of $A_1(\eta)$ and $A_2(\eta)$ can be integrated numerically with η as a parameter because of the functional relationship Eq. (2.81) between χ and W_0. Chapman and Cowling listed the values of $A_1(\eta)$ and $A_2(\eta)$ for several values of

2 SOME BASIC CONCEPTS OF KINETIC THEORY

η (see Table 2.1, cited from [1], p.172). Substituting Eq. (2.86) into Eq. (2.71) yields the viscosity for the inverse power law molecule

$$\mu = \frac{5m(RT/\pi)^{1/2}(2mRT/\kappa)^{\frac{2}{\eta-1}}}{8A_2(\eta)\Gamma\left[4-\frac{2}{(\eta-1)}\right]}. \tag{2.89}$$

The advantage of the inverse power law is that the deflection angle χ is dependent only on one impact parameter W_0 (or b) and the power in the dependence of μ on T can be adjusted to be the same as in the real gas by choosing appropriate η (see Eq. (2.89)). The shortcoming is that the specification of σ_T and hence the collision frequency and mean free path depends on the arbitrary stipulation of the value of W_0 which is inconvenient. In the direct simulation of gas flows the inverse power law model was in use (see [4]), but now is replaced by the variable hard sphere (VHS) model (see 2.4.4).

2.4.3 MAXWELL MODEL

The special case of the inverse power law model with $\eta = 5$ is called the *Maxwellian model*, and W_1 in Eq. (2.81) can be found in explicit form as

$$W_1 = W_0^2 \left[\left(1 + \frac{2}{W_0^4}\right)^{1/2} - 1 \right]^{1/2}.$$

And the deflection angle χ can be found through quadrature

Table 2.1 Several values of $A_1(\eta)$ and $A_2(\eta)$

η	$A_1(\eta)$	$A_2(\eta)$
5	0.422	0.436
7	0.385	0.357
9	0.382	0.332
11	0.383	0.319
15	0.393	0.309
21	-	0.307
25	-	0.306
∞	0.5	0.333

2.4 COLLISION CROSS-SECTIONS AND MOLECULE MODELS 77

$$\chi = \pi - \frac{2}{\left[1+\dfrac{2}{W_0^4}\right]^{\frac{1}{4}}} K\left[\frac{1}{2} - \frac{1}{2}\left(1+\frac{2}{W_0^4}\right)^{-\frac{1}{2}}\right]. \tag{2.90}$$

$K(\alpha)$ is the first kind complete elliptic integral

$$K(\alpha) = \int_0^{\pi/2} \left(1 - \alpha \sin^2 y\right)^{-1/2} dy. \tag{2.91}$$

The expression Eq. (2.82) of the differential cross-section takes the form

$$\sigma d\Omega = \frac{W_0}{c_r}\left(\frac{\kappa}{m_r}\right)^{1/2} dW_0 d\varepsilon, \tag{2.92}$$

that is, the collision cross-section is inversely proportional to the relative velocity c_r. In the Introduction we have shown that the collision frequency is proportional to the product of the cross-section and the relative velocity, thus the collision frequency in the gas of Maxwell molecules is independent of c_r. This brings about the possibility of simplification in the analytical methods (see the calculation of the collision integrals in section 2.8), so the Maxwellian molecule is in wide application in analytical works. But the viscosity of the gas of Maxwellian molecules is linearly proportional to the temperature. In fact, in Eq. (2.89) let $\eta = 5$, one has

$$\mu = \frac{2kT}{3\pi A_2(5)}\left(\frac{m}{2k}\right)^{1/2}. \tag{2.93}$$

This is unrealistic, and thus the Maxwell molecule has significant limitations.

2.4.4 VARIABLE HARD SPHERE (VHS) MODEL

The finite truncation of the miss distance b or the deflection angle χ for the inverse power law is a practical necessity, but this inevitably introduces arbitrariness, followed by the arbitrary determination of the molecular diameter and the mean free path. We have seen, that by comparison of the Chapman-Enskog theoretical expression of the viscosity for hard sphere model and the viscosity measured in real gas the molecular diameter can be derived (see Eq. (2.77)). And the isotropic distribution of the direction of the post-collision relative velocity of the hard sphere molecule is another advantage of this model. Unfortunately the hard sphere molecule has fixed cross-section, but the collision cross-section of real gas changes with the relative velocity (decreases with the increase of c_r). The difference of various molecular models manifests in the different dependence of the viscosity μ on the temperature T. For hard sphere, μ is proportional to T in the power of 1/2, but in real gas (for oxygen, nitrogen and air) this power is near 0.75. The problem is that any molecular model in classical kinetic theory is unable to simultaneously possess a finite collision cross-section and reveal the correct dependence of the viscosity on the temperature. Bird put forward the *variable hard sphere* (*VHS*) model [5] and resolved this problem successfully. In this model the molecules possess isotropic scattering probability as the hard sphere (see Eq. (2.73) and the explanation following it), but the cross-section is a function of the relative velocity between the molecules like the inverse power law molecules (see Eq. (2.84)). Possession of the same scattering law as the hard sphere makes the sampling of the distribution of the direction of the post-collision velocity very simple, and analysis and the practical application of the mathematical investigation show, that the way of scattering of molecules has little consequences on the flow field. Choosing the collision cross-section (or the diameter) of the molecules to be proportional to a certain inverse power of the relative velocity of the molecules can lead to that the derived viscosity varies as a certain power of the temperature, say, lead to that μ to be proportional approximately to the 0.75 power of T, i.e. in the same way as in the real gas. The practical applications show, that the influence of this dependence of the collision cross-section on the properties of the flow field is essential.

The *VHS model or the Variable Hard Sphere model* [5] stipulates that the collision cross-section is proportional to the inverse power of the relative velocity. The reference values $\sigma_{Tref}, d_{ref}, c_{rref}$ and ε_{tref} are introduced for the total cross-section σ_T, molecular diameter d, relative velocity c_r and relative translational energy ε_t, where σ_{Tref} and d_{ref} are the values when the relative velocity is c_{rref}. The VHS model can be defined as follows

$$(\sigma_T / \sigma_{Tref}) = (d / d_{ref})^2 = (c_r / c_{rref})^{-2\xi} = (\varepsilon_t / \varepsilon_{tref})^{-\xi}. \qquad (2.94)$$

2.4 COLLISION CROSS-SECTIONS AND MOLECULE MODELS 79

ξ by definition is the (negative) power of ε_t in the dependence of collision cross-section σ_T on the translational energy ε_t.

The deflection angle is the same as the hard sphere model (see Eq. (2.72))

$$b = d\cos(\chi/2) . \qquad (2.95)$$

That is, the deflection angle χ does not depend on c_r explicitly. Thus, in the VHS model in the integral(Eq. (2.69) of the viscosity the collision cross-section σ as in the case of hard sphere can be taken out of the integral and yield the Equation (2.75). Now making use of Eq. (2.94), writing σ_T as $\left(c_r/c_{r\,ref}\right)^{-2\xi}\sigma_{Tref}$ and substituting into Eq. (2.71), one obtains

$$\mu = \frac{\frac{15}{8}(\pi mk)^{1/2}(4k/m)^{\xi} T^{1/2+\xi}}{\Gamma(4-\xi)\sigma_{Tref} c_{r\,ref}^{2\xi}} . \qquad (2.96)$$

Note, for the case of $\xi = 1/2$ one has $(\Gamma(7/2) = (15/8)\sqrt{\pi})$

$$\mu = \frac{2kT}{\sigma_{Tref} c_{rref}} . \qquad (2.96)'$$

From the comparison of Eq. (2.76) and Eq. (2.96) it is seen, that the viscosity of the hard sphere model is proportional to the 1/2 power of the temperature, but in the VHS model μ is proportional to the power $(1/2+\xi)$ of T. By adjusting the power ξ in the definition formula Eq. (2.94), one can attain the required power ω in the dependence of viscosity on temperature

$$\mu \propto T^{\omega} . \qquad (2.97)$$

For this one only needs to put

$$\frac{1}{2}+\xi = \omega . \qquad (2.98)$$

We also list the relations of ω and ξ with the power η in the inverse power law Eq. (2.78). From the Comparison of Eq. (2.97), Eq. (2.96) and Eq. (2.89) it is easy to see

$$\xi = \frac{2}{\eta-1}, \quad \omega = \frac{1}{2}\frac{\eta+3}{\eta-1} . \qquad (2.99)$$

For the hard sphere (HS) model one has $\omega = 1/2$, $\eta = \infty$ and $\xi = 0$, for the variable hard sphere (VHS and also the inverse power law IPL) molecule with $\omega = 0.75$, one has $\eta = 9$ and $\xi = 0.25$, and for the Maxwellian molecule one has $\omega = 1$, $\eta = 5$ and $\xi = 1/2$. The hard sphere model with $\eta = \infty$ is the most hard molecule, and the Maxwellian molecule is the most soft among the molecular models under study.

The relation of the molecule diameter in dependence of the relative translational energy ε_t in a gas with viscosity law of Eq. (2.97) and having μ_{ref} at temperature T_{ref} can be obtained by making use of Eq.(2.94), Eq. (2.96) and Eq. (2.98)

$$d_{VHS} = \left[\frac{(15/8)(m/\pi)^{1/2}(kT_{ref})^{\omega}}{\Gamma(9/2-\omega)\mu_{ref}\varepsilon_t^{\omega-\frac{1}{2}}} \right]^{1/2}. \tag{2.100}$$

2.4.5 VARIABLE SOFT SPHERE (VSS) MODEL

For the hard sphere (HS) and the VHS models we have the relation of the viscosity collision cross-section σ_μ with the total cross-section σ_T (see Eq. (2.75))

$$\sigma_\mu = \frac{2}{3}\sigma_T.$$

Similarly, from the expression Eq. (2.70) of σ_D by using Eq. (2.73) we have

$$\sigma_D = \sigma_T. \tag{2.101}$$

For VHS and HS models one has constant ratios of σ_μ and σ_D to σ_T. But for inverse power law molecule the ratio of σ_D to σ_μ is $A_1(\eta)/A_2(\eta)$ (see Eq.(2.85), Eq.(2.86)), and for realistic values of η, $A_1(\eta)/A_2(\eta)$ is quite different from the value $\sigma_D/\sigma_\mu = 1.5$ for HS and VHS models (see Table 2.1 when $\eta = 9$ and 11). This results in quite large deviation from practice of the VHS model in considering multi-component gas mixture with diffusion playing essential role. Koura and Matsumoto [6, 7] introduced the *variable soft sphere* (*VSS*) model and overcame this shortcoming of the VHS model. They suggest that the total cross-section (or the molecular diameter) depends on the energy during the collision as in the VHS model, i.e., Eq. (2.94) still holds, but instead of

$$b = d\cos(\chi/2),$$

the scattering law of the VSS model is

$$b = d\cos^\alpha(\chi/2), \tag{2.102}$$

where d and α both depend on the relative energy in the collision, α being the power of the cosine of the deflection angle. As d is variable and for $\alpha > 1$ (it is the case for real gas, vide post) the deflection angel χ_{VSS} is less than the deflection angle χ_{VHS} of the HS and VHS model, so it is named the variable soft sphere model.

For the VSS model substituting Eq. (2.102) into Eqs. (2.68), (2.69) and (2.70) it is easy to obtain

$$\sigma_T = 2\pi \int b\,db = \pi d^2, \tag{2.103}$$

$$\sigma_\mu = 2\pi d^2 \int (1-\cos^2\chi)\left(\cos\frac{\chi}{2}\right)^\alpha d\left(\cos\frac{\chi}{2}\right)^\alpha = S_\mu\left(\frac{2}{3}\pi d^2\right) = S_\mu \frac{2}{3}\sigma_T, \tag{2.104}$$

$$\sigma_D = 2\pi d^2 \int (1-\cos\chi)\left(\cos\frac{\chi}{2}\right)^\alpha d\left(\frac{\chi}{2}\right)^\alpha = S_D \pi d^2 = S_D \sigma_T, \tag{2.105}$$

where

$$S_\mu = \frac{6\alpha}{(\alpha+1)(\alpha+2)}, \tag{2.106}$$

$$S_D = \frac{2}{\alpha+1} \tag{2.107}$$

are called soft coefficients of viscosity and diffusivity collision cross-sections, respectively. Equalizing the ratio $\sigma_\mu/\sigma_D = (2/3)S_\mu/S_D$ (from Eqs. (2.104) and (2.105)) for the VSS model to the ratio $\sigma_\mu/\sigma_D = A_2(\eta)/A_1(\eta)$ (from Eqs. (2.85) and (2.86)) for the inverse power law model, the value of α is obtained

$$\alpha = \left[\frac{A_1(\eta)}{A_2(\eta)} - \frac{1}{2}\right]^{-1}, \tag{2.108}$$

under which the ratio σ_μ/σ_D of the VSS model would vary according to the same law of the inverse power law model. The values of α and the corresponding S_μ and S_D for various gases can be determined from $A_1(\eta)$ and $A_2(\eta)$ corresponding to η (and the value of ξ is also determined according to Eq. (2.99)). The values of α for some gases determined in this way are listed in the 6th column of Table 2.2.

The viscosity μ under the VSS model is obtained by substitution of Eq. (2.104) into Eq. (2.71)

$$\mu_{VSS} = \mu_{VHS}/S_\mu, \qquad (2.109)$$

where μ_{VHS} is given by Eq.(2.96), and

$$\mu_{VSS} = \frac{5(\alpha+1)(\alpha+2)(\pi m k)^{1/2}(4k/m)^\xi T^{\frac{1}{2}+\xi}}{16\alpha\Gamma(4-\xi)\sigma_{T_{ref}}c_{rref}^{2\xi}}. \qquad (2.109)'$$

Correspondingly, the diameter of the molecule determined by the viscosity under the VSS model is

$$d_{VSS} = d_{VHS}/S_\mu^{1/2}, \qquad (2.110)$$

where d_{VHS} is given by Eq. (2.100), i.e.,

$$d_{VSS} = \left[\frac{5(\alpha+1)(\alpha+2)(m/\pi)^{1/2}(kT_{ref})^\omega}{16\alpha\Gamma(9/2-\omega)\mu_{ref}\varepsilon_t^{\omega-\frac{1}{2}}}\right]^{1/2}. \qquad (2.110)'$$

This expression for d_{VSS} (which yields the expression Eq. (2.100) of d_{VHS} when $\alpha = 1$) gives the relation of the molecular diameter with the collision translational energy of the VSS (and VHS) model, and can be used to implement the simulation of VSS(VHS) molecules in DSMC. Obviously, the dependence on ε_t can be written in a dimensionless form as

$$\left[(kT_{ref}/\varepsilon_t)^{\omega-\frac{1}{2}}\right]^{\frac{1}{2}}.$$

And the right hand side of Eq. (2.110)' can be written as

$$d_*\left[(kT_{ref}/\varepsilon_t)^{\omega-\frac{1}{2}}\right]^{\frac{1}{2}},$$

2.4 COLLISION CROSS-SECTIONS AND MOLECULE MODELS 83

where d_* is a value having the dimension of length and depending on μ_{ref} and T_{ref}. This is convenient. We defer the introduction of d_{ref} (which differs from d_* by a coefficient) till section 2.11, for it is most convenient to define d_{ref} by using the average value of $c_r^{2\omega-1}$ in the equilibrium state(see Eq. (2. 235)).

The way of determination of α is not unique. It could be determined also by the fitting of the experimental data of the viscosity of various gases with the expressions like Eq. (2.109)' (and simultaneously determined also is the value of ξ), in reference [7] the values of ξ and α are determined in this way in the temperature ranges $20K \sim 300K$ and $300K \sim 2000K$. And Bird in reference [8] determined the value of α using the experimental data of the self-diffusion coefficient of various gases. The values of α determined by these two means are also listed in Table 2.2. It can be seen that the value of α determined by the experimental data of the self-diffusion coefficient agrees quite well with that determined by the viscosity data (in the temperature range of $300K \sim 2000K$). The values of Bird [8] are recommended (see Appendix II).

The VHS and VSS collision models are peculiarly devised for being used in DSMC simulations. The implementation of collision for the VSS model molecules in DSMC simulation is more complicated than for the VHS molecules. For all HS, VHS and VSS models there is the invariance of the relative velocity before and after collision (see Eq. (2.44)). For HS and VHS molecules there is the result of the isotropic distribution of the post-collision velocity c_r^* over all directions (see Eq. (2.73) and the explanation after it). In a polar coordinate system with θ as the

Table 2.2 The values of α for some gases

	η	ξ	$A_1(\eta)$	$A_2(\eta)$	α [6]	α [7] 20~300	α [7] 300~2000	α [8]
O_2	8.4	0.27	0.381	0.3360	1.5775	1.92	1.429	1.4
N_2	9.4	0.238	0.3810	0.3273	1.5059	1.784	1.362	1.36
Air	8.4	0.27	0.3810	0.3360	1.5775	1.878	1.492	
H_2	12.9	0.168	0.3878	0.3130	1.3532	1.578	1.396	1.35
He	13.7	0.157	0.3898	0.3115	1.3309	1.349	1.431	1.26
Ar	7.5	0.308	0.383	0.3477	1.6624	1.856	1.425	1.40

zenith angle and ϕ as the azimuth angle, the probability of c_r^* pointing into an element of solid angle $d\omega = \sin\theta d\theta d\phi$ is uniform. As $d\omega = -d(\cos\theta)d\phi$, so the probability over $\cos\theta$ and ϕ is uniform. ϕ is uniformly distributed between 0 and 2π, and $\cos\theta$ is uniformly distributed between -1 and 1. From Appendix III the sampling from a variate uniformly distributed between $a = -1$ and $b = 1$ is easily obtained (see Eq. (III-5)), and the values of $\cos\theta$ and ϕ into which c_r^* is directed can be sampled as

$$\cos\theta = 2ranf - 1,$$

$$\phi = 2\pi ranf, \quad (2.111)$$

where $ranf$ is an uniform variate in the interval $(1,0)$. The three components of c_r^* are easily obtained as $c_r^* \sin\theta \cos\phi$, $c_r^* \sin\theta \sin\phi$ and $c_r^* \cos\theta$. The components of the post-collision velocities of the two collision partners can be found from Eqs. (2.41) as

$$u_1^* = \frac{m_1 u_1 + m_2 u_2}{m_1 + m_2} + \frac{m_2}{m_1 + m_2}\sin\theta\cos\phi c_r^*,$$

$$u_2^* = \frac{m_1 u_1 + m_2 u_2}{m_1 + m_2} - \frac{m_1}{m_1 + m_2}\sin\theta\cos\phi c_r^*,$$

$$v_1^* = \frac{m_1 v_1 + m_2 v_2}{m_1 + m_2} + \frac{m_2}{m_1 + m_2}\sin\theta\sin\phi c_r^*,$$

$$v_2^* = \frac{m_1 v_1 + m_2 v_2}{m_1 + m_2} - \frac{m_1}{m_1 + m_2}\sin\theta\sin\phi c_r^*,$$

$$w_1^* = \frac{m_1 w_1 + m_2 w_2}{m_1 + m_2} + \frac{m_2}{m_1 + m_2}\cos\theta c_r^*,$$

$$w_2^* = \frac{m_1 w_1 + m_2 w_2}{m_1 + m_2} - \frac{m_1}{m_1 + m_2}\cos\theta c_r^*. \quad (2.112)$$

For the VSS model the scattering law is Eq. (2.102), and one has

$$\cos\chi = 2\left[\left(\frac{b}{d}\right)^2\right]^{1/\alpha} - 1. \quad (2.113)$$

Because $(b/d)^2$ is uniformly distributed between 0 and 1, so $\cos\chi$ is sampled as follows

$$\cos\chi = 2(ranf)^{1/\alpha} - 1. \quad (2.114)$$

Though this is analogous with the sampling formula Eq. (2.111) of the VHS model, but the three components of the post-collision relative velocity c_r^* in the VHS model are obtained easily through multiplication by $\cos\theta, \sin\theta\cos\phi$ and $\sin\theta\sin\phi$. But in the VSS model the components of the post-collision relative velocity are to be obtained by using Eqs. (2.59)', (2.61)' and (2.63)'.

2.4.6 GENERALIZED HARD SPHERE (GHS) MODEL

In the VHS model the power ξ (in the dependence of collision cross-section σ_T on the translational energy ε_t) is related to the power η in the inverse power law (Eq. (2.78)) and the power ω in the dependence of viscosity on the temperature (Eq. (2.97)) (see Eqs (2.98) and (2.99)). The inverse power law describes the intermolecular action as a pure repulsion force. For flow fields around re-entry vehicles the temperature variation range is very large, the dependence of viscosity on the temperature can not be represented entirely by an one-exponent power law. The interaction between molecules except the short distance repulsion force reveals the attraction character at large intermolecular distances. A molecular model capable of reproducing attractive-repulsive potential is desirable. The *generalized hard sphere* (*GHS*) introduced by Hassan and Hash [9], being a generalization of the concepts of VHS model and VSS model, contains the intermolecular force that possesses both attraction and repulsion. Its scattering law is similar with the hard sphere model (or the soft sphere model), but the relation of its total cross-section σ_T with the relative kinetic energy ε_t allows the reproduction of the attractive–repulsive potential. More concretely, the GHS model suggests that σ_t can be written

$$\sigma_T / \rho^2 = \sum \alpha_j \left(\frac{\varepsilon_t}{\varepsilon}\right)^{-\xi_j}, \qquad (2.115)$$

where ε has the dimension of energy, ρ has the dimension of length and is a parameter depending on the gas component, ε_t is, as before,

$$\varepsilon_t = \frac{1}{2} m_r c_r^2. \qquad (2.116)$$

If take only one term in Eq. (2.115), it can be seen to be the same as the definition formula Eq. (2.94) of the VHS model, with only different definition of the constants. Aiming at the expressions usually used in practice for the viscosity and interaction potential, two terms in Eq. (2.115) are taken as follows

$$\sigma_T / \rho^2 = \alpha_1 \left(\frac{\varepsilon_I}{\varepsilon}\right)^{-\xi_1} + \alpha_2 \left(\frac{\varepsilon_I}{\varepsilon}\right)^{-\xi_2}. \tag{2.117}$$

If the scattering law is taken as the hard sphere, the relation between σ_μ and σ_T is Eq. (2.75):

$$\sigma_\mu = \frac{2}{3}\sigma_T.$$

Substituting Eqs. (2.115), (2.116) and (2.75) into the expression Eq. (2.71) of the viscosity, one obtains

$$\mu = \frac{\frac{15}{8}(\pi m k T)^{1/2} / \rho^2}{\sum \alpha_j \Gamma(4-\xi_j)(kT/\varepsilon)^{-\xi_j}}. \tag{2.118}$$

If the two-term expression Eq. (2.117) is used, then

$$\mu = \frac{\mu_0}{1 + ST^{\xi_2 - \xi_1}},$$

$$\mu_0 = \frac{15(\pi m k T)^{1/2}}{8} \frac{(kT)^{\xi_1}}{\alpha_1 \Gamma(4-\xi_1)\rho^2}, \quad S = \frac{\alpha_2}{\alpha_1} \frac{\Gamma(4-\xi_2)}{\Gamma(4-\xi_1)} \left(\frac{k}{\varepsilon}\right)^{\xi_1 - \xi_2}. \tag{2.119}$$

In Eq. (2.119) let $\xi_1 = 0$, $\xi_2 = 1$, the viscosity law of Sutherland is obtained. For the Lennard-Jones interaction potential

$$F = \frac{K_{12}}{r^\eta} - \frac{K'_{12}}{r^{\eta'}}, \quad \eta > \eta' \tag{2.120}$$

Chapman and Cowing gives the following expression of the viscosity ([1], p.182)

$$\mu = \mu_{IPL} / \left(1 + ST^{-\frac{\eta - \eta'}{\eta - 1}}\right), \tag{2.121}$$

where μ_{IPL} is the expression of the viscosity under inverse power law (see Eq. (2.89)). Comparison of Eq. (2.119) and Eq. (2.121) yields

$$\xi_1 = \frac{2}{\eta - 1}, \quad \xi_2 = \frac{2 - \eta' + \eta}{\eta - 1}. \tag{2.122}$$

For the Lennard-Jones 6-12 ($\eta = 13, \eta' = 7$) model, one has $\xi_1 = 1/6$, $\xi_2 = 2/3$.

2.4.7 GENERALIZED SOFT SPHERE (GSS) MODEL

Fan in [10] suggested the simultaneous adoption of the GHS dependence of σ_T on ε_t (Eq. (2.115)) and the VSS molecular scattering law (Eq. (2.102)) which naturally lead to the *generalized soft sphere (GSS)* model. In the GSS model σ_t is calculated explicitly from its expression in dependence of c_r (see Eq. (2.117)), where the values of ε and ρ for Lennard-Jones 6-12 model have been tabled in reference [11] (see Table 73, in section 4 of Chapter 8, and Table 1 of Appendix I of [11], also Table 4 of Appendix I in the present book). The deflection angles of the GSS molecules in collisions are determined by Eq. (2.102) or Eq. (2.113). To implement the model the power α of the cosine of the deflection angle and the parameters $\alpha_1, \alpha_2, \xi_1$ and ξ_2 in Eq. (2.117) need to be determined. For this purpose the expression Eq. (2.71) for the viscosity is written as [11]

$$\mu = \frac{5}{16}\left(\frac{\sqrt{\pi m k T}}{\pi \sigma^2 \Omega^{(2,2)^*}}\right), \tag{2.123}$$

where

$$\Omega^{(2,2)^*} = \frac{1}{2\pi\sigma^2} \int_0^\infty \exp(-\zeta^2)\zeta^7 \sigma_\mu d\zeta, \tag{2.124}$$

$$\zeta = c_r / \sqrt{2\frac{k}{m_r}T}. \tag{2.125}$$

Substituting Eq. (2.104) and Eq. (2.117) into Eq. (2.124), one obtains the viscosity integral for GSS model

$$\Omega^{(2,2)^*} = \frac{\alpha}{\pi(\alpha+1)(\alpha+2)}\left[\alpha_1\Gamma(4-\xi_1)T_*^{-\xi_1} + \alpha_2\Gamma(4-\xi_2)T_*^{-\xi_2}\right], \tag{2.126}$$

where $T_* = kT'/\varepsilon$. The parameters in Eq. (2.126) can be determined by comparison of its right hand side with those of Eq. (2.124), if the latter for certain potential can be expressed explicitly as function of T_*. In fact, the integral $\Omega^{(2,2)*}$ for Lennard-Jones 6-12 potential has been calculated through numerical quadrature by Hirschfelder, Curtiss and Bird and tabulated against T_* from 0.3 to 400 (see [11], Table XII in the Appendix). As direct comparison shows, the following fitting formula

$$\Omega^{(2,2)*}_{fit} = a_1 T_*^{-c_1} + a_2 T_*^{-c_2}, \tag{2.127}$$

with $a_1 = 1.1$, $a_2 = 0.4$, $c_1 = 0.133$ and $c_2 = 1.25$ gives a good fit to the exact tabulated data of [11]. Comparison of Eq. (2.126) and Eq. (2.127) determines all parameters required by the GSS model:

$$\alpha = 1.5, \quad \alpha_1 = 3.962, \quad \alpha_2 = 4.558, \quad \xi_1 = 0.133 \text{ and } \xi_2 = 1.25. \tag{2.128}$$

By modification of the coefficients a_1 and a_2 in Eq. (2.127) this fitting formula can also be used to fit the exact numerical quadrature of $\Omega^{(2,2)*}$ for the Stockmayer potential [12], which takes into account the electrostatic interaction between polar molecules (for details see [10]). The viscosity, self-diffusivity and diffusivity calculated by the GSS model yield good agreement with the experimental data in the entire range of variation of temperature. The GSS model is of meaning especially for the case of low temperature and polar molecules.

2.5 THE EIGHT VELOCITY GAS MODEL

In the present section a simplified discrete gas model, the *eight velocity gas model* introduced by Broadwell [13], is considered before the derivation of the Boltzmann equation in the next section. The eight velocity gas model is the first gas model in researching gas flow problems by the discrete velocity method. A brief consideration of this simple model also has the heuristic meaning for the easier comprehension of the derivation of the Boltzmann equation.

In the eight velocity gas model the velocity of a molecule can be taken only as one of eight velocities $c_1, c_2, \cdots c_8$, formed by pointing from the center of a cube (see Fig.2.8) to its eight corner points (1, 2, ..,8). The magnitude of every velocity is the same and equals to $\bar{c}$. The molecule with velocity c_i is called molecule

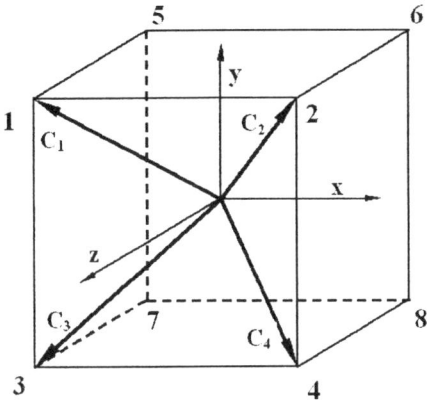

Fig. 2.8 The eight-velocity gas model, the velocities of the gas can attain only $c_1, c_2, ..., c_8$, in the figure only c_1, c_2, c_3, c_4 are shown

of class i. Obviously, the magnitudes q of the components of every velocity are the same and $q = \bar{c}/\sqrt{3}$. The momentum conservation requires the constancy of the combination of signs of the velocity components of molecular pairs. The value c_{r12} of the relative velocity between molecules 1 and 2 is $2q = 2\bar{c}/\sqrt{3}$, the values of $c_{r13}, c_{r14}, ..., c_{r18}$ can be written easily as well.

Let n_i represent the number density of molecules of class i, and in general it is a function of time and space coordinates. Giving the whole set (c_1, n_1), (c_2, n_2), ..., (c_8, n_8) is equivalent to the fixing of the distribution function f. In equilibrium state the magnitudes of various n_i equal each other, and $n_i = n/8$, where n is the density of the gas. Obviously, the collision frequency ν of the molecule can be calculated:

$$\nu = \sum_{i=1}^{8} c_{ri} \sigma_T n_i = \frac{10.36}{8} \bar{c} \sigma_T n = 1.295 \bar{c} \sigma_T n. \tag{2.129}$$

σ_T is the collision cross-section of the molecule, the mean free path is obtained

$$\lambda = \frac{\bar{c}}{\nu} = 0.772/(\sigma_T n). \tag{2.130}$$

Now consider in the eight velocity gas the *depletion* of the number of molecules of class 1 caused by collisions with molecules of class 4. The number of collisions of the type (1-4) that cause the reduction of molecules of class 1 in unit

90 2 SOME BASIC CONCEPTS OF KINETIC THEORY

time in unit volume is $c_{r14}\sigma_T p_{14} n_1 n_4$, where p_{14} is the probability that the result of the (1-4) collision is the depletion of molecule of class 1. As the (1-4) collision has two consequences of equal probability, (1-4)→(4-1), and (1-4)→(2-3) (please refer to Fig. 2.8), so one has

$$p_{14} = \frac{1}{2}. \tag{2.131}$$

Molecules of class 1 deplete also because of the collisions with molecules of class 6, class 7 and class 8. So the number of the depleted (lost) molecules of class 1 as the consequence of collisions in unit time in unit volume is $\left(p_{16} = p_{17} = 1/2 = p_{14}\right)$.

$$L_1 = c_{r14}\sigma_T p_{14}(n_1 n_4 + n_1 n_6 + n_1 n_7) + c_{r18}\sigma_T p_{18} n_1 n_8, \tag{2.132}$$

where

$$p_{18} = \frac{3}{4}, \tag{2.133}$$

as the (1-8) collision has four consequences of equal probability: (1-8)→(8-1), (1-8)→(2-7), (1-8)→(3-6) and (1-8)→(4-5), with three leading to the depletion of molecules of class 1.

For planar two-dimensional flow the flow characteristics do not depend on z, and the symmetry condition holds

$$n_1 = n_5, n_2 = n_6, n_3 = n_7, n_4 = n_8. \tag{2.134}$$

Then Eq. (2.132) can be written as

$$L_1 = \bar{c}\sigma_T \left[(2/3)^{1/2} (n_1 n_4 + n_1 n_2 + n_1 n_3) + \frac{3}{2} n_1 n_4 \right]. \tag{2.135}$$

Analogous reasoning yields the number of replenished (gained) molecules of class 1 as the consequence of collisions in unit time in unit volume is

$$G_1 = c_{r23}\sigma_T p_{23}(n_2 n_3 + n_2 n_5 + n_3 n_5) + c_{r27}\sigma_T p_{27}(n_2 n_7 + n_3 n_6 + n_4 n_5). \tag{2.136}$$

where

$$p_{23} = \frac{1}{2}, \quad p_{27} = \frac{1}{4}. \tag{2.137}$$

2.5 THE EIGHT VELOCITY GAS MODEL

For planar two-dimensional flow case by using the symmetry condition Eq. (2.134) one has

$$G_1 = \bar{c}\sigma_T \left[(2/3)^{1/2} (n_2 n_3 + n_2 n_1 + n_3 n_1) + \frac{1}{2}(n_2 n_3 + n_2 n_3 + n_4 n_1) \right]. \quad (2.138)$$

The fact that the change rate in time $dn_1/dt = \partial n_1/\partial t + u_i(\partial n_1/\partial x_i)$ of the density of molecules of class 1 is caused by the depleting and replenishing collisions of molecules of class 1 can be written

$$\frac{\partial n_1}{\partial t} + u_1 \frac{\partial n_1}{\partial x} + v_1 \frac{\partial n_1}{\partial y} + w_1 \frac{\partial n_1}{\partial z} = \left(\frac{\partial n_1}{\partial t}\right)_c = G_1 - L_1, \quad (2.139)$$

where the terms G_1 and L_1 are given in Eq. (2.138) and Eq. (2.135). This is one of the *discrete velocity Boltzmann equations* for the eight velocity gas. It is appropriate here to write the Boltzmann equation without the external forces to be derived in the next section

$$\frac{\partial f}{\partial t} + u\frac{\partial f}{\partial x} + v\frac{\partial f}{\partial y} + w\frac{\partial f}{\partial z} = \left(\frac{\partial f}{\partial t}\right)_c. \quad (2.140)$$

It is noted that the whole set $(c_1, n_1), (c_2, n_2), \cdots, (c_8 n_8)$ is equivalent to the distribution function f. Analogous equations for $n_2, n_3, \cdots, n_8$ need to be added to Eq. (2.139) to have the whole formulation of the problem which will be done later in section 6.6.

Eq. (2.139) can be further simplified

$$\frac{\partial n_1}{\partial t} - q\frac{\partial n_1}{\partial x} + q\frac{\partial n_1}{\partial y} = \left(1 + (2/3)^2\right)\bar{c}\sigma_T (n_2 n_3 - n_1 n_4).$$

or

$$\frac{\partial n_1}{\partial t} - q\frac{\partial n_1}{\partial x} + q\frac{\partial n_1}{\partial y} = 2\theta (n_2 n_3 - n_1 n_4)/n, \quad (2.141)$$

where

$$\theta = \frac{1}{2}\left(1 + (2/3)^2\right)\bar{c}\sigma_T n = 0.70\nu. \quad (2.142)$$

2.6 BOLTZMANN EQUATION

In the present chapter the concept of the velocity distribution function has been introduced, various macroscopic quantities have been expressed through the velocity distribution function, thus showing how the distribution function gives the statistical description of the gas on the molecular level. Now the problem is how to find this velocity distribution function. An equation first derived by Ludwig Boltzmann in 1872 (see [14]) has given answer to this question and is known as the *Boltzmann equation*, it presents the evolution of the distribution function in dependence of space position and time.

For N particle distribution function $F^{(N)}$ (Eq. (2.1)), Liouville equation is the basic statistical equation, this is a conservation equation of $F^{(N)}$ in $6N$ dimensional phase space. Successive integration of the Liouville equation leads to a hierarchy of equations, the BBGKY equations named after the five authors Bogoliubov, Born, Green, Kirkwood and Yvon who first suggested these equations independently (for the derivation of BBGKY equations and the original references see e.g. [15]). Each chain of this hierarchy is an equation for certain reduced distribution function $F^{(R)}$, but involving higher order distribution function $F^{(R+1)}$. The last chain of this hierarchy is an equation for single particle distribution function $F^{(1)}$, involving also the two particle distribution function $F^{(2)}$. If at this stage introduce the assumption of *molecular chaos*

$$F^{(2)}(c_1,c_2,r_1,r_2,t) = F^{(1)}(c_1,r_1,t) \cdot F^{(1)}(c_2,r_2,t), \qquad (2.143)$$

a closed equation for $F^{(1)}$ is obtained. As there is a relationship Eq. (2.3) between f and $F^{(1)}$, the Boltzmann equation follows. But we would not go this road but give a derivation in analogy with the conservation equation for f. Such a derivation has the advantage of independence on the Liouville equation and simplicity, and also the ease in gaining clear comprehension of the physical meaning of various terms of the equation. Note, the unknown f searched by us is the probabilistic number of molecules in the vicinity of certain velocity c, so unlike ordinary conservation, the unknown variable f changes as the sequence of the collision. This is a peculiarity of the derivation of the Boltzmann equation in difference of the derivation of ordinary conservation equations. The counting of the change of number of molecules of class 1 as the result of collisions with other molecules described in the previous section for the eight velocity gas illustrates this situation.

Consider $f(c,r,t)$, the number of molecules at time instant t, in the physical space volume element $dr \equiv dxdydz$, with velocity in the velocity space element $dc \equiv dudvdw$ near c is $fdcdr$ (see Eq. (2.4) and Fig. 2.1). In the fixed phase space element the rate of change of the number of molecules is

$$\frac{\partial}{\partial t} f(c,r,t) dcdr \ . \tag{2.144}$$

This change is caused by the following three processes: 1) The convection of molecules across the surface of dr as they move with velocity c. 2) The passing through of the molecules across the surface of dc in the phase space as the result of the action of external force. 3) The depletion (loss) and replenishment (gain) of molecules of class c as a result of intermolecular collisions.

First consider the convection of molecules across the side surfaces of cubic element $dr \equiv dxdydz$, (see Fig. 2.9). The number of molecules interring dr at x in time dt across surface $dydz$ perpendicular to axis x is $fdc \cdot udtdydz$. The number of molecules leaving dr at $x+dx$ crossing $dydz$ in time dt is $\left[uf + (\partial(uf)/\partial x)dx\right]dcdtdydz$. Thus, the gain of number of molecules of class c in time t caused by the flux across the two side surface perpendicular to axis x is

$$-\frac{\partial}{\partial x}(uf)dcdrdt \ .$$

And the gain of number of molecules of class c across all side surfaces of dr is

$$-\left[\frac{\partial}{\partial x}(uf) + \frac{\partial}{\partial y}(vf) + \frac{\partial}{\partial z}(wf)\right]dcdrdt \ .$$

As $u,v,w,$ and x,y,z are equal arguments of f, this expression can be written as

$$-\left(u\frac{\partial}{\partial x} + v\frac{\partial}{\partial y} + w\frac{\partial}{\partial z}\right)fdcdrdt = -c \cdot \frac{\partial f}{\partial r}dcdrdt \ . \tag{2.145}$$

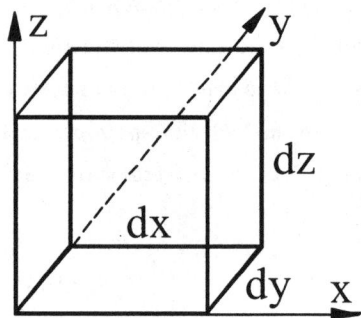

Fig. 2.9 Volume element $d\mathbf{r}$ in the physical space

Next let us consider the convection of molecules across various side surfaces of $d\mathbf{c}$. Suppose there is external force acting on the gas molecules, characterized quantitatively by the external force per unit mass $\mathbf{F}$. $\mathbf{F}$ makes the molecules to accelerate, its action in changing molecules in the phase space $\mathbf{c}$ is analogous to the action of velocity $\mathbf{c}$ in changing molecules in the physical space. Taking advantage of the analogy between $\mathbf{F},\mathbf{c}$ and $\mathbf{c},\mathbf{r}$ one can write the change of molecules of class $\mathbf{c}$ in time dt passing through all side surfaces of $d\mathbf{c}$.

$$-\left(F_x \frac{\partial}{\partial u}+F_y \frac{\partial}{\partial v}+F_z \frac{\partial}{\partial w}\right) f \cdot d\mathbf{c} d\mathbf{r} dt = -\mathbf{F} \cdot \frac{\partial f}{\partial \mathbf{c}} d\mathbf{c} d\mathbf{r} dt. \qquad (2.146)$$

Now consider the change of the molecules of class $\mathbf{c}$ as the consequence of the collisions in the element $d\mathbf{c} d\mathbf{r}$ in time dt : $(\partial f/\partial t)_c d\mathbf{c} d\mathbf{r} dt$. We consider the collision of molecules of class $\mathbf{c}$ with molecules of class $\mathbf{c}_1$, their velocities become $\mathbf{c}^*$ and $\mathbf{c}_1^*$ after collision. A molecule of class $\mathbf{c}$ is considered as moving with relative velocity $\mathbf{c}_r$ in mass of molecules of class $\mathbf{c}_1$. In time dt the volume swept out by this molecule of class $\mathbf{c}$ is $c_r \sigma d\Omega dt$ (note, that the differential cross–section is employed here), and the number density of molecules of class $\mathbf{c}_1$ is $f_1 d\mathbf{c}_1$. Obviously, the number of collisions of this molecule of class $\mathbf{c}$ with molecules of class $\mathbf{c}_1$ in dt is

$$c_r f_1 \sigma d\Omega d\mathbf{c}_1 dt.$$

But in the phase space element $dcdr$ there are $fdcdr$ molecules of class c, so in time dt in the phase space element the number of collisions between molecules of class c and molecules of class c_1 is

$$c_r f f_1 \sigma d\Omega dc dc_1 r dt .\tag{2.147}$$

This is the number of collisions of type $(c, c_1 \to c^*, c_1^*)$ in time dt in phase space element $dcdr$. In section 2.3 it is shown, that for the direct collision of type $(c, c_1 \to c^*, c_1^*)$ there exists the inverse collision of type $(c^*, c_1^* \to c, c_1)$ (see Fig.2.3 and the paragraph following Eq. (2.46)). Denote the velocity distribution functions of molecules of class c^* and of class c_1^* by f^* and f_1^*, according to the analogy with Eq. (2.147), one can write the number of collisions of type $(c^*, c_1^* \to c, c_1)$ in time dt in phase space element $dcdr$ as

$$c_r^* f^* f_1^* (\sigma d\Omega)^* dc^* dc_1^* dr dt .\tag{2.148}$$

Equation (2.44) shows $c_r^* = c_r$, and the symmetry of the direct and inverse collisions guarantees that the Jacobian of the transformation between the values of the product of the differential cross-section and phase space elements $\sigma d\Omega dcdr$ for the direct and the inverse collisions equals unity.

$$|(\sigma d\Omega) dc dc_1| = |(\sigma d\Omega)^* dc^* dc_1^*| .\tag{2.149}$$

So Eq. (2.148) can be written as

$$c_r f^* f_1^* \sigma d\Omega dc dc_1 r dt .\tag{2.148}'$$

Noting that the direct collision $(c, c_1 \to c^*, c_1^*)$ causes the depletion of molecules of class c, while the reverse collision $(c^*, c_1^* \to c, c_1)$ causes the replenishment of molecules of class c, one obtains the gain of the number of molecules of class c in time t as the result of collisions with molecules of class c_1 in phase space element as

$$(f^* f_1^* - f f_1) c_r \sigma d\Omega dc dc_1 r dt .\tag{2.150}$$

When counting the volume swept out by the molecule of class c in time dt, the differential cross-section $\sigma d\Omega \equiv b db d\varepsilon$ has been used, so in order to obtain the total number of the increase of molecules of class c in time dt in $dcdr$, integration over the whole cross-section is needed, and as the collisions of molecules of class c with all molecules are to be taken into account, integration over entire space of c_1 is needed. Thus, in the phase space $dcdr$, the total increase of the number of molecules of class c caused by collisions is

$$\left(\frac{\partial f}{\partial t}\right)_c dcdrdt = \left(\int_{-\infty}^{\infty}\int_0^{4\pi}(f^*f_1^* - ff_1)c_r\sigma d\Omega dc_1\right)dcdrdt. \qquad (2.151)$$

Dividing the sum of Eqs.(2.145), (2.146) and (2.151) by dt, the rate of change of the number of molecules of class c in the phase space $dcdr$ as the result of convection and collision is obtained. Balancing Eq. (2.144) with this rate of change, moving the convection terms to the left hand side of the equation, and canceling the common factor $dcdr$, the *Boltzmann equation* for the single-component gas is obtained

$$\frac{\partial f}{\partial t} + c\cdot\frac{\partial f}{\partial r} + F\cdot\frac{\partial f}{\partial c} = \left(\frac{\partial f}{\partial t}\right)_c \equiv \int_{-\infty}^{\infty}\int_0^{4\pi}(f^*f_1^* - ff_1)c_r\sigma d\Omega dc_1. \qquad (2.152)$$

The Boltzmann equation has two important prerequisites which must be kept in mind when using this equation. First, it is valid only for *dilute gas*, for only binary collisions have been taken into account (see analysis of the condition under which the three body collision is not important given in the beginning of section 2.3). Then in counting the number Eq. (2.147) of collisions between molecules of class c and molecules of class c_1, the *molecular chaos* assumption has been adopted, that is, the distribution function f of molecules of class c has been assumed to be uncorrelated with the distribution function f_1 of molecules of class c_1. The necessity of the molecular chaos for the validity of the Boltzmann equation is explicitly seen in the derivation of it from the BBGKY equations (see Eq. (2.143)).

Boltzmann equation as written in the form of Eq. (2.152) is only valid for single component monatomic gas. But in nature most gases are polyatomic gases with rotational and vibrational degrees of freedom and possibly with excited electronic levels. In the polyatomic gases there are the exchanges of translational and internal energies and more transport coefficients are to be introduced. Wang Chengshu (known in the literature as Wang Chang, Chang is her husband's family name) and Uhlenbeck [16] proposed to treat the collision term in the Boltzmann equation for the polyatomic gas semi-classically, i.e., to treat the translational energy classically and the rotational and vibrational energies quantum-mechanically. Wang's method of treatment and her method of the solution are both of value not only for polyatomic gases but for treating other problems. The Boltzmann equation for the gas mixture is given in section 2.12.

Boltzmann equation is the basic equation of molecular gas dynamics and plays a central role in rarefied gas dynamics. In the free molecular flow regime the collisionless Boltzmann equation and sometimes the equilibrium solution of the Boltzmann equation, i.e. the Maxwellian distribution, is employed. In the slip flow regime the first approximation, the Navier-Stokes equation, or the second approximation, the Burnett equation, of the Chapman-Enskog expansion in solving the Boltzmann equation is employed, with slip boundary condition derived also basing on the Chapman-Enskog expansion, or, the fluid dynamics equations obtained from the systematic asymptotic expansion in solving the Boltzmann equation are employed. In the transitional regime the Boltzmann equation or some equivalent methods must be used to solve the problems of gas flows.

To obtain a certain solution of the Boltzmann equation the initial and boundary conditions must be prescribed. Certain function of c and r can be assigned as the initial distribution function at the initial time. The problems related to the prescription of the boundary conditions on body surfaces will be discussed separately in Chapter 3.

Boltzmann equation is a differential-integral equation. Its right hand side term is called the *collision integral term* or *collision term*. If divide it into two terms, every term is a function of c, r and t. In the second term (the depleting term) $f(c)$ can be taken out of the integral, for c and c_1 are independent, but for the first term (the replenishing term) there is not such a simplification f^*, f_1^* are the distribution functions with the post-collision velocities c^* and c_1^* as their arguments, and c^* and c_1^* can be expressed as functions of c and c_1 and of angle χ and angle ε (see Eq. (2.41), Eq. (2.59)*, Eq. (2.61)*, Eq. (2.63)*, Eq. (2.59)**, Eq. (2.61)**, Eq. (2.63)** and Eq. (2.39)). In calculating the collision integral or its moments the expressions of c^* and c_1^* through c and c_1 are used. One can imagine what enormous difficulty the presence of the collision term brings to the solution of the Boltzmann equation.

Except the tremendous difficulty brought by the collision term, the Boltzmann equation has the peculiarity of possessing many arguments. The advantage of the simplicity of having only one unknown function f becomes not so essential compared with the complexity brought by the large number of arguments. In general (three-dimensional and unsteady flow case) f depends on seven independent scalar arguments, and the velocity space needs to be extended to quite large extent. This sets difficulty of arranging the grid points in phase space for numerical calculations, and in general case (complex geometry or large disturbances) it is impossible to obtain analytical solutions.

2.7 COLLISION INTEGRAL AND TOTAL NUMBER OF COLLISIONS

The multiplication of the right hand side of the Boltzmann equation Eq. (2.152) by a function $Q(c)$ associated with molecular velocity c and integration over the entire velocity space c leads to an expression called *collision integral*

$$\Delta[Q] = \int_{-\infty}^{\infty}\int_{-\infty}^{\infty}\int_{0}^{4\pi} Q(f^*f_1^* - ff_1) c_r \sigma d\Omega dc_1 dc . \tag{2.153}$$

For the convenience of further application we discuss some modified forms of this integral.

In the integral Eq. (2.153) let us interchange the places of c and c_1 and replace $Q \equiv Q(c)$ by $Q_1 \equiv Q(c_1)$, the value of integral is unchanged, for this is simply the change of the notations of the two molecules, so one has

$$\Delta[Q] = \int_{-\infty}^{\infty}\int_{-\infty}^{\infty}\int_{0}^{4\pi} Q_1(f_1^*f^* - f_1 f) c_r \sigma d\Omega dc dc_1 . \tag{2.154}$$

In this integral interchange c, c_1, f, f_1 with c^*, c_1^*, f^*, f_1^*, and replace $Q_1 \equiv Q(c_1)$ by $Q_1^* \equiv Q(c_1^*)$, the value of integral is unchanged. This is because there are inverse collisions, any collision is the inverse collision of another collision, and the result of integration over c and c_1, is the same as that of integration over c^* and c_1^*. Here the fact that the Jacobian of the transformation between the pre-collision and

2.7 COLLISION INTEGRAL AND TOTAL NUMBER OF COLLISIONS 99

post-collision values of the product $\sigma d\Omega dc dr$ (see Eq. (2.149)) is unity has been used

$$\Delta[Q] = \int_{-\infty}^{\infty}\int_{-\infty}^{\infty}\int_{0}^{4\pi} Q_1^*(f_1 f - f_1^* f^*) c_r \sigma d\Omega dc dc_1 \ . \tag{2.155}$$

In this integral once more change the notations of c and c_1, one has

$$\Delta[Q] = \int_{-\infty}^{\infty}\int_{-\infty}^{\infty}\int_{0}^{4\pi} Q^*(f f_1 - f^* f_1^*) c_r \sigma d\Omega dc dc_1 \ . \tag{2.156}$$

From Eqs. (2.153), (2.154), (2.155) and (2.156) one finds easily

$$\Delta[Q] = \frac{1}{4}\int_{-\infty}^{\infty}\int_{-\infty}^{\infty}\int_{0}^{4\pi} (Q + Q_1 - Q^* - Q_1^*)(f^* f_1^* - f f_1) c_r \sigma d\Omega dc dc_1 \ . \tag{2.157}$$

Equation (2.153) can be written in a form that brings out its physical meaning in full relief. For this first write it like

$$\Delta[Q] = \int_{-\infty}^{\infty}\int_{-\infty}^{\infty}\int_{0}^{4\pi} Qf^* f_1^* c_r \sigma d\Omega dc dc_1 - \int_{-\infty}^{\infty}\int_{0}^{\infty}\int_{0}^{4\pi} Q f f_1 c_r \sigma d\Omega dc_1 dc \ .$$

To the first term of the right hand side apply the reasoning that led to Eq. (2.155), i.e., take advantage of the existence of the inverse collisions, the term $Q f^* f_1^*$ can be replaced by $Q^* f f_1$, so one has

$$\Delta[Q] = \int_{-\infty}^{\infty}\int_{-\infty}^{\infty}\int_{0}^{4\pi} (Q^* - Q) f_1 f c_r \sigma d\Omega dc_1 dc \ . \tag{2.158}$$

In this equation simple change of the notations of c and c_1 leads to

$$\Delta[Q] = \int_{-\infty}^{\infty}\int_{-\infty}^{\infty}\int_{0}^{4\pi} (Q_1^* - Q_1) f_1 f c_r \sigma d\Omega dc dc_1 \ . \tag{2.159}$$

Add the above two equations and divide the sum by 2, one has

$$\Delta[Q] = \frac{1}{2}\int_{-\infty}^{\infty}\int_{-\infty}^{\infty}\int_{0}^{4\pi} (Q^* + Q_1^* - Q - Q_1) f_1 f c_r \sigma d\Omega dc dc_1 \ . \tag{2.160}$$

This equation shows that $\Delta[Q]$ is the change in the quantity Q summed over all collisions. The appearance of 1/2 is to take into account the double counting of collisions when integrating. The change of Q in a $(c,c_1 \to c^*, c_1^*)$ collision is $(Q^* + Q_1^* - Q - Q_1)$. It is easily seen, that the integral

$$\frac{1}{2}\int_{-\infty}^{\infty}\int_{-\infty}^{\infty}\int_0^{4\pi} f_1 f c_r \sigma d\Omega dc dc_1 = \frac{1}{2}\int_{-\infty}^{\infty}\int_{-\infty}^{\infty} \sigma_T c_r f_1 f dc_1 dc \qquad (2.161)$$

is the number of collisions in the gas in unit volume in unit time. In the derivation of the Boltzmann equation we have seen, that $f_1 f c_r \sigma d\Omega dc d\mathbf{c}_1 d\mathbf{r} dt$ (see Eq. (2.147)) is the number of collisions of type $(c,c_1 \to c^*, c_1^*)$ in time dt in phase space $dcdr$. Integrating it over c and c_1, and over the entire cross-section, one obtains the total number of collisions between various molecules in time dt in space dr counted twice. Divide it by $dtdr$ and by 2, the above expressions is obtained, it represents the total number N_c of molecular collisions in unit time in unit volume

$$N_c = \frac{1}{2}\int_{-\infty}^{\infty}\int_{-\infty}^{\infty} \sigma_T c_r f_1 f dc_1 dc . \qquad (2.162)$$

Similarly with the definition Eq. (2.9) of the mean value $\bar{Q}$ of a quantity Q associated with molecular velocity in gas, the mean value of a quantity associated with the relative velocity is defined as

$$\overline{Q(c_r)} = \frac{1}{n^2}\iint Q(c_r) f_1 f dc_1 dc . \qquad (2.163)$$

And Eq. (2.162) can be written

$$N_c = \frac{1}{2} n^2 \overline{\sigma_T c_r} . \qquad (2.164)$$

2.8 EVALUATION OF COLLISION INTEGRALS

In the last section we have seen that the collision integral $\Delta[Q]$ is the summation of the change of the quantity Q over all collisions (see Eq. 2.160). In this section we discuss the evaluation of this integral in some cases.

2.8 EVALUATION OF COLLISION INTEGRALS

When Q is either of the invariants — the mass m, the momentum mc or the kinetic energy $(1/2)mc^2$ — in the collision, the conservation of mass, momentum and energy yields

$$Q + Q_1 - Q^* - Q_1^* = 0.\qquad(2.165)$$

Then from Eq. (2.160), one has, obviously

$$\Delta[Q] = 0, \quad Q = (1, u, v, w, c^2).\qquad(2.166)$$

The quantity Q that satisfies condition Eq. (2.165) is called the *summational invariant*. It can be proved, that the collision invariants $(1, u, v, w, c^2)$ or their linear combinations are the only summational invariants, that is, $\Delta[Q] = 0$, when and only when

$$Q = Ac^2 + \boldsymbol{B} \cdot \boldsymbol{c} + C.\qquad(2.167)$$

According to the definition of the collision integral one has, obviously, when a and b are independent of c

$$\Delta[aQ_1 + bQ_2] = a\Delta[Q_1] + b\Delta[Q_2].\qquad(2.168)$$

According to this property and the definition of the thermal velocity c' it is easy to prove

$$\Delta[c_i c_j] = \Delta[c'_i c'_j],\qquad(2.169)$$

$$\Delta[c_i c^2] = \Delta[c'_i c'^2] + 2c_{0j}\Delta[c'_i c'_j].\qquad(2.170)$$

As an example of evaluation of $\Delta[Q]$, the case of $Q = u^2$ and the Maxwellian molecules is chosen. For the Maxwellian molecules the differential cross-section $\sigma d\Omega$ can be written as Eq. (2.92), i.e., is inversely proportional to c_r, thus the expression $c_r \sigma d\Omega$ in the collision integral (like Eq. (2.153)) does not involve the relative velocity

$$c_r \sigma d\Omega = \left(\frac{2\kappa}{m}\right)^{1/2} W_0 d W_0 d\varepsilon.\qquad(2.171)$$

Making use of Eq. (2.158), one has

$$\Delta[u^2] = \int_{-\infty}^{\infty}\int_{-\infty}^{\infty}\int_{0}^{2\pi}\int_{0}^{\infty}(u^{*2}-u^2)ff_1\left(\frac{2\kappa}{m}\right)^{1/2}W_d W_0 d\varepsilon dc dc_1. \quad (2.172)$$

For further evaluation the expression Eq. (2.59)* of the post-collision velocity through the pre-collision velocity is needed. Write $u^{*2} - u^2$ as $(u^* - u)^2 + 2u(u^* - u)$. From Eqs. (2.40) and (2.41) one has $c_1^* - c_1 = (1/2)(c_r^* - c_r)$. To keep the notation consistent with that used in Eq. (2.172), c and c_1 need to be used instead of c_1 and c_2 in Eq. (2.40), Eq. (2.41) (and also in Eq. (2.39) to be used in the following). Thus, one has $(u^* - u) = (1/2)(u_r^* - u_r)$. Making use of Eq. (2.59)', one has

$$u^* - u = -\frac{1}{2}(u_r - u_r^*) = -\frac{1}{2}\left[(1-\cos\chi)u_r - \sin\chi\sin\varepsilon(v_r^2 - w_r^2)^{1/2}\right],$$

from where

$$u^{*2} - u^2 = \frac{1}{4}\left[(1-\cos\chi)^2 u_r^2 - \right.$$

$$2(1-\cos\chi)\sin\chi\sin\varepsilon u_r(v_r^2 + w_r^2)^{1/2} + \sin^2\chi\sin^2\varepsilon(v_r^2 + w_r^2)] -$$

$$u\left[(1-\cos\chi)u_r - \sin\chi\sin\varepsilon(v_r^2 + w_r^2)^{1/2}\right].$$

Substituting this expression into Eq. (2.172) and noting that

$$\int_0^{2\pi}\sin\varepsilon d\varepsilon = 0, \int_0^{2\pi}\sin^2\varepsilon d\varepsilon = \pi, \quad (1-\cos\chi)^2 = 2(1-\cos\chi) - \sin^2\chi,$$

the integral over ε can be written as

$$\int_0^{2\pi}(u^{*2} - u^2) d\varepsilon = \pi(u_r^2 - 2uu_r)(1-\cos\chi) - \frac{1}{4}\pi(3u_r^2 - c_r^2)\sin^2\chi.$$

Equation (2.39) gives $u_r^2 = (u-u_1)^2 = u_1^2 - u^2 + 2u^2 - 2uu_1 = u_1^2 - u^2 + 2uu_r$, from where

$$u_r^2 - 2uu_r = u_1^2 - u^2.$$

When this term is being integrated in Eq. (2.172) over the velocity space c, c_1, because of that f and f_1 is the same distribution function, u_1^2 and u^2 gives the same result which cancels each other. According to the property Eq. (2.169), when integrating, thermal velocity can be used to express $u_r : u_r^2 = u_1'^2 - 2u_1'u' + u'^2$, and the integration over c, c_1 space gives $2n^2\overline{u'^2}$. Similarly, $c_r^2 = u_r^2 + v_r^2 + w_r^2$, the integral for v_r and w_r has the same result as for u_r and yields $2n^2\overline{u'^2}$. Thus, the result of the whole collision integral is

$$\Delta[u^2] = -\frac{3\pi}{2}\left(\frac{2\kappa}{m}\right)^{1/2} n^2 \left(\overline{u'^2} - \frac{\overline{c'^2}}{3}\right)\int_0^\infty \sin^2\chi W_1 dW_0 . \tag{2.173}$$

According to the definition of the shear stress τ_{xx} (Eq. (2.24), also see Eq. (2.25)) and the definition of $A_2(\eta)$, Eq. (2.88), the collision integral being explored can be written as

$$\Delta[u^2] = \frac{3\pi}{2} A_2(\eta)\left(\frac{2\kappa}{m}\right)^{1/2} \frac{n}{m}\tau_{xx} . \tag{2.174}$$

For Maxwellian molecules $\eta = 5, A_2(5) = 0.436$ (see Table 2.1), and the viscosity μ is expressed by Eq. (2.93), and as $nkT = p$, so we have

$$\Delta[u^2] = \frac{p\,\tau_{xx}}{m\,\mu} . \tag{2.175}$$

Similarly, one can write, for the Maxwell molecules

$$\Delta[c_i c_j] = \frac{3\pi}{2} A_2(5)\left(\frac{2\kappa}{m}\right)^{1/2} \frac{n}{m}\tau_{ij} = \frac{p\,\tau_{ij}}{m\,\mu} , \tag{2.176}$$

$$\Delta[c_i c^2] = 3\pi A_2(5)\left(\frac{2\kappa}{m}\right)^{1/2} \frac{n}{m}\left(\tau_{ij}c_{0j} - \frac{2}{3}q_i\right) = \frac{2p}{m\mu}\left(\tau_{ij}c_{0j} - \frac{2}{3}q_i\right) . \tag{2.177}$$

The collision integrals $\Delta[u^2]$, $\Delta[c_i c_j]$ and $\Delta[c_i c^2]$ can also be evaluated for a special kind of VHS molecules. That is, when their collision cross-section is inversely proportional to c_r, as in the Maxwellian molecular model, $\zeta = 1/2$ (see Eq. (2.94)), and $c_r \sigma d\Omega$ is also independent of c_r. The differential cross-section σ according to the hard sphere expression Eq. (2.73) is written as $d^2/4$, and then according to Eq. (2.94) is written as $d_{ref}^2 c_{r,ref}/4c_r$. Writing $d\Omega$ as $\sin\chi d\chi d\varepsilon$, the collision integral can be written as

$$\Delta[u^2] = \int_{-\infty}^{\infty}\int\int_0^{2\pi}\int_0^{\pi}(u^{*2} - u^2)ff_1\frac{d_{ref}^2 c_{r,ref}}{4}\sin\chi d\chi d\varepsilon \, dc\, dc_j ,$$

Transforming $\left(u^{*2} - u\right)$ as above and integrating first over ε, one obtains

$$\Delta[u^2] = -\frac{3\pi}{2}n^2\left(\overline{u'^2} - \frac{\overline{c'^2}}{3}\right)\frac{d_{ref}^2 c_{rref}}{4}\int_0^\pi \sin^3\chi\, d\chi = \frac{\pi}{2}d_{ref}^2 \varsigma_{ref}\frac{n}{m}\tau_{xx}. \qquad (2.178)$$

Making use of the expression Eq. (2.96)' of the viscosity for Maxwell gas model with $\xi = 1/2$, one obtains

$$\Delta[u^2] = \frac{p}{m}\frac{\tau_{xx}}{\mu},$$

i.e. the same result Eq. (2.175) as for the Maxwell molecules. Similarly, for the collision integrals $\Delta[c_i c_j]$ and $\Delta[c_i c^2]$, using the VHS model and taking $\xi = 1/2$, the same results as for the Maxwell molecules are obtained (see Eqs. (2.176) and (2.177)).

2.9 THE MAXWELL TRANSPORT EQUATION – THE MOMENT EQUATION

In section 2.2 it is seen, that the average value of quantity Q (being a constant or a function of c) associated with a molecule in the gas can be obtained by finding moment of the distribution function (see Eq. (2.9). Multiplying the Boltzmann equation by Q and integrating each term over the entire velocity space, one obtains the moment of the Boltzmann equation. This is the *moment equation* of the Boltzmann equation. Historically, this equation was derived by Maxwell in 1866 before the derivation of the Boltzmann equation in 1872, so we also call it the *Maxwell transport equation*. Maxwell by using this equation and with the help of the Maxwellian molecular model ($\eta = 5$, and Eq. (2.92)) obtained the expressions of various transport coefficients of the gas. Now the natural way of finding the moment from the Boltzmann equation is used to derive the moment equation.

Multiplying the Boltzmann equation (2.152) by Q leads to

$$Q\frac{\partial f}{\partial t} + Qc\cdot\frac{\partial f}{\partial r} + QF\cdot\frac{\partial f}{\partial c} = Q\int_{-\infty}^{\infty}\int_0^{4\pi}(f^*f_1^* - ff_1)\varsigma_r\sigma\, d\Omega dc_1. \qquad (2.179)$$

2.9 THE MAXWELL TRANSPORT EQUATION – THE MOMENT EQUATION

To obtain the moment equation one must take integral of each term of this equation over the entire velocity space c. As Q is a function only of c, the result of integration of the first term can be written as

$$\int_{-\infty}^{\infty} \frac{\partial}{\partial t}(Qf)dc = \frac{\partial}{\partial t}\left[\int_{0}^{\infty}(Qf)dc\right],$$

or, with the help of Eq. (2.9), as

$$\frac{\partial}{\partial t}(n\overline{Q}). \qquad (2.180)$$

In the integral of the second term of Eq. (2.179) cQ can be taken into the symbol of differential quotient, for $cQ(c)$ is function of c, and c and r are equal arguments of f. But after multiplication by $f(r,c,t)$ and integration over c, $\overline{cQ}$ is dependent of the space coordinates, as the quantity $\overline{Q}$ in Eq.(2.180) is a function dependent of t. Thus, the integral of the second term of Eq. (2.179) can be written as

$$\int_{-\infty}^{\infty} \nabla \cdot (cQf)dc = \nabla \cdot n\overline{cQ}. \qquad (2.181)$$

The integral of the third term of Eq. (2.179)

$$\int_{-\infty}^{\infty} QF \cdot \frac{\partial f}{\partial c}dc,$$

can be written as

$$\int_{-\infty}^{\infty} F \cdot \frac{\partial}{\partial c}(Qf)dc - \int_{-\infty}^{\infty} F \cdot \frac{\partial Q}{\partial c}fdc.$$

Here the first integral is zero, for F is independent of c, and $f \to 0$, when $c \to \infty$:

$$\int_{-\infty}^{\infty} F_1 \frac{\partial}{\partial c_1}(Qf)dc_1 dc_2 dc_3 = \int_{-\infty}^{\infty} F_1(Qf)\Big|_{c_1=-\infty}^{c_1=\infty} dc_2 dc_3.$$

And the other integral can be written as

$$-n\mathbf{F} \cdot \overline{\frac{\partial Q}{\partial \mathbf{c}}}.\qquad(2.182)$$

The integral of the right hand side of Eq. (2.179) is the collision integral Eq. (2.153) already discussed is section 2.7. Thus, the result of integration of Eq. (2.179) can be written (making use of Eq. (2.180), Eq. (2.181), Eq. (2.182) and Eq. (2.153))

$$\frac{\partial}{\partial t}(n\overline{Q}) + \nabla \cdot \overline{n\mathbf{c}Q} - n\mathbf{F} \cdot \overline{\frac{\partial Q}{\partial \mathbf{c}}} = \Delta[Q].\qquad(2.183)$$

This is the moment equation for Q. If the collision integral $\Delta[Q]$ is written in the form of Eq. (2.160), thus it represents explicitly the change of quantity Q in the collisions, and with such right hand side the equation (2.183) could have been derived not from the Boltzmann equation. What Maxwell derived independently was just Eq. (2.183) with $\Delta[Q]$ expressed in the form of Eq. (2.160). Such equation is called the Maxwell transport equation.

The terms on the left hand side of Eq. (2.183) can be expressed by the macroscopic quantities of the gas, and $\Delta Q = 0$, when $Q = m$, $m\mathbf{c}$ or $(1/2)mc^2$. The equations thus obtained are the conservation equations of gas dynamics. We shall discuss in detail the various cases of the conservation equations in section 5.2 of Chapter 5.

2.10 MAXWELL DISTRIBUTION

An uniform in space single-component gas of monatomic molecules in the case of absence of external force yields a solution of the Boltzmann equation, i.e., the *distribution in equilibrium state*. Historically, this distribution was obtained by Maxwell in 1860 [17] before the establishment of the Boltzmann equation, and is thus called Maxwell distribution. *Maxwell distribution*, i.e. the velocity distribution function under the equilibrium state of the gas, is

$$f_0 = n\left(\frac{m}{2\pi kT}\right)^{3/2} \exp\left(-\frac{m}{2kT}c'^2\right) =\qquad(2.184)$$

$$n\left(\frac{m}{2\pi kT}\right)^{3/2} \exp\left[-\frac{m}{2kT}(u'^2 + v'^2 + w'^2)\right].$$

The original derivation of Maxwell is not sufficiently strict, but is simple and direct. It is of sense to know how historically this was done first, before cognition of the mathematically strict derivation based on the Boltzmann's H-theorem [14]. As the three components u', v' and w' of the molecular thermal velocity are perpendicular to each other, Maxwell assumed that the distribution of certain component is independent of the distributions of the other components. If the probability that the value of the velocity component in the x direction is between u' and $u'+du'$ is denoted by $F(u')du'$, then according to the assumption $F(u')$ is independent of v' and w'. Similarly, the probabilities that the values of the velocity components in the y and z directions are between v' and $v'+dv'$ and w' and $w'+dw'$ are denoted by $F(v')dv'$ and $F(w')dw'$, respectively. According to the definition of the distribution function

$$f(u',v',w')dc'$$

is the probabilistic number of molecules in unit volume with velocity in the element $dc' \equiv du'dv'dw'$ near c', and in the assumption of independence of the velocity component distribution function on each other is to be equal to

$$f(u',v',w')du'dv'dw' = nF(u')F(v')F(w')du'dv'dw'.$$

For the quiescent gas the distribution function $f(u',v',w')$ can not distinguish any peculiar direction, it can depend only on the value of the thermal velocity $c' \equiv \sqrt{u'^2 + v'^2 + w'^2}$. Thus, basing on the consideration of the isotropy in the quiescent gas, one should have

$$nF(u')F(v')F(w') = f(u',v',w') = \phi(u'^2 + v'^2 + w'^2).$$

The solution of this functional equation is

$$F(u') = ae^{bu'^2},$$
$$f(u',v',w') = \phi(u'^2 + v'^2 + w'^2) = na^3 \exp(bc'^2). \tag{2.185}$$

Here a and b are arbitrary constants, appropriate choosing them can make Eq. (2.185) identical with Eq. (2.184).

The strict mathematical derivation of the Maxwell distribution based on the Boltzmann's H-theorem is as follows. For the quiescent uniform gas without the action of the external force $\partial/\partial r$ is zero and F is zero. Then the Boltzmann equation is

$$\frac{\partial f}{\partial t} = \int_{-\infty}^{\infty}\int_{0}^{4\pi}(f^*f_1^* - ff_1)c_r\sigma d\Omega dc_1. \tag{2.186}$$

The *H function* introduced by Boltzmann is (see the definition of the average value Eq. (2.9))

$$H = n\overline{\ln f} = \int f \ln f dc. \tag{2.187}$$

Taking $Q = \ln f$, making use of the moment equation (2.183) and writing the collision integral ΔQ in the form of Eq.(2.157), one can write

$$\frac{\partial H}{\partial t} = \frac{1}{4}\int_{-\infty}^{\infty}\int_{-\infty}^{\infty}\int_{0}^{4\pi}(\ln f + \ln f_1 - \ln f^* - \ln f_1^*) \cdot$$

$$(f^*f_1^* - ff_1)c_r\sigma d\Omega dc_1 dc =$$

$$\frac{1}{4}\int_{-\infty}^{\infty}\int_{-\infty}^{\infty}\int_{0}^{4\pi}\ln(ff_1/f^*f_1^*)(f^*f_1^* - ff_1)c_r\sigma d\Omega dc_1 dc. \tag{2.188}$$

Obviously, if $\ln(ff_1/f^*f_1^*)$ is positive, then $f^*f_1^* - ff_1$ must be negative, if $\ln(ff_1/f^*f_1^*)$ is negative, then $f^*f_1^* - ff_1$ must be positive. So the integral in Eq. (2.188) must be negative or zero, that is, H can not increase

$$\frac{\partial H}{\partial t} \leq 0. \tag{2.189}$$

This is the *Boltzmann's H theorem*.

It is seen from the expression Eq. (2.187) of H, that H may diverse, as one has $f \to 0$ and $\ln f \to -\infty$, when $c \to \infty$. There arises a question: whether H would decrease all the way to $-\infty$, or approach a finite value? To answer this question one may compare the expression Eq. (2.187) of H with the following integral

$$\int_{-\infty}^{\infty} fc^2 dc.$$

This integral represents the total translational energy of the gas so is convergent. Suppose that $H = \int f \ln f dc$ diverges, then $-\ln f$ must tend to infinity more rapidly than c^2, when $c \to \infty$, or

$$\ln f < -c^2,$$

i.e.

$$f < \exp(-c^2).$$

That is to say, f tends to zero more rapidly, than $\exp(-c^2)$. As $c^n \exp(-c^2)\big|_{c \to \infty} \to 0$ for any n, $H = \int f \ln f dc$ is surely convergent. So the assumption that H diverges is not valid. H decreases with time monotonically and tends to a finite bound. This corresponds to the state of the gas with

$$\frac{\partial H}{\partial t} = 0.$$

From Eq. (2.188) it is seen, that this is possible when and only when the following equality holds

$$f^* f_1^* - f f_1 = 0, \tag{2.190}$$

or when and only when the following equality holds

$$\ln f + \ln f_1 = \ln f^* + \ln f_1^*. \tag{2.191}$$

It is seen from Eq. (2.186), that $\partial H / \partial t = 0$ leads to $\partial f / \partial t = 0$, that is, in equilibrium the probabilistic number of molecules in any velocity space element remains unchanged with time.

The condition Eq. (2.191) shows, that in the state of equilibrium the quantity $Q = \ln f$ satisfies the condition $Q + Q_1 - Q^* - Q_1^* = 0$, i.e., Q is the summational invariant.. It is known, that the collision invariants — the mass m, the momentum $m\mathbf{c}$ and the kinetic energy $(1/2) mc^2$ of the molecule — satisfies this condition. It has been proved that the collision invariants and their linear combinations are the only summational invariants (see [2], pp. 42~45). So one has

$$\ln f = Ac^2 + \mathbf{B} \cdot \mathbf{c} + C. \tag{2.192}$$

If introduce the thermal velocity c' according to Eq. (2.10), then the above equation can be written as

$$\ln f = Ac'^2 + (2Ac_0 + B)\cdot c' + Ac_0^2 + B\cdot c_0 + C.$$

As it is impossible to distinguish any peculiar direction in an equilibrium gas, the distribution function should be isotropic and should not depend explicitly on c'. So one has

$$B = -2Ac_0,$$

and

$$\ln f = Ac'^2 - Ac_0^2 + C,$$

i.e.

$$f = \exp(Ac'^2 - Ac_0^2 + C).$$

New constants can be introduced, let $A = -\beta^2$, the negative sign is introduced to satisfy the condition that f should be bounded and the coefficient of c'^2 should be negative. And for the sake of simplicity let $\exp(C + \beta^2 c_0^2) = \alpha$, one has

$$f = \alpha \exp(-\beta^2 c'^2). \tag{2.193}$$

The constants α and β are determined from the expressions Eqs (2.5) and (2.31) of the number density n and temperature T. First making use of Eq. (2.5)

$$n = \int f dc = \alpha \int_{-\infty}^{\infty} \exp(-\beta^2 c'^2) dc' =$$

$$\alpha \int_{-\infty}^{\infty}\int_{-\infty}^{\infty}\int_{-\infty}^{\infty} \exp\{-\beta^2(u'^2 + v'^2 + w'^2)\} du' dv' dw' =$$

$$\alpha \left(\int_{-\infty}^{\infty} \exp(-\beta^2 x^2) dx\right)^3.$$

The integral Eq. (II-11) in the Appendix II gives the value of the integral in the last formula as $\sqrt{\pi}/\beta$. So one has

$$n = \alpha(\sqrt{\pi}/\beta)^3,$$

i.e.

$$\alpha = n(\beta/\sqrt{\pi})^3. \tag{2.194}$$

Making use of Eq. (2.31)

$$\frac{3}{2}\frac{k}{m}T = \frac{\alpha}{2n}\int_{-\infty}^{\infty} c'^2 \exp(-\beta^2 c'^2) dc',$$

The value of the integral according to Eq. (II.11) and Eq. (II.13) in the Appendix II can be evaluated as $(3/2)\pi^{\frac{3}{2}}/\beta^5$. The value β^2 is obtained after substituting the value of α according to Eq. (2.194)

$$\beta^2 = \left(2\frac{k}{m}T\right)^{-1} = (2RT)^{-1}. \tag{2.195}$$

Substituting the value of α and β according to Eq. (2.194) and Eq. (2.195) into Eq. (2.193), the expression for the *equilibrium* or *Maxwell distribution* is obtained as (Eq. (2.184))

$$f_0 = n\left(\frac{m}{2\pi kT}\right)^{3/2} \exp\left(-\frac{m}{2kT}c'^2\right).$$

Sometimes it is expressed through β:

$$f_0 = n\left(\frac{\beta}{\pi^{1/2}}\right)^3 \exp(-\beta^2 c'^2), \beta = (2RT)^{-1/2}. \tag{2.196}$$

Comparison with the most probable thermal velocity c_m introduced in the next section shows, that β is just the reciprocal of c_m.

The Maxwell distribution Eq. (2.184) may be considered as a basic physical law, it stipulates the most probable velocity distribution of the gas in the quiescent equilibrium state. The molecular beam technique can measure with sufficient accuracy the velocity distribution of molecules in an equilibrium, quiescent gas and confirm the validity of the Maxwellian distribution.

The Boltzmann's *H*-theorem has proven that the Maxwellian distribution is the sufficient and necessary condition for the equilibrium state. Without the introduction of $H = \int f \ln f dc$, it is impossible to prove, that Eq. (2.191) or Eq. (2.192) is the necessary condition of $\partial f/\partial t = 0$, for from Eq. (2.186) it is seen, that to make the right hand side integral vanish, it is possible to maintain $(f^* f_1 - ff_1)$ somewhere positive and some where negative in the integration domain. The above proof was aimed at the single component, monatomic molecule gas. But the Maxwellian distribution applies also to the velocities of gas mixture and of diatomic and polyatomic molecules. In Chapter 1 we introduced the equilibrium energy distribution derived by the method of statistical mechanics, i.e., the *Boltzmann distribution* (see Eq. (1.68)), which is the more general distribution than the Maxwellain distribution derived from the kinetic theory. Maxwellian distribution can be considered as a particular example of the Boltzmann distribution with $\varepsilon_j = (1/2)mc'^2$.

2.11 EQUILIBRIUM STATE OF GASES

Many useful properties of gas in equilibrium state can be obtained from the Maxwellian distribution, and these will be discussed in the following sections.

2.11.1 SOME PECULIAR SPEEDS OF GAS

Here we start from the introduction of the concepts of the most probable molecular thermal speed c'_m, the average thermal speed $\overline{c'}$ and the root mean thermal speed c'_s.

For this in the velocity space we introduce the spherical coordinates (c',θ,φ), the volume of the element in the velocity space is

$$dc' = c'^2 \sin\theta \, d\theta \, d\varphi \, dc'.$$

According to the definition of the velocity distribution function the fraction of molecules with velocity of value between c' and $c' + dc'$, of the polar angle between θ and $\theta + d\theta$, and of the azimuth angle between φ and $\varphi + d\varphi$ is (see Eq. (2.4) and Eq. (2.196))

$$\frac{dn}{n} = (\beta^3/\pi^{3/2})c'^2 \exp(-\beta^2 c'^2)\sin\theta \, d\theta \, d\varphi \, dc'.$$

The fraction of molecules with speed in the velocity space between the sphere of radius c' and the sphere of radius $c' + dc'$ obviously can be obtained from the integration of the above equation over φ from 0 to 2π, and over θ from 0 to π

2.11 EQUILIBRIUM STATE OF GASES

$$\frac{dn}{n} = \left(4/\sqrt{\pi}\right)\beta^3 c'^2 \exp\left(-\beta^2 c'^2\right) dc'. \tag{2.197}$$

A distribution function $\chi(c')$ of the value of the velocity can be defined, that the fraction of the molecules with speed between c' and $c' + dc'$ is $\chi(c') dc'$. Obviously

$$\chi(c') = \left(4/\sqrt{\pi}\right)\beta^3 c'^2 \exp\left(-\beta^2 c'^2\right) = 4\pi \left(\frac{m}{2\pi kT}\right)^{3/2} c'^2 \exp\left(-\frac{mc'^2}{2kT}\right). \tag{2.198}$$

The dimensionless distribution function $\chi/\beta \equiv (2kT/m)^{1/2}\chi$ is shown in Fig. 2.10, the abscissa is also represented by the dimensionless quantity $\beta c' \equiv c'/(2kT/m)^{1/2}$.

The peculiar speeds mentioned in the beginning of this section can be evaluated from the distribution function $\chi(c')$.

The most probable molecular thermal speed c'_m is the value of c' where $\chi(c')$ reaches the maximum, from the differentiation $\partial\chi/\partial c' = 0$ it is easy to obtain

$$c'_m = 1/\beta = \left(\frac{2kT}{m}\right)^{1/2}. \tag{2.199}$$

The average thermal speed $\overline{c'}$ is defined by the following equation

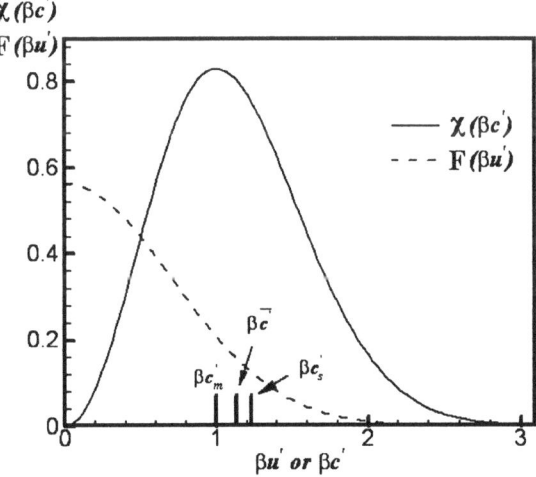

Fig. 2.10 Equilibrium distribution function $\chi(\beta c')$ of the molecular speed and the equilibrium distribution function $F(\beta u')$ of the component of the thermal velocity

$$\bar{c'} = \int_0^\infty c' \chi(c') dc',$$

making use of the integral Eq. (II.14) in appendix II, it is easy to obtain

$$\bar{c'} = 2/(\pi^{1/2}\beta) = \left(\frac{8kT}{\pi m}\right)^{1/2}. \tag{2.200}$$

The root mean thermal speed c'_s is defined as $c'_s \equiv (\overline{c'^2})^{1/2}$, and $\overline{c'^2}$ has been defined in Eq (2.28) and can be expressed through $\chi(c')$ as

$$\overline{c'^2} = \int_0^\infty c'^2 \chi(c') dc',$$

the result of integration (with the help f Eq. (II.15)) gives

$$c'_s \equiv \left(\overline{c'^2}\right)^{1/2} = \left(\frac{3}{2}\right)^{1/2}/\beta = \left(\frac{3kT}{m}\right)^{1/2}. \tag{2.201}$$

The values of c'_m, $\bar{c'}$ and c'_s are marked on the abscissa of Fig. 2.10 in the successively increasing order in the proportion of $\sqrt{2}$, $\sqrt{8/\pi}$ and $\sqrt{3}$. This order is caused by the high speed tail of the equilibrium distribution function.

From the Maxwellian distribution function Eq (2.196) the *distribution function for a component of the thermal velocity* can be obtained. The fraction of molecules with velocity having a component in certain range can be obtained through the integration of E. (2.196) over the other two components. For example, the fraction of molecules with the x component of the thermal velocity lying between u' and $u' + du'$ is

$$\left(\beta/\pi^{1/2}\right)^3 \exp\left(-\beta^2 u'^2\right) \left(\int_{-\infty}^{\infty}\int_{-\infty}^{\infty} \exp\left[-\beta^2\left(v'^2 + w'^2\right)\right] dv' dw'\right) du',$$

or, making use of the table of integrals in Appendix II-2, it can be written as

$$\beta/\pi^{1/2} \exp\left(-\beta^2 u'^2\right) du'. \tag{2.202}$$

And the distribution function for the u component of the thermal velocity is

$$F(u') = \beta/\pi^{1/2} \exp\left(-\beta^2 u'^2\right) = \left(\frac{m}{2\pi kT}\right)^{1/2} \exp\left(-\frac{mu'^2}{2kT}\right). \tag{2.203}$$

2.11 EQUILIBRIUM STATE OF GASES

This result can be reached by writing the Maxwellian distribution function Eq. (2.196) as the product of the number density n and three equal distribution functions $F(u')$, $F(v')$ and $F(w')$ of the components u', v' and w'. Here the probabilities in the three directions have been supposed to be independent on each other, and this is just the assumption introduced by Maxwell when deriving the equilibrium distribution. The dimensionless distribution function $F/\beta \equiv (2kT/m)^{1/2} F$ is also shown in Fig. 2.10. From the figure it is seen, the most probable value of the component of the thermal velocity is zero. The average value of the u component of the thermal velocity in the positive x direction is (by using Eq. (II.12) in Appendix II)

$$\int_0^\infty u' F(u') du' / \left(\int_0^\infty F(u') du' \right) =$$

$$\left(\beta / \left(\pi^{1/2} \int_0^\infty u' \exp(-\beta^2 u'^2) du' \right) \right) \cdot 2 = 1/(\pi^{1/2} \beta) = \frac{1}{2} \overline{c'} .$$
(2.204)

The dominator on the left hand side of this formula is the normalizing factor, its value is 1/2, because the molecules moving in the positive x direction obviously constitutes half the entirety of molecules.

2.11.2 MOLECULAR COLLISION FREQUENCY AND THE MEAN FREE PATH

In the following the Maxellian distribution function is used to obtain the molecular collision frequency v and the mean free path λ. In section 2.7 the total collision number of molecules in the gas in unit volume in unit time was obtained (Eq. (2.162))

$$N_c = \frac{1}{2} \int_{-\infty}^{\infty} \int_{-\infty}^{\infty} \sigma_T c_r f_1 \, dc_1 dc .$$

The relation of N_c with the number density n and the *collision frequency* v of *each molecule* is as follows

$$N_c = \frac{1}{2} nv .$$
(2.205)

Here the symmetric factor 1/2 is introduced because every collision involves two molecules. From Eq. (2.162) and Eq. (2.205) the expression for v is obtained as

$$v = \frac{1}{n}\int_{-\infty}^{\infty}\int_{-\infty}^{\infty} \sigma_T c_r f_1 f dc_1 dc \; . \tag{2.206}$$

Or, using the definition of the mean value Eq. (2.163), it can be written as

$$v = n\overline{\sigma_T c_r} \; . \tag{2.207}$$

In the discussion of the collision cross-section in section 2.4 it is seen, that the total collision cross-section σ_T in general varies in dependence of the relative velocity c_r, for the inverse power law model $\sigma_r \sim c_r^{-4/(n-1)}$, see Eq. (2.84), for the VHS model and the VSS model $\sigma_r \sim c_r^{-2\xi}$, see Eq. (2.94). Consequently, to calculate v, it needs to calculate the mean value of some power of the relative velocity c_r (for the definition of the mean value see Eq. (2.163)).

So we proceed to calculate the following integral

$$\overline{c_r^j} = \frac{1}{n^2}\int_{-\infty}^{\infty}\int_{-\infty}^{\infty} c_r^j f_1 f_2 \, dc_1 dc_2 \; ,$$

where $c_r = c_1 - c_2$, f_1 and f_2 given by Eq. (2.184)

$$\overline{c_r^j} = \frac{(m_1 m_2)^{3/2}}{(2\pi kT)^3}\int_{-\infty}^{\infty}\int_{-\infty}^{\infty} c_r^j \exp\left[-\left(m_1 c_1^2 + m_2 c_2^2\right)/(2kT)\right] dc_1 dc_2 \; .$$

To calculate this integral the most convenient way is to transform the integration variable c_1, c_2 to c_r, c_m. The Jacobian of the transformation

$$\frac{\partial(u_1, v_1, w_1, u_2, v_2, w_2)}{\partial(u_r, v_r, w_r, u_m, v_m, w_m)}$$

is to be constructed by Eq. (2.40). From the form of Eq. (2.40) it is seen, that it is sufficient to calculate the reduced one-dimensional Jacobian:

$$\frac{\partial(u_1, u_2)}{\partial(u_r, u_m)} = \begin{vmatrix} \dfrac{\partial u_1}{\partial u_r} & \dfrac{\partial u_1}{\partial u_m} \\ \dfrac{\partial u_2}{\partial u_r} & \dfrac{\partial u_2}{\partial u_m} \end{vmatrix} = \begin{vmatrix} m_2/(m_1+m_2) & 1 \\ -m_1/(m_1+m_2) & 1 \end{vmatrix} = 1 \; ,$$

from where it is concluded that the whole Jacobian is 1. The expression $\left(m_1 c_1^2 + m_2 c_2^2\right)$ in the integral can be substituted according to Eq. (2.42). Then in the velocity space c_r, c_m introducing the spherical polar coordinates, and because of the existence of the spherical symmetry integrating first over the polar angle and the azimuth angle, we have

$$dc_r = 4\pi c_r^2 dc_r, \quad dc_m = 4\pi c_m^2 dc_m.$$

There for

$$\overline{c_r^j} = \frac{2(m_1 m_2)^{3/2}}{\pi(kT)^3} \int_0^\infty \int_0^\infty c_r^{j+2} c_m^2 \exp\left[-\frac{(m_1+m_2)c_m^2}{2kT} - \frac{m_r c_r^2}{2kT}\right] dc_m dc_r,$$

or

$$\overline{c_r^j} = \frac{(2m_1 m_2)^{3/2}}{\pi(kT)^3} \int_0^\infty c_m^2 \exp\left[-\frac{(m_1+m_2)c_m^2}{2kT}\right] dc_m \cdot \int_0^\infty c_r^{j+2} \exp\left(-\frac{m_r c_r^2}{2kT}\right) dc_r. \quad (2.208)$$

Equation (2.208) can be expressed through distribution functions $\phi(c_m)$ and $\psi(c_r)$ of c_m and c_r,

$$\overline{c_r^j} = \int_0^\infty \phi(c_m) dc_m \int_0^\infty c_r^j \psi(c_r) dc_r, \quad (2.209)$$

where

$$\phi(c_m) = \frac{4(m_1+m_2)^{3/2}}{\pi^{1/2}(2kT)^{3/2}} c_m^2 \exp\left[-\frac{(m_1+m_2)c_m^2}{2kT}\right], \quad (2.210)$$

$$\psi(c_r) = \frac{4m_r^{3/2}}{\pi^{1/2}(2kT)^{3/2}} c_r^2 \exp\left(-\frac{m_r c_r^2}{2kT}\right). \quad (2.211)$$

Note, that $\phi(c_m)$ and $\psi(c_r)$ formally are identical with the distribution function $\chi(c')$ of the value of the thermal velocity (see Eq. (2.198)), only the mass m is to be replaced by $(m_1 + m_2)$ and m_r, respectively. Obviously, $\phi(c_m)$ and $\psi(c_r)$ have been normalized (see Eq. (II.13) of Appendix)

$$\int_0^\infty \phi(c_m) dc_m = 1, \quad \int_0^\infty \psi(c_r) dc_r = 1.$$

The second integral in Eq. (2.209) can be expressed through gamma function (see Eq. (II.1)), thus one obtains

$$\overline{c_r^j} = \frac{2}{\pi^{1/2}}\Gamma\bigl[(j+3)/2\bigr](2kT/m_r)^{j/2}. \tag{2.212}$$

The most direct application of Eq. (2.212) is to obtain the mean value $\overline{c_r}$ of the relative velocity by letting $j=1$

$$\overline{c_r} = (2/\pi^{1/2})(2kT/m_r)^{1/2}. \tag{2.213}$$

For a single-component gas $m_r = m/2$, with the help of Eq. (2.200) one obtains the expression of $\overline{c_r}$ through the mean thermal speed $\overline{c'}$

$$\overline{c_r} = 2^{3/2}/(\pi^{1/2}\beta) = 2^{1/2}\overline{c'}. \tag{2.214}$$

In obtaining Eq. (0.5) in the Introduction we have used this result.

Now we use Eq. (2.207) to calculate the *collision frequency* of the VHS model molecules. The collision cross-section σ_T is expressed according to Eq. (2.94), and with the help of Eq. (2.98) one has

$$\sigma_T = \sigma_{Tref} c_{rref}^{2\xi} c_r^{-2\xi} = \sigma_{Tref} c_{rref}^{2\omega-1} c_r^{1-2\omega}.$$

Substituting it into Eq. (2.207) and making use of the result for $\overline{c_r^j}$, Eq. (2.212) with $j = 2 - 2\omega$, one obtains

$$v_0 = n\sigma_{Tref} c_{,ref}^{2\omega-1} \frac{2}{\pi^{1/2}} \Gamma\left(\frac{5}{2}-\omega\right)\left(\frac{2kT}{m_r}\right)^{1-\omega}. \tag{2.215}$$

This result is valid also for the VSS model, for the relation of the collision cross-section for the VSS model with the relative velocity is the same as for the VHS model, only with different coefficient of proportionality (see Eq. (2.110))

In fact, we can make use of the expression Eq. (2.109)' of the viscosity for the VSS model to express the collision cross-section v_0 (Eq. (2.215)). Note that $\Gamma(4-\xi)$ in Eq. (2.109)', because of Eq. (2.98), can be written as

$$\Gamma(\frac{9}{2}-\omega) = (\frac{7}{2}-\omega)(\frac{5}{2}-\omega)\Gamma(\frac{5}{2}-\omega),$$

so one has

$$v_0 = \frac{5(\alpha+1)(\alpha+2)}{\alpha(7-2\omega)(5-2\omega)}\frac{np}{\mu}. \tag{2.216}$$

Making use of the definition of the molecular mean free path $\lambda = \bar{c}/\nu$ (see Eq. (0.4)) and the expression of $\bar{c}$ (Eq. (2.200)), one obtains the *mean free path* λ_0 for VSS model in the equilibrium state

$$\lambda_0 = \frac{4\alpha(7-2\omega)(5-2\omega)}{5(\alpha+1)(\alpha+2)} \left(\frac{m}{2\pi kT}\right)^{1/2} \frac{\mu}{\rho}. \tag{2.217}$$

In this equation let $\alpha = 1$, one obtains the result for the VHS model. For the VHS model with the collision cross-section in reverse proportion with the relative velocity (as for the Maxwellian molecule $\eta = 5$, $\omega = 1$), one has

$$\lambda_0 = 2\left(\frac{m}{2\pi kT}\right)^{1/2} \frac{\mu}{\rho}. \tag{2.217}'$$

For simple hard sphere model, $\omega = 1/2$, $\sigma_{Tref} = \sigma_T$, $m_r = m/2$, one has

$$\nu_0 = \left(4/\pi^{\frac{1}{2}}\right) n\sigma_T (kT/m)^{1/2} = 2^{1/2} n\sigma_T \bar{c}. \tag{2.218}$$

This result can be obtained also from the derivation given in the Introduction: for this one just need to substitute the expression Eq. (2.214) of $\bar{c}_r$ through $\bar{c}'$ into Eq. (0.3). The number of collisions in the gas per unit volume per unit time is obtained from Eq. (2.205) as

$$N_{c,0} = 2^{-1/2} n^2 \sigma_T \bar{c}. \tag{2.219}$$

The mean free path λ_0 of molecule in equilibrium is obtained from Eq. (0.4) and Eq. (2.218) as

$$\lambda_0 = \left(2^{1/2} n\sigma_T\right)^{-1} = \left(2^{1/2} n\pi d^2\right)^{-1}. \tag{2.220}$$

If use the expression Eq. (2.76) of the viscosity for the hard sphere model, λ_0 can also be written as

$$\lambda_0 = \frac{16}{5}\left(\frac{m}{2\pi kT}\right)^{1/2} \frac{\mu}{\rho}, \tag{2.221}$$

$$\lambda_0 = \frac{32}{5\pi} \frac{\mu}{\bar{c}\rho} \approx 2.037\mu/(\bar{c}\rho). \tag{2.222}$$

It is evident from the comparison of Eq. (2.221) and Eq. (2.217) that multiplication of the mean free path of the usual HS model by a factor yields the result for the VSS model. The value of this factor for more realistic values $\omega = 0.75$ and $\alpha = 1.5$ is 0.825, for $\omega = 0.75$ and $\alpha = 1$ of the VHS model is approximately 0.8021, for the VHS model ($\alpha = 1$) with Maxwellian character ($\omega = 1$) this factor equals $5/8 = 0.625$.

2.11.3 THE MEAN VALUE OF COLLISION QUANTITIES

The mean value of a quantity Q that is a function only of the relative velocity c_r averaged over all collisions is a quantity that will be needed in the following. According to the definition this mean value is obtained by multiplying the expression in the integral of the collision frequency ν (Eq. (2.206)) by Q and then normalization of thus obtained integral by ν. Or alternatively, by multiplying the expression in the integral of $\overline{c_r^j}$ (Eq. (2.208)) by Q and then normalization of thus obtained integral by the value of $\overline{c_r^j}$ (Eq. (2.212)), where the value of j should chosen equal to the power in the dependence of $\sigma_T c_r$ on c_r

$$\overline{Q} = 2\left(\frac{m_r}{2kT}\right)^{(j+3)/2} \frac{1}{\Gamma[(j+3)/2]} \int_0^\infty Q c_r^{j+2} \exp\left(-\frac{m_r c_r^2}{2kT}\right) dc_r . \qquad (2.223)$$

For the VHS model, according to Eq. (2.94), $\sigma_T c_r \sim c_r^{1-2\xi}$, or according to Eq (2.98), $j = 2 - 2\omega$. So for the VHS model molecules we have

$$\overline{Q} = \frac{2}{\Gamma\left(\frac{5}{2}-\omega\right)} \left(\frac{m_r}{2kT}\right)^{5/2-\omega} \int_0^\infty Q c_r^{2(2-\omega)} \exp\left(-\frac{m_r c_r^2}{2kT}\right) dc_r . \qquad (2.224)$$

Take the example of the translational kinetic energy in the frame of reference of the center of mass $\varepsilon_t = (1/2) m_r c_r^2$. The substitution of $Q = (1/2) m_r c_r^2$ in the above equation yields the *mean translational kinetic energy in collisions*

$$\overline{\varepsilon_t} = \left(\frac{5}{2}-\omega\right) kT . \qquad (2.225)$$

For the HS model $\omega = 1/2$ (see Eq. (2.97))

2.11 EQUILIBRIUM STATE OF GASES

$$\overline{\varepsilon}_t = 2kT. \tag{2.226}$$

The fraction of collisions in which the translational kinetic energy $\varepsilon_t = (1/2)m_r c_r^2$ exceeds a certain prescribed value ε_0 is an important characteristic quantity and will be of use in discussing chemical reactions. To find this quantity one can put $Q=1$ and set the lower limit for the integration to $c_r = (2\varepsilon_0/m_r)^{1/2}$ and obtain

$$\frac{\Delta N}{N} = \frac{2}{\Gamma\left(\frac{5}{2}-\omega\right)} \left(\frac{m_r}{2kT}\right)^{5/2-\omega} \int_{(2\varepsilon_0/m_r)^{1/2}}^{\infty} c_r^{2(2-\omega)} \exp\left(-\frac{m_r c_r^2}{2kT}\right) dc_r. \tag{2.227}$$

The integral in the above expression is the incomplete gamma function (see Eq. (II.4)), one has

$$\frac{\Delta N}{N} = \frac{\Gamma(5/2-\omega, \varepsilon_0/kT)}{\Gamma(5/2-\omega)}. \tag{2.228}$$

For the HS model, $\omega = 1/2$, and the reduction formula Eq. (II.5) in Appendix II gives

$$\Delta N/N = \exp(-\varepsilon_0/(kT))\left(\frac{\varepsilon_0}{kT}+1\right). \tag{2.229}$$

Sometimes we are not interested in the total translational kinetic energy in the collisions but rather in the kinetic energy corresponding to the component of the relative velocity along the apse line of the collision. That is to say, we need to calculate the fraction of molecules in collisions in which the following condition

$$\frac{1}{2}m_r c_r^2 \cos^2\theta_A > \varepsilon_0, \tag{2.230}$$

is satisfied (see Fig. 2.7)). The differential cross-section for HS and VHS molecules is (see Eq. (2.66) and Eq. (2.69))

$$\sigma d\Omega = \left(\frac{d_{12}^2}{4}\right)\sin\chi d\chi d\varepsilon = \frac{\sigma_T}{4\pi}\sin\chi d\chi d\varepsilon.$$

And Eq. (2.56) gives

$$\sin\chi d\chi = -4\sin\theta_A \cos\theta_A d\theta_A.$$

The condition Eq. (2.230) determines the range of values of θ_A

$$0 < \theta_A < \cos^{-1}\left[\left(\frac{2\varepsilon_0}{m_r}\right)^{1/2}/c_r\right].$$

Thus the effective total cross-section σ_T satisfying the condition Eq. (2.230) should be

$$\sigma_\varepsilon = -2\sigma_T \int_0^{\cos^{-1}\left[(2\varepsilon_0/m_r)^{1/2}/c_r\right]} \sin\theta_A \cos\theta_A d\theta_A,$$

i.e.

$$\sigma_\varepsilon = \sigma_T\left(1 - \frac{2\varepsilon_0}{m_r c_r^2}\right). \tag{2.231}$$

Thus to obtain the fraction of collisions in which the condition Eq. (2.230) is satisfied, one should multiply the expression in the integral of Eq. (2.227) by a factor $\left[1 - 2\varepsilon_0/(m_r c_r^2)\right]$

$$\frac{\Delta N}{N} = \frac{2}{\Gamma(5/2-\omega)}\left(\frac{m_r}{2kT}\right)^{5/2-\omega} \int_{(2\varepsilon_0/m_r)^{1/2}}^\infty c_r^{2(2-\omega)}\left(1 - \frac{2\varepsilon_0}{m_r c_r^2}\right)\exp\left(-\frac{m_r c_r^2}{2kT}\right)dc_r.$$

Introducing again the incomplete gamma function according to Eq. (II.4) in Appendix II, we obtain

$$\frac{\Delta N}{N} = \frac{1}{\Gamma(5/2-\omega)}\left[\Gamma\left(\frac{5}{2}-\omega, \frac{\varepsilon_0}{kT}\right) - \frac{\varepsilon_0}{kT}\Gamma\left(\frac{3}{2}-\omega, \frac{\varepsilon_0}{kT}\right)\right]. \tag{2.232}$$

For the hard sphere molecules, $\omega = 1/2$, the fraction of collisions in which condition Eq. (2.230) is satisfied is extremely simple

$$\frac{\Delta N}{N} = \exp\left(-\frac{\varepsilon_0}{kT}\right). \tag{2.233}$$

2.11.4 THE REFERENCE DIAMETER OF THE VSS MODEL AND THE VHS MODEL

When the relation Eq. (2.110)' between the molecular diameter d_{VSS} in the VSS model and the translational kinetic energy ε_t was being obtained, it was pointed out that d_{VSS} could be written in the form of

$$d_* \left[(kT/\varepsilon_t)^{\omega - \frac{1}{2}} \right]^{\frac{1}{2}}.$$

In fact, d_{VSS} (and d_{VHS}, when $\alpha = 1$) can be written as

$$d_{VSS} = d_{ref} \left[(kT_{ref}/\varepsilon_t)^{\omega - \frac{1}{2}} \right]^{\frac{1}{2}} / \Gamma(5/2 - \omega), \qquad (2.234)$$

where

$$d_{ref} = \left[\frac{5(\alpha+1)(\alpha+2)(mkT_{ref}/\pi)^{1/2}}{4\alpha(5-2\omega)(7-2\omega)\mu_{ref}} \right]^{1/2}. \qquad (2.235)$$

The d_{ref} defined by Eq. (2.235) is the reference diameter of the VSS model (and VHS model, with $\alpha = 1$). The reason that there is a factor of $1/\Gamma(5/2 - \omega)$ being separated is because that the definition of d_{ref} is obtained from the relation with the value of $\overline{c_r^{2\omega - 1}}$ in the equilibrium state. In fact, according to Eq. (2.94) and Eq. (2.98) the total cross-section $\sigma_T = \pi d^2$ can be written as

$$\pi d^2 = \sigma_{T ref} c_{rref}^{2\omega - 1} / c_r^{2\omega - 1}. \qquad (2.236)$$

And Bird defines d_{ref} as

$$\pi d_{ref}^2 = \sigma_{T(ef)} c_{rr(ej)}^{2\omega - 1} / \overline{c_r^{2\omega - 1}}, \qquad (2.237)$$

where $\overline{c_r^{2\omega - 1}}$ is the mean value obtained in the assumption that the gas is in equilibrium. In Eq. (2.224) by setting $Q = c_r^{2\omega - 1}$ one obtains

$$\overline{c_r^{2\omega - 1}} = (2kT/m_r)^{\omega - \frac{1}{2}} / \Gamma\left(\frac{5}{2} - \omega\right). \qquad (2.238)$$

Substituting Eq. (2.238) into Eq. (2.237), then expressing $\sigma_{Tref} c_{r,ref}^{2\omega-1}$ by the μ_{ref} of the VSS model through Eq. (2.109)', one obtains the expression of d_{ref}, Eq. (2.235).

The reference diameter d_{ref} is easily calculated at certain reference temperature T_{ref}, and it is convenient to calculate the molecular diameters of the VSS model and VHS model ($\alpha = 1$) according to Eq. (2.234). In Table 2 of Appendix I the molecular weight m, the value of μ at the reference temperature T_{ref}, the viscosity power ω, the power α of the cosine of the deflection angle in scattering (see Eq. (2.102)) and the reference diameter d_{ref} of the VSS and VHS ($\alpha = 1$) models calculated according to Eq. (2.235) are listed for some typical gases.

2.12 GAS MIXTURE

The previous exposition and derivation of this Chapter is aimed at the single-component gas. Now consider a gas mixture composed of s species, the quantity of each species is denoted by the subscript p or q, p and q can take the values from 1 to s. The concepts and results concerning the expression of the macroscopic quantities, the Boltzmann equation, the collision frequencies and the mean free path etc. can easily be extended to the case of gas mixture.

2.12.1 THE MACROSCOPIC PROPERTIES

It is easy to extend the expressions of the macroscopic quantities of the single – component gas through the distribution function to the case of the gas mixture (see Eqs. (2.5), (2.7), (2.8), (2.21), (2.24), (2.25), (2.27) and (2.31)):

component number density

$$n_p = \int f_p dc_p , \tag{2.239}$$

total number density

$$n = \sum_{p=1}^{s} n_p , \tag{2.240}$$

density

$$\rho = \sum_{p=1}^{s} m_p n_p \equiv \sum_{p=1}^{s} \rho_p, \qquad (2.241)$$

average molecular weight

$$\rho = \overline{m}n, \overline{m} = \sum_{p=1}^{s} m_p (n_p/n), \qquad (2.242)$$

mass average velocity

$$c_0 = \frac{1}{\rho}\sum_{p=1}^{s} m_p n_p \overline{c}_p = \frac{1}{\rho}\sum_{p=1}^{s} m_p \int c_p f_p dc_p, \qquad (2.243)$$

thermal velocity

$$c' = c - c_0, \qquad (2.244)$$

species mean thermal velocity

$$\overline{c'}_p = \overline{c}_p - c_0, \qquad (2.245)$$

species diffusion velocity

$$c_p \equiv \overline{c'}_p, \qquad (2.246)$$

pressure tensor

$$P_{ij} = \sum_{p=1}^{s} \rho_p \overline{c'_{ip} c'_{jp}} = \sum_{p=1}^{s} n_p m_p \overline{c'_{ip} c'_{jp}}, \qquad (2.247)$$

stress tensor

$$\tau_{ij} = -\left(\sum_{p=1}^{s} \rho_p \overline{c'_{ip} c'_{jp}} - \delta_{ij} p\right), \qquad (2.248)$$

scalar pressure

$$p = \frac{1}{3}\sum_{p=1}^{s} \rho_p \overline{c'^2_p}, \qquad (2.249)$$

heat flux

$$q_i = \sum \left[\int c'_{ip}\left(\frac{1}{2}m_p c'^2_p\right) f_p dc_p + \int c'_{ip} \varepsilon_{ip} f_p dc_p\right], \qquad (2.250)$$

translational temperature

$$\frac{3}{2}kT = \frac{1}{n}\sum_{p=1}^{s}\int f_p \frac{1}{2}m_p c_p'^2 dc_p .$$ (2.251)

2.12.2 THE BOLTZMANN EQUATIONS

To treat the gas mixture separate distribution functions f_i for the components of the mixture must be introduced, the behavior of which is governed by a system of simultaneous Boltzman equations, the collision term of the ith equation for species i is a sum over s collision integrals, with the jth term representing the collisions of molecules of component i with molecules of component j. The Boltzmann equation for species i is written as (compare with Eq. (2.152))

$$\frac{\partial f_i}{\partial t} + c_i \cdot \frac{\partial f_i}{\partial r} + F \frac{\partial f_i}{\partial c} = \left(\frac{\partial f_i}{\partial t}\right)_c \equiv \sum_{j=1}^{s}\int_{-\infty}^{\infty}\int_{0}^{4\pi} (f_i^* f_{1j}^* - f_i f_{1j}) c_{rij} \sigma_{ij} d\Omega dc_{1j} .$$ (2.252)

2.12.3 NUMBER OF COLLISIONS, COLLISION FREQUENCY AND MEAN FREE PATH

The total number of collisions of molecules of species p with molecules of species q in unit time and unit volume is (see Eq. (2.162), as the collisions here are collisions between different species, so the factor $1/2$ appeared there to avoid double counting is not needed).

$$N_{cpq} = \iint \sigma_{Tpq} c_{rpq} f_p f_q dc_p dc_q .$$ (2.253)

Making use of the definition Eq. (2.163) of the mean value in a collision, one can write

$$\overline{\sigma_{Tpq} c_{rpq}} = \frac{1}{n_p n_q}\iint \sigma_{Tpq} c_{rpq} f_p f_q dc_p dc_q ,$$

consequently

$$N_{cpq} = n_p n_q \overline{\sigma_{Tpq} c_{rpq}} = n_p v_{pq} ,$$ (2.254)

where v_{pq} is the mean collision frequency between the p molecules and the q molecules:

$$v_{pq} = n_q \overline{\sigma_{Tpq} c_{rpq}}. \tag{2.255}$$

The mean collision frequency of p molecules with all species is

$$v_p = \sum_{q=1}^{s} \left(n_q \overline{\sigma_{Tpq} c_{rpq}} \right). \tag{2.256}$$

The total number of collisions of molecules of all species in unit time in unit volume is

$$N_c = \frac{1}{2} \sum_{p=1}^{s} n_p v_p. \tag{2.257}$$

If define the mean collision frequency v of molecules in the gas mixture as the weighted average of the species collision frequencies

$$v = \sum_{p=1}^{s} v_p (n_p/n), \tag{2.258}$$

then N_c can be written if the form of Eq. (2.205).

The mean free path of specie p can be written as

$$\lambda_p = c_p'/v_p, \tag{2.259}$$

and the mean free path in the gas mixture is

$$\lambda = \sum_{p=1}^{s} \lambda_p (n_p/n). \tag{2.260}$$

2.12.4 COLLISION FREQUENCY OF A MOLECULE OF SPECIES A WITH MOLECULES OF SPECIES B IN GAS MIXTURE OF VSS (OR VHS) MOLECULES

The value of the collision frequency of a molecule of species A with molecules of species B in equilibrium state will be needed in the context of simulation of chemical reactions. According to Eq. (2.255) this collision frequency is

$$v_{AB} = n_B \overline{\sigma_{TAB} c_r}. \tag{2.261}$$

For VSS (or VHS) model Eq. (2.94) with the help of Eq. (2.98) gives

$$\sigma_{TAB}/(\sigma_{T.ref})_{AB} = (c_r/c_{rref})^{1-\omega_{AB}}.$$ (2.262)

Substituting into Eq. (2.261) and calculating the integral according to Eq. (2.212) with

$$j = 2 - 2\omega_{AB},$$ (2.263)

one obtains

$$v_{AB} = n_B\left(\sigma_{Tref}\right)_{AB} c_{rref}^{2\omega_{AB}-1} \frac{2}{\pi^{1/2}} \Gamma\left(\frac{2}{5}-\omega_{AB}\right)\left(\frac{2kT}{m_r}\right)^{1-\omega_{AB}}.$$ (2.264)

Expressing $\left(\sigma_{Tref}\right)_{AB} c_{rref}^{2\omega_{AB}-1}$ with the help of Eq. (2.237) and using Eq. (2.238) yields the searched *collision frequency of a molecule of species A with molecules of species B*

$$v_{AB} = 2\pi^{1/2}\left(d_{ref}\right)_{AB}^2 n_B\left[T/\left(T_{ref}\right)_{AB}\right]^{-\omega_{AB}}\left[2k\left(T_{ref}\right)_{AB}/m_r\right]^{1/2}.$$ (2.265)

REFERENCES

1. Chapman S and Cowling TG (1970) The Mathematical Theory of Non-uniform Gases. 3rd Edition, Cambridge University Press
2. Kennard EH. (1938) Kinetic Theory of Gases. McGraw-hill
3. Vincenti WG and Kruger GH (1965) Introduction to Physical Gas Dynamics. John Wiley and Sons, New York
4. Bird GA, (1976) Molecular Gas Dynamics. Clarendon press
5. Bird GA. (1981) Monte Garlo simulation in an engineering context. Progress in Astro Aero, 74, Proceedings of International Symposium on Rarefied Gas Dynamics, 239-255
6. Koura K and Matsumoto H (1991) Variable soft sphere model for inverse power law or Lennard-Jones potential. Phys of Fluids A3: 2459-2465
7. Koura K and Matsumoto H (1992) Variable soft sphere model for air species. Phys of Fluids A4: 1083-1085
8. Bird GA, (1994) Molecular Gas Dynamics and the Direct Simulation of Gas Flows. Clarendon press, Oxford
9. Hassan HA and Hash DB. (1993) A generalized hard sphere model for Monte Carlo simulations. Phys of Fluids A5: 738-744,

10. Fan J (2002) A generalized soft sphere model for Monte Carlo simulation. Phys. of Fluids, 14:4399-4405
11. Hirschfelder JO, Curtiss CF and Bird RB (1954) Molecular Theory of Gases and Liquids. John Wiley and Sons
12. Stockmayer WH, J. (1941). Chem. Phys, 9: 398
13. Broadwell JE. (1964) Study of rarefied shear flow by the discrete velocity method. J of Fluid Mechanics, 19: 401-414
14. Boltzmann, Wien (1872) Sitz., 66:275
15. Green MS, (1956) Boltzmann equation from the statistical mechanical point of view. J Chem Phys, 25: 835-855
16. Wang Chang CS and Uhlenbeck GE (1964) Transport phenomena in polyatomic molecules. Univ. of Michigan, CM 681; Wang Chang CS Ublenbeck, de Boer J (1964) The heat conductivity and viscosity of polyatomic gases. in Studies in Statist. Mech., 2: 247-277, North-Holland, Amsterdam
17. Maxwell (1860) Phil. Mag. (4), 19:22; Collected Works, 1:377

3. INTERACTION OF MOLECULES WITH SOLID SURFACE

3.1 INTRODUCTION

Boltzmann equation (2.152) gives the rate of change of the distribution function with space position and time. When considering the flows of gas around bodies, for obtaining the solution of the Boltzmann equation certain boundary conditions must be proposed, so it is necessary to know, how do the molecules incident with certain velocity onto the body surface reflex from it. The interaction of the gas molecules with the body surface is also the origin of the drag, the lift, the force moment and the heat transfer subjected by the body from the gas flow. So the proposal and application of the gas-surface interaction model in accordance with the reality is the basis of correct prediction of the force and heat actions subjected by the body in rarefied gas flows. Because of the physical complexity of this problem both the theoretical and experimental research on it is far from attaining perfection. In aerospace and other engineering practice for a long period of time the so called complete diffuse reflection model has been in use, or alternatively, the Maxwellian type boundary condition has been adopted, i.e., on the surface the molecules are supposed to partially reflect in a fashion of diffuse reflection model, and the remainder reflect specularly. Such simple models and the definitions of the accommodation coefficients and reflection coefficients will be introduced in section 3.2. Recently Cercignani and Lampis introduced a phenomenological model which was extended by Lord to implement it in DSMC calculations. This is the so called CLL (Cercignani, Lampis, Lord) reflection model. This model satisfies the reciprocity principle, a general rule the molecules reflected from the surface should obey, and can produce a number of distributions of the reflected molecules that are physically reasonable with the introduction of parameters of direct physical meaning. It is easy to be implemented in the DSMC method and can include the practically important cases such as the diffuse reflection with incomplete energy accommodation and the reflection of molecules with discrete energy distributions. In section 3.3 a simple version of the proof of the reciprocity principle will be given. In section 3.4 the CLL reflection model, the diffuse reflection with incomplete energy accommodation and the reflection of molecules with discrete energy distributions will be elucidated.

3.2 SPECULAR AND DIFFUSE REFLECTION

In the description of the interaction of the gas molecules with solid surface there are two simple models proposed by Maxwell [1] in 1879 : the specular reflection model and the diffuse reflection model.

The word 'specular' originated from specularis in Latin which means 'mirror', the word diffuse originated from diffusus which in Latin means 'spread abroad', 'scatter'. The *specular reflection* model assumes that the incident molecules reflect on the body surface as the elastic spheres reflect on the entirely elastic surface, i.e., the normal to the surface component of the relative velocity reverses its direction while the parallel to the surface components remain unchanged. Thus the normal pressure originated from the reflected molecules equals to that originated from the incident molecules; the sheer stress subjected by the surface from the reflected molecules has the opposite sign to that from the incident molecules and the net sheer stress is zero; the total energy exchange with the surface is zero. The *complete diffuse reflection* assumes that the molecules leaving the surface scatter with an equilibrium Maxwellian distribution, the condition of equilibrium being the equality of the surface temperature, the temperature in the Maxwellain distribution and the static temperature of the incident flow. As the distribution function of the reflected molecules is known, so the momentum flux and energy flux carried by the reflected molecules are easily calculated.

Usually to characterize the scatter of the molecules on the surface the so called *scatter kernel* or the *scatter probability* $R(c_i, c_r, r, t)$ of the molecules on the surface is introduced, where c_i denotes the velocity of the incident molecules, c_r denotes the velocity of the reflected molecules. And

$$R(c_i, c_r, r, t) dc_r \qquad (3.1)$$

represents the probability that the molecules with the incident velocity between c_i and $c_i + dc_i$ reflect with velocity between c_r and $c_r + dc_r$. For simplicity, the following discussions will be limited to cases in which the reflection probability is not dependent on the position r of the surface and time t. Thus in R only c_i and c_r are retained as the arguments. If suppose that the incident molecules are not absorbed on the surface, and this is usually the case, the *normalization condition of the scatter kernel*

$$\int_{c_r \cdot n > 0} R(c_i, c_r) dc_r = 1 \qquad (3.2)$$

follows, where n is the external normal of the surface.

We use $f^-(c_i)$ and $f^+(c_r)$ to denote the velocity distribution functions of the incident molecules $(c_i \cdot n < 0)$ and the reflected molecules $(c_r \cdot n > 0)$, respectively. $f^-(c_i) c_i \cdot n \, dc_i$ is the number of incident molecules striking on unit area of the surface in unit time (the number flux of the incident molecules). Multiplying it by Eq. (3.1) and integrating over the whole half space $c_i \cdot n < 0$, one gets the number flux of the reflected molecules

$$f^+(c_r) c_r \cdot n \, dc_r = - \int_{c_i \cdot n < 0} R(c_i, c_r) c_i \cdot n f^-(c_i) dc_i \cdot dc_r$$

$$(3.3)$$

$$(c_r \cdot n > 0).$$

According to the definition, obviously the distribution function f^+ of the reflected molecules under specular reflection is obtained from the distribution function f^- of the incident molecules according to the following expression

$$f^+(c_r) = f^-(c_i - 2n(c_i \cdot n)) \qquad (c_r \cdot n > 0). \qquad (3.4)$$

The scatter kernel of the specular reflection is

$$R_s(c_i,c_r)=\delta\left(c_i-c_r+2n[n\cdot c_r]\right), \tag{3.5}$$

where $\delta(r)$ is the Dirac δ function, according to the definition

$$\int \delta(r-a)\phi(r)dr = \phi(a). \tag{3.6}$$

Substituting Eq. (3.5) into Eq. (3.3), with the help of Eq. (3.6), one obtains the velocity distribution Eq. (3.4) of the reflected molecules under specular reflection.

Similarly, the distribution function f^+ of the reflected molecules under diffuse reflection is the Maxwellian distribution (see Eq. (2.184)) at temperature T_r

$$f^+(c_r) = n_r\left(\frac{m}{2\pi k T_r}\right)^{3/2} \exp\left(-\frac{m}{2kT_r}c_r^2\right). \tag{3.7}$$

The scatter kernel of the diffuse reflection is

$$R_d(c_i,c_r) = \frac{1}{2\pi}\left(\frac{m}{kT_r}\right)^2 \exp\left(-\frac{m}{2kT_r}c_r^2\right). \tag{3.8}$$

When Eq. (3.8) is substituted into Eq. (3.3), as the above expression is not dependent on c_i it can be taken out of the integral, the remained integral over c_i

$$\int_{c_i\cdot n} c_i\cdot n f^-(c_i) dc_i = N_i, \tag{3.9}$$

is just the total number flux of the incident molecules. As there is not accumulation nor leakage of the molecules on the surface, the number flows of the incident and reflected molecules equal each other, i.e.

$$N_i = N_r = n_r\sqrt{\frac{kT_r}{2\pi m}}. \tag{3.10}$$

N_r is the number flow of the molecules scattered in the Maxwell distribution, the last equality of Eq. (3.10) has made use of the result of Chapter 4 See Eq (4.17). Thus from Eq. (3.8) and Eq. (3.3) one obtains Eq. (3.7).

3.2 SPECULAR AND DIFFUSE REFLECTION

It is easy to verify that both the specular reflection model and the diffuse reflection model satisfy the normalization condition Eq. (3.2).

Now let us consider the *implementation of the specular reflection and the diffuse reflection* in the DSMC simulations. For the specular reflection this is a simple and easy matter. At the surface the normal velocity component changes sign, and the molecule moves with the changed velocity in the remainder of the time step into new position. For the diffuse reflection the velocity distribution of the scattered molecules is Maxwelian (see Eq. (2.196)). Here the distribution function divided by n, $f_K = f/n$, is employed (see Eq. (2.6))

$$f_K = \left(\frac{\beta}{\pi^{1/2}}\right)^3 \exp(-\beta^2 c_r^2), \qquad (3.11)$$

where

$$\beta = \left(2\frac{k}{m}T_r\right)^{-1/2}. \qquad (3.12)$$

Let c_r be represented as (u,v,w), where u is the normal to the surface component, v, w are the tangential components. The distribution functions of the v, w components are (see Eq. (2.203))

$$F(v) = \left(\beta/\pi^{1/2}\right)\exp(-\beta^2 v^2)$$
$$F(w) = \left(\beta/\pi^{1/2}\right)\exp(-\beta^2 w^2). \qquad (3.13)$$

So we have

$$F(v)F(w)\,dvdw = \frac{\beta^2}{\pi}\exp(-\beta^2(v^2+w^2))\,dvdw. \qquad (3.14)$$

If transform from the coordinates v, w to the coordinates V, θ

$$v = V\cos\theta, \quad w = V\sin\theta, \qquad (3.15)$$

one will have

$$F(v)F(w)\,dvdw = \frac{\beta^2}{\pi}\exp(-\beta^2 V^2)VdVd\theta =$$

$$\exp(-\beta^2 V^2)d(\beta^2 V^2)d(\theta/2\pi). \tag{3.16}$$

$\beta^2 V^2$ is distributed between 0 and ∞, and has the distribution function:

$$F_{\beta^2 V^2} = \exp(-\beta^2 V^2), \tag{3.17}$$

and the cumulative distribution function (see Appendix III)

$$\Phi_{\beta^2 V^2} = 1 - \exp(-\beta^2 V^2).$$

According to the *sampling method based on inversion of the cumulative distribution* (see (III.1)), by letting $\Phi_{\beta^2 V^2}$ equal to a random fraction *ranf* uniformly distributed between 0, 1, solving the equality relative to $\beta^2 V^2$, the sampled V is obtained. As *ranf* and $(1 - ranf)$ are equivalent, so V is obtained as

$$V = (-\ln(ranf))^{1/2}/\beta. \tag{3.18}$$

From Eq. (3.16) it is known that θ is uniformly distributed between 0 and 2π

$$\theta = 2\pi \, ranf. \tag{3.19}$$

From Eq. (3.18) and Eq. (3.19) the samples of V and θ are obtained through generating the random fractions, then by using Eq. (3.15) two components of the velocity of the scattered molecules according to Maxwellian distribution are obtained. From the derivation given here it is found by the way that the *sampling of the components of the thermal velocity in an equilibrium gas* can also be run as follows: the resultant of the velocity is given by Eq. (3.18), and three values of θ are successively generated according to Eq. (3.19), by taking sines of them and multiplying the resultant velocity the searched values of the components are obtained.

To obtain the sampling of the normal component u of the diffusely reflected molecules, it must be taken into account that it is the product of $u = c \cdot l$ and the

3.2 SPECULAR AND DIFFUSE REFLECTION

distribution function f that appears in the flux of the gas properties as distribution function, see Eq. (2.13) and Eq. (2.14). So the distribution function of the normal component u of the molecules scattered from the surface diffusely is

$$F(u) = Cu\exp(-\beta^2 u^2).$$

The normalization condition $\int_0^\infty F(u)du = 1$ gives

$$F(u) = 2\beta^2 u\exp(-\beta^2 u^2).$$

So the distribution function of $\beta^2 u^2$ is

$$F_{\beta^2 u^2} = \exp(-\beta^2 u^2).$$

This is identical with the distribution Eq. (3.17), and the sampling of u can obtained according to

$$u = -\left(\ln(ranf)\right)^{1/2}/\beta, \qquad (3.20)$$

i.e., the sampling of the normal component u in the diffuse reflection is identical with the sampling of the resultant of the tangential components v and w.

The experimental research on the reflection of molecules on the body surfaces reveals that pure specular reflection or complete diffuse reflection can not describe properly the real situation of the scatter of molecules on the surfaces. Maxwell put forward a model composed of these two reflection models [2]. He suggested that the σ portion of the incident molecules reflects diffusely and the other $(1-\sigma)$ portion reflects specularly, such a reflection model is called *Maxwellian type boundary condition*.

In the ordinary problems of rarefied gas dynamics usually the flow of the gas relative to the bodies is considered, in such case even when the molecules are scattered from the surface according to diffuse reflection model, the distribution of the gas near the body surface would not remain in equilibrium. For the stagnation temperature of the gas flow is different from the static temperature, the temperature in correspondence with the Maxwell distribution of the reflected molecules

138 3. INTERACTION OF MOLECULES WITH SOLID SURFACE

would be different at least with one of the above two temperatures. So the distribution function of the reflected molecules generally is different from that of the incident molecules, so there is generally no equilibrium distribution function near the body surface. Knudsen [3] introduced the concept of the thermal accommodation coefficient α to characterize in what degree the temperature of the reflected molecules has accommodated (adjusted) to the situation (temperature) of the surface

$$\alpha = \frac{q_i - q_r}{q_i - q_w}. \tag{3.21}$$

In the above expression q_i, q_r are the energy fluxes of the incident and reflected molecules; q_w is the energy flux under complete diffuse reflection when the temperature of the reflected flow equal to the temperature T_w of the body; α characterizes the degree to which the reflected molecules has adjusted to the temperature of the body surface. $\alpha = 1$ corresponds to the case of complete thermal accommodation, the molecules reflect with the Maxwellian distribution under temperature $T_r = T_w$. $\alpha = 0$ corresponds to the case when the incident molecules are entirely not adjusted to the conditions of the surface, $q_r = q_i$.

By analogy with the introduction of the thermal accommodation coefficient α one can introduce the *accommodation coefficients* σ' and σ of the *normal* and *tangential momentum* components

$$\sigma' = \frac{p_i - p_r}{p_i - p_w}, \tag{3.22}$$

$$\sigma = \frac{\tau_i - \tau_r}{\tau_i - \tau_w} = \frac{\tau_i - \tau_r}{\tau_i}. \tag{3.23}$$

p and τ represent the normal and tangential components of the momentum flux; i and r denote the values of incident and reflected flows respectively; p_w and τ_w are the normal and tangential momentum fluxes in the case of complete diffuse reflection with $T_r = T_w$, obviously one has $\tau_w = 0$. In the case of complete specular reflection $\sigma' = \sigma = 0$, in the case of complete diffuse reflection $\sigma' = \sigma = 1$.

3.2 SPECULAR AND DIFFUSE REFLECTION

The accommodation coefficients α, σ' and σ are defined through macroscopic physical quantities that can be measured and are parameters describing macroscopically the reflection mechanism. By measuring the heat flux and the force actions of the gas flows on the body surfaces their values can be determined separately. Provided that their values are known, in the regime of free molecular flow the aerodynamic forces and heat transfer acted on the body can be calculated. But they can not offer the quantitative information about the distribution function of the reflected molecules, and in the transition regime they do not constitute the boundary conditions of the Boltzmann equation for solving the problem. Besides, when introducing the definitions Eq. (3.21), Eq. (3.22) and Eq. (3.23) of the accommodation coefficients, it was implicitly supposed that they are independent of the magnitude of the incident velocity, the angle between the incident velocity and the normal to the surface and the body temperature, and that they are independent of each other. But the experimental results show that what happens in the reality is the opposite. All these impose restrictions on the application of the accommodation coefficients, especially the normal and tangential momentum coefficients.

For surfaces processed by usual industrial means at normal atmospheric temperature and with not too high incident energy of the oncoming flow, the diffuse reflection model with complete energy accommodation can serve as fairly good approximation of the interaction of molecules with the body surface. But experimental investigation shows that for the clean, precisely processed surfaces at high temperature and in the high vacuum environment, when the incident energy is high (for example, the energy flux corresponding to satellite speed), there is apparent reduction of the degree of accommodation of the energy on the surface, α is much smaller than 1. The molecular beam experiments also show that the distribution of the reflected molecules has the leaf blade form in certain direction. All these imply that more realistic molecular surface interaction model must be put forward to describe the scattering of the molecules on the body surface. In the next section it will be seen that besides the normalization condition Eq. (3.2) the scatter kernel should also satisfy the reciprocity principle, which is the basis of construction of the physically realistic reflection model.

3.3 THE RECIPROCITY PRINCIPLE

The requirement of equilibrium between the incident molecules and the surface imposes a restriction condition on the interaction between them, i.e., the *reciprocity principle* or *the principle of the detailed balance* [4, 5, 6]. This manifests itself as the limitation relation on the scatter kernel $R(c_i, c_r)$ of molecules on the surface introduced in section 3.2. For the positive interaction scattering the molecules from the range $(c_i, c_i + dc_i)$ into the range $(c_r, c_r + dc_r)$ there is the reverse interaction scattering the incident velocity $(-c_r, -c_r - dc_r)$ into the range of $(-c_i, -c_i - dc_i)$. The reciprocity principle is a relation between the scatter kernel $R(c_i, c_r)$ and the scatter kernel $R(-c_r, -c_i)$

$$c_r \cdot nR(-c_r, -c_i)\exp\left[-\varepsilon_{c_r}/(kT_w)\right] = -c_i \cdot nR(c_i, c_r)\exp\left[-\varepsilon_q/(kT_w)\right]. \qquad (3.24)$$

In the above formula ε_{c_r}, ε_{c_i} are the energy of the reflected and the incident molecules.

Assume that the surface consists of a number of independent, identical constituent elements (atoms, molecules or certain crystal units), and each element interacts with the gas molecule only once. The constituent element of the surface has momentum p_i and energy ε_{pi} before the collision with the gas molecule (i denotes the initial status), after the collision the momentum and energy become p_f and energy ε_{pf} (f denotes the final status). Correspondingly, the momentum and energy of the gas molecule become from mc_i and ε_{ci} to mc_r and ε_{cr}. The following formula

$$P\left[c_i, p_i \to c_r, p_f\right]dc_r dp_f \qquad (3.25)$$

is used to denote the probability of transformation in unit time for the velocity of the molecule to transform from $(c_i, c_i + dc_i)$ into $(c_r, c_r + dc_r)$ and the momentum of the surface element to transform from $(p_i, p_i + dp_i)$ into $(p_f, p_f + dp_f)$. In the interaction of the molecules with the surface the energy conservation is satisfied

3.3 THE RECIPROCITY PRINCIPLE

$$\varepsilon_{c_i} + \varepsilon_{p_i} = \varepsilon_{c_r} + \varepsilon_{p_f}. \tag{3.26}$$

Summing up the probability of transformation Eq. (3.25) under the condition Eq. (3.26) over the initial and final statuses yields obviously the transform probability in unit time for the velocity of the gas molecule to transform from $(c_i, c_i + dc_i)$ into $(c_r, c_r + dc_r)$:

$$-c_i \cdot nR(c_i, c_r) dc_r = \int\int_{p_i\, p_f} P(c_i, p_i \to c_r, p_f) \cdot$$

$$n(p_i)\delta\left(\varepsilon_{c_i} + \varepsilon_{p_i} - \varepsilon_{c_r} - \varepsilon_{p_f}\right) dp_i\, dp_f\, dc_r. \tag{3.27}$$

In the above formula $n(p_i)$ denotes the density of the constituent elements of the surface; the presence of the δ function guarantees the energy conservation. As the laws governing the change of the probability of transformation P (they are essentially quantum mechanical) obey the invariant principle with time reversal, so the value of P remains unchanged when changing places of the initial and final statuses and reversing the sign of the momentum

$$P(c_i, p_i \to c_r, p_f) = P(-c_r, -p_f \to -c_i, -p_i). \tag{3.28}$$

Assume that the distribution of the orientations of the constituent elements of the surface is random and the energy obeys the normal distribution characterized by the wall temperature (see Eq. (1.68)' in Chapter 1)

$$n(p_i) = C\exp\left(-\varepsilon_{p_i}/kT_w\right). \tag{3.29}$$

Substituting Eq. (3.29) into Eq. (3.27) gives

$$-c_i \cdot nR(c_i, c_r) dc_r =$$

$$\int\int_{p_i\, p_f} P(c_i, p_i \to c_r, p_f) C\exp\left(-\varepsilon_{p_i}/kT_w\right) \cdot \tag{3.30}$$

$$\delta\left(\varepsilon_{ci} + \varepsilon_{pi} - \varepsilon_{cr} - \varepsilon_{pf}\right) dp_i \, dp_f \, dc_r.$$

Changing the places of the initial and the final statuses and reversing the signs, one can rewrite the above equation as

$$c_r \cdot nR(-c_r, -c_i) dc_i =$$

$$\iint_{p_i \, p_f} P(-c_r, -p_f \to -c_i, -p_i) C \exp(-\varepsilon_{pf} / kT_w) \cdot$$

$$\delta\left(\varepsilon_{cr} + \varepsilon_{pf} - \varepsilon_{ci} - \varepsilon_{pi}\right) dp_f \, dp_i \, dc_i. \tag{3.31}$$

Substituting the time reverse invariant principle Eq. (3.28) and the energy conservation Eq. (3.26) into Eq. (3.31) yields

$$c_r \cdot nR(-c_r, -c_i) dc_i =$$

$$\iint_{p_i \, p_f} P(c_i, p_i \to c_r, p_f) C \exp(-\varepsilon_{pi} / kT_w) \cdot$$

$$\delta\left(\varepsilon_{cr} + \varepsilon_{pf} - \varepsilon_{ci} - \varepsilon_{pi}\right) dp_f \, dp_i \, \exp\left[\left(\varepsilon_{ci} - \varepsilon_{cr}\right) / kT_w\right] dc_i. \tag{3.32}$$

Comparison of Eq. (3.32) and Eq. (3.30) gives the reciprocity principle given in the beginning the present section

$$c_r \cdot nR(-c_r, -c_i) \exp(-\varepsilon_{cr} / kT_w) = -c_i \cdot nR(c_i, c_r) \exp(-\varepsilon_{ci} / kT_w). \tag{3.24}$$

It is easy to verify that both the specular and the diffuse reflection models satisfy the reciprocity principle.

3.4 THE CLL GAS SURFACE INTERACTION MODEL

Cercignani and Lampis [7] constructed a scatter kernel that satisfies the normalization condition Eq. (3.2), the reciprocity principle Eq. (3.24) and is positive, it involves two adjustable parameters, the energy accommodation coefficients of the tangential velocity component and the normal velocity component. The number of

3.4 THE CLL GAS SURFACE INTERACTION MODEL

molecules scattered into given directions calculated by using this model is found to be in good agreement with the experimental result of the molecular beam measurement [8],[9]. Lord [10, 11] developed this model, implemented it in the DSMC method and extended it to involve the cases such as the internal energy exchange, the diffuse reflection with incomplete energy accommodation and the discrete energy exchange.

In this *CLL (Cercignani-Lampis-Lord)* model the scatter kernels of the normal velocity component u and the tangential velocity components v, w are independent. The *scatter kernel of the tangential component* v normalized by $\sqrt{2RT_w}$ is

$$R(v_i, v_r) = \frac{1}{\sqrt{\pi \sigma_t (2-\sigma_t)}} \exp{-\frac{(v_r - (1-\sigma_t)v_i)^2}{\sigma_t (2-\sigma_t)}}. \tag{3.33}$$

σ_t is the tangential reflection coefficient in the CLL model. Obviously, Eq. (3.33) satisfies the normalization condition (see Eq. (3.2)) of the following form

$$\int_{-\infty}^{\infty} R(v_i, v_r) dv_r = 1. \tag{3.34}$$

For v the reciprocity principle Eq. (3.24) takes the following form

$$R(-v_r, -v_i)\exp(-v_r^2) = R(v_i, v_r)\exp(-v_i^2). \tag{3.35}$$

Equation (3.33) satisfies apparently also this condition.

The scatter kernel Eq. (3.33) stipulates the mean value $\overline{v_r}$ of the reflected tangential velocity component

$$\overline{v_r} = \int v_r R(v_i, v_r) dv_r =$$

$$\int_{-\infty}^{\infty} v_r \frac{1}{\sqrt{\pi \sigma_t (2-\sigma_t)}} \exp\left\{-\frac{(v_r - (1-\sigma_t)v_i)^2}{\sigma_t (2-\sigma_t)}\right\} dv_r = (1-\sigma_t)v_i. \tag{3.36}$$

144   3. INTERACTION OF MOLECULES WITH SOLID SURFACE

Thus, the meaning of σ_t introduced in Eq. (3.33) becomes apparent, it is identical with accommodation coefficient σ of the tangential velocity component introduced according to Eq. (3.23)

$$\sigma = \frac{v_i - \overline{v_r}}{v_i} = \sigma_t. \qquad (3.37)$$

And the accommodation coefficient $\alpha_t = \left(v_i^2 - \overline{v_r^2}\right)/v_i^2$ of the kinetic energy of the tangential velocity component equals

$$\alpha_t = \sigma_t(2 - \sigma_t). \qquad (3.38)$$

Obviously

$$1 - \sigma_t = (1 - \alpha_t)^{1/2}. \qquad (3.38)'$$

Thus, the scatter kernel of v, Eq. (3.33), can also be written as

$$R(v_i, v_r) = \frac{1}{\sqrt{\pi \alpha_t}} \exp\left[-\frac{\left(v_r - (1-\alpha_t)^{1/2} v_i\right)^2}{\alpha_t}\right]. \qquad (3.39)$$

For the *scatter kernel of the normal velocity component u*, the reciprocity principle (Eq. (3.24)) is written as (u is normalized still by $\sqrt{2RT_w}$)

$$u_r R(-u_r, -u_i)\exp(-u_r^2) = -u_i R(u_i, u_r)\exp(-u_i^2). \qquad (3.40)$$

And as u_r is always positive, the normalization condition Eq. (3.2) is expressed in a form different from Eq. (3.34)

$$\int_0^\infty R(u_i, u_r)du_r = 1. \qquad (3.41)$$

The scatter kernel of the normal velocity component u is different from those of the tangential components v, w, just as the sampling of u is different from that of v, w in the diffuse reflection model, there the sampling of u is the same as the sampling of the resultant V of v and w. In the CLL model the scatter ker-

3.4 THE CLL GAS SURFACE INTERACTION MODEL   145

kernel of u is also constructed by using the resultant velocity V of v and w whose scatter kernels are known. In the plane v, w introduce the polar coordinates V, θ

$$v = V\cos\theta, \quad w = V\sin\theta, \quad dvdw = VdVd\theta .$$

Without violating generality the coordinate system can be chosen according to the direction of the incident velocity so that $v_i = V_i$ and $w_i = 0$. The probability of scattering from V_i into V_r $(V_i \rightarrow V_r)$ is obviously the product of the probability of $(v_i \equiv V_i \rightarrow v_r)$ and the probability of $(w_i \equiv 0 \rightarrow w_r)$ integrated over all directions, i.e.

$$R(V_i, V_r) = \int R(V_i, v_r) R(0, w_r) V_r d\theta =$$

$$\int_0^{2\pi} \frac{1}{\sqrt{\pi\alpha}} \exp\left[-\frac{\left(V_r \cos\theta - (1-\alpha)^{\frac{1}{2}} V_i\right)^2}{\alpha}\right].$$

$$\frac{1}{\sqrt{\pi\alpha}} \exp\left[-\frac{V_r^2 \sin^2\theta}{\alpha}\right] V_r d\theta =$$

$$\frac{V_r}{\pi\alpha} \exp\left[-\frac{V_r^2 + (1-\alpha)V_i^2}{\alpha}\right] \int_0^{2\pi} \exp\left[\frac{2(1-\alpha)^{\frac{1}{2}} V_i V_r \cos\theta}{\alpha}\right] d\theta =$$

$$\frac{2V_r}{\alpha} \exp\left[-\frac{V_r^2 + (1-\alpha)V_i^2}{\alpha}\right] I_0\left(\frac{2(1-\alpha)^{1/2} V_r V_i}{\alpha}\right) \quad (3.42)$$

I_0 is the first kind zero order modified Bessel function

$$I_0(z) = \frac{1}{\pi} \int_0^\pi \exp(z\cos\theta) d\theta . \quad (3.43)$$

The expression Eq. (3.42) is taken as the *scatter kernel of the normal velocity component*

146   3. INTERACTION OF MOLECULES WITH SOLID SURFACE

$$R(u_i,u_r) = \frac{2u_r}{\alpha_n} \exp\left[-\frac{u_r^2 + (1-\alpha_n)u_i^2}{\alpha_n}\right] I_0\left[\frac{2(1-\alpha_n)^{\frac{1}{2}} u_r u_i}{\alpha_n}\right], \qquad (3.44)$$

where α_n is the energy accommodation coefficient of the normal velocity component u. It can be seen, the scatter kernel (Eq. (3.44)) satisfies the reciprocity principle Eq. (3.40). To prove the normalization condition (Eq. (3.41)) Eq. (3.42) is used, the integration of $R(V_i,V_r)$ over V_r from 0 to ∞ is transformed to the integration with arguments changing from V_r, θ into v_r, w_r and range of integration changing from $-\infty$ to $+\infty$. Because of the normalization conditions of $R(V_i,v_r)$ and $R(w_i,w_r)$ the normalization of $R(V_i,V_r)$ is verified.

The experiment of the molecular beam measurement is to direct a molecular beam with certain distribution function to strike onto the body surface and to determine the distribution function of the reflected molecules through the counting of the number of molecules scattered in certain direction. For the ideal case when the incident molecular beam is collimated and has single-velocity U, the probability that the reflected molecules scattered into unit solid angle in the direction θ, φ in unit time is

$$N(\theta,\varphi) = \int_0^\infty R(U,c)|c \cdot n|c^2 dc^2 . \qquad (3.45)$$

The scatter kernel $R(U,c)$ is easily constructed from the scatter kernel of the tangential velocity component Eq. (3.33) and the scatter kernel of the normal velocity component Eq. (3.44). Thus the quoted probability, or the ratio of the number of molecules scattered into unit solid angle in the direction θ,φ in unit time to the total number of the scattered molecules, can be obtained by simple quadrature. For more realistic collimated thermal molecular beams the distribution function of the incident beam can be calculated and additional integration over U is needed. The experimental result [8] of the argon beam of temperature $295K$ scattered on a platinum surface at temperature $1081K$ is given in Fig. 3.1. The experimental data for cases of incident angles of $15°$, $22.5°$, $30°$ and $45°$ are given by small circles together with the calculated results by using the kernels Eq. (3.33), Eq. (3.44) with $\sigma_t = 0.1$ and $\alpha_n = 0.3$ [9]. One can see that the agree-

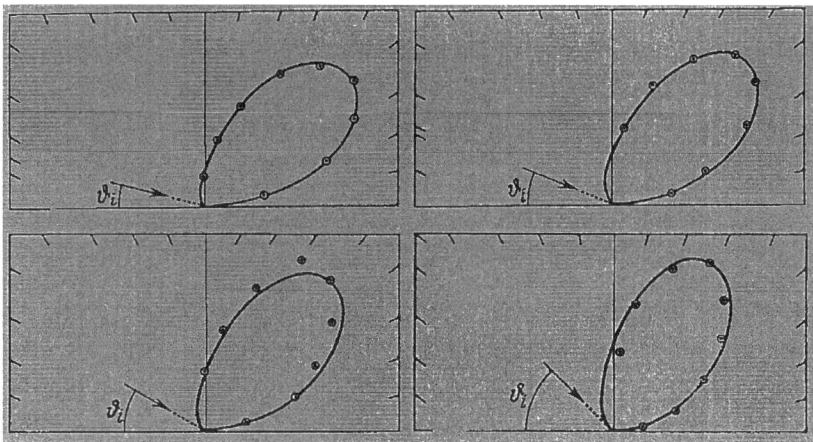

Fig. 3.1 Comparison of the experimental data of [8] with the calculated results by using the kernel Eq. (3.33) and (3.44), $\sigma_t = 0.1$, $\alpha_n = 0.3$, $\vartheta_i = 15°$, $22.5°$, $30°$ and $45°$ [9].

ment is quite good for all incident angles with the same values of the accommodation coefficients $\sigma_t = 0.1$ and $\alpha_n = 0.3$. It is noted that both the experimental and the predicted by CLL model distribution of the reflected molecules has the leaf blade form in certain direction, whereas the specular reflection would have a narrow peak in certain direction and the diffuse reflection would have a circular (spherical) form indicatrix (see the discussion on the spatial distribution of the diffuse reflection at the end of this section).

Now we proceed to the discussion of the *implementation of the CLL reflection model in DSMC simulations*. According to Eq. (3.33), Eq. (3.36) and Eq. (3.38) the probability that an incident molecule with tangential velocity components $(v_i, 0)$ scatters into (v_r, w_r) is

$$R(v_i, v_r) R(0, w_r) dv_r dw_r = \frac{1}{\pi \alpha} \exp\left(-\frac{(v_r - \overline{v_r})^2}{\alpha}\right) \exp\left(-\frac{w_r^2}{\alpha}\right) dv_r dw_r.$$

Now transform from the coordinates (v_r, w_r) into the polar coordinates (r, θ) with pole at point $Q(\overline{v_r}, 0)$, where $r = \sqrt{(v_r - \overline{v_r})^2 + w_r^2}$ (see Fig 3.2)). As $dv_r dw_r = r dr d\theta$, the transform probability can be written as

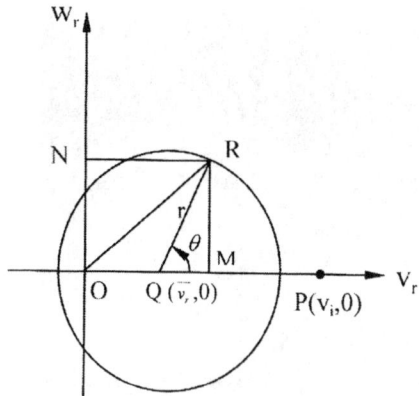

Fig. 3.2 The diagram showing the obtaining of the reflected velocity of the scattered molecule

$$\frac{1}{\pi\alpha}e^{-\frac{r^2}{\alpha}}rdrd\theta, \qquad (3.46)$$

showing that the transform probability is uniformly distributed over θ. Integrating over θ from 0 to 2π yields the distribution probability of r

$$f(r)dr = \frac{2r}{\alpha}\exp\left(\frac{-r^2}{\alpha}\right)dr. \qquad (3.47)$$

The cumulative distribution function is

$$F(r) = \int_0^r f(r)dr = 1 - \exp\left(-\frac{r^2}{\alpha}\right). \qquad (3.48)$$

Again according to the inverse cumulative method (Appendix III.1) and because of the equivalence of $ranf$ and $(1-ranf)$, the sample of r is obtained from the following formula

$$ranf_1 = e^{-\frac{r^2}{\alpha}}. \qquad (3.49)$$

As the searched probability is uniform for θ in $(0, 2\pi)$

$$ranf_2 = \theta/2\pi .\tag{3.50}$$

So the sampling of r and θ is obtained from the inversion of the above formulae

$$r = \sqrt{-\alpha \ln ranf_1}, \quad \theta = 2\pi ranf_2 .\tag{3.51}$$

Consequently the sampling of v_r and w_r is obtained

$$v_r = \overline{v_r} + r\cos\theta, w_r = r\sin\theta ,\tag{3.52}$$

where $\overline{v_r}$ is given by Eq. (3.36).

As for the normal velocity component u_r, as it is considered as the resultant velocity V_r of v_r and w_r, the sampling of it in DSMC implementation is the same as the above method. Only the pole of the polar coordinates is taken as $(\overline{u_r}, 0)$, where $\overline{u_r}$ is the average normal velocity of the scattered molecules

$$\overline{u_r} = \sqrt{(1-\alpha_n)}u_i .\tag{3.53}$$

The sampling of u_r uses the resultant of v_r and w_r, Eq. (3.52), only with $\overline{v_r}$ replaced by $\overline{u_r}$

$$u_r = \sqrt{\overline{u_r}^2 + 2r\overline{u_r}\cos\theta + r^2} .\tag{3.54}$$

Now it is easy to explain the *diagram of the CLL model* (see Fig. 3.2) [10]. Point P represents the oncoming velocity (v_i or u_i) of the incident molecule, in the case of tangential velocity, it is assumed that $w_i = 0$, so the framework can be rotated so that v_i is in the direction of the incident tangential velocity. Point Q represents the average state of the scattered molecules, $\overline{v_r}$ or $\overline{u_r}$, its value is given by Eq. (3.36) or Eq. (3.53), so one has

$$OQ/OP = \sqrt{(1-\alpha)} .\tag{3.55}$$

Point R represents the actually attained reflected velocity, its distribution along the circle with Q as the center and r as the radius is given by Eq. (3.52). ON and OM represent v_r and w_r, and OR represents u_r or $\sqrt{v_r^2 + w_r^2}$.

Now we discuss the issue of *exchange of the internal mode energy of molecules with the surface*. From the derivation of the reciprocity principle Eq. (3.24) and the construction of the above scattering model it is seen that the model is not only suitable for the velocity components but is also suitable for the description of the *exchange of the angular velocity component* of a polyatomic molecule about the principle axis of inertia. The scattering kernel Eq. (3.33) of the tangential velocity component can be considered as a scattering kernel of an angular velocity component ω_v, only this time the normalization factor of ω_r is $\sqrt{2kT/I_v}$, I_v being the rotational inertia of the molecule about the axis v. In the case of diatomic molecule with two rotational degrees of freedom the two transverse moments of inertia equal each other and the scattering probability of the resultant angular velocity ω is described by Eq. (3.44) derived for the normal velocity u. It certainly satisfies the reciprocity principle Eq. (3.40) and the normalization condition Eq. (3.41), where u is replaced by ω and α_n replaced by the accommodation coefficient of the rotational energy.

If set $E = u^2$, from Eq. (3.44) the distribution of the kinetic energy of the normal velocity of the scattered molecules can be written

$$f(E_r)dE_r = \frac{1}{\alpha}\exp\left[-\frac{E_r + (1-\alpha)E_i}{\alpha}\right]I_0\left[\frac{2\{(1-\alpha)E_r E_i\}^{1/2}}{\alpha}\right]dE_r. \qquad (3.56)$$

This distribution at the same time is also the distribution function of the kinetic energy $E = V^2$ of the resultant tangential velocity and of the kinetic energy $E = \omega^2$ of the angular velocity. In different cases the value of α should be taken as α_n, α_t and α_r correspondingly. According to the proved normalization of $R(u_i, u_r)$ obviously follows

$$\frac{1}{\alpha}\int_0^\infty \exp\left[-\frac{E_r + (1-\alpha)E_i}{\alpha}\right]I_0\left[\frac{2\{(1-\alpha)E_r E_i\}^{1/2}}{\alpha}\right]dE_r = 1. \qquad (3.57)$$

3.4 THE CLL GAS SURFACE INTERACTION MODEL

The CLL model can be extended to the case of *discrete energy levels of the internal energy* of molecules [11]. The above classic method in treating the internal degrees of freedom is suitable for the case of rotational degree of freedom at ordinary and fairly high temperatures, for the rotational characteristic temperature is low and the spacing of the rotational energy levels is small. The spacing of the vibrational energy levels is much larger, such levels are suitable to be treated by the discrete energy method (see the discussion in section 1.1).

The harmonic vibrator model of the diatomic molecule gives the discrete vibrational energy levels (Eq. (1.16))

$$\varepsilon_{v,n} = \left(n + \frac{1}{2}\right) h\nu, \quad n = 0,1,2,3,\cdots.$$

Let us try to construct the transition probability

$$R(m_i, n_r) \tag{3.58}$$

for a molecule to transit from an initial state m_i, $E_i = (m_i + 1/2) E_0$, into the final state n_r, $E_r = (n_r + 1/2) E_0$, where $E_0 = h\nu$, m_i and n_r are the vibrational quantum number of the molecule at incidence and at scattering. The reciprocity principle and the normalization condition $R(m_i, n_r)$ should satisfy for discrete energy levels can be expressed as

$$R(n_r, m_i) \exp\left[-\left(n_r + \frac{1}{2}\right) E_0\right] = R(m_i, n_r) \exp\left[-\left(m_i + \frac{1}{2}\right) E_0\right], \tag{3.59}$$

$$\sum_{n_r=0}^{\infty} R(m_i, n_r) = 1. \tag{3.60}$$

It is known that the energy distribution for scattered molecules having two degrees of freedom is Eq. (3.56), and now the vibrational energy of the incident molecules are stipulated to be discrete, we can try to quantize the energy E_r in Eq. (3.56), relating all energy in the range between $n_r E_0$ and $(n_r + 1) E_0$ to the n_r th quantum state, thus obtaining a preliminary scattering kernel

$$R_p(m_i, n_r) = \frac{1}{\alpha} \int_{n_r E_0}^{(n_r+1)E_0} \exp\left[-\frac{E_r + (1-\alpha)E_i}{\alpha}\right] I_0\left[\frac{2\{(1-\alpha)E_r E_i\}^{1/2}}{\alpha}\right] dE_r, \quad (3.61)$$

where $E_i = (m_i + 1/2) E_0$. R_p satisfies the normalization condition Eq. (3.60) (in view of Eq. (3.57)), but can not satisfy the reciprocity condition Eq. (3.59). We attempt to modify $R_p(m_i, n_r)$ so that the reconstructed scattering kernel is symmetric with m_i and n_r, or, which is the same, is symmetric with E_i and E_r. It is easily seen that the expression in the integral should be multiplied by a factor $\exp(-E_i)$, and the energy level m_i should be formally obtained according to the rule that relating all energy in the range between $m_i E_0$ and $(m_i + 1) E_0$ to the m_ith quantum state. Thus the searched scattering kernel is written in the following form

$$R(m_i, n_r) = \frac{\exp(m_i E_0)}{1 - \exp(-E_0)} \int_{m_i E_0}^{(m_i+1)E_0} \exp(-E_i) R_p(m_i, n_r) dE_i =$$

$$\frac{\exp(m_i E_0)}{1 - \exp(-E_0)} \frac{1}{\alpha} \int_{m_i E_0}^{(m_i+1)E_0} \exp(-E_i) \times$$

$$\int_{n_r E_0}^{(n_r+1)E_0} \exp\left[-\frac{E_r + (1-\alpha)E_i}{\alpha}\right] I_0\left[\frac{2\{(1-\alpha)E_r E_i\}^{1/2}}{\alpha}\right] \cdot dE_r dE_i. \quad (3.62)$$

The scattering kernel Eq. (3.62) now satisfies the reciprocity principle Eq. (3.59), for one has

$$(1 - \exp(-E_0)) \cdot \exp\left[-\left(m_i + \frac{1}{2}\right) E_0\right] R(m_i, n_r) =$$

$$\frac{1}{\alpha} \int_{m_i E_0}^{(m_i+1)E_0} \int_{n_r E_0}^{(n_r+1)E_0} \exp\left[-\frac{E_r + E_i + \frac{1}{2}\alpha E_0}{\alpha}\right] I_0\left[\frac{2\{(1-\alpha)E_r E_i\}^{1/2}}{\alpha}\right] \quad (3.63)$$

$$dE_r dE_i = (1-\exp(-E_0))\exp\left[-\left(n_r+\frac{1}{2}\right)E_0\right]R(n_r,m_i).$$

It still satisfies the normalization condition Eq. (3.60), for when summing over n_r the integral over E_r (after multiplying by the factor $1/\alpha$) yields 1 because of Eq. (3.57), and the integration over E_i gives

$$\int_{m_i E_0}^{(m_i+1)E_0} \exp-E_i dE_i = (1-\exp(-E_0))\exp(-m_i E_0). \tag{3.64}$$

The *implementation of the scattering kernel (Eq. (3.62)) in the DESMC simulation* is to replace the vibrational energy $E_i = (m_i + 1/2)E_0$ by $m_i E_0 + E^*$. As the probability density of E^* between 0 and E_0 is an exponential function, the sampling of E^* is obtained by using the inverse cumulative method (see Appendix III.1, let the cumulative distribution of E^*, $(1-e^{-E^*})/(1-e^{-E_0})$, equal *ranf*)

$$E^* = -\ln[1-ranf + ranf \exp(-E_0)], \tag{3.65}$$

where *ranf* is a random fraction uniformly distributed between 0 and 1. Then the known method of sampling from the incident normal velocity $u_i = E_i^{1/2}$ is used to get the sampling of u_r, and consequently $E_r = u_r^2$, the final quantum number is $n_r = INT(E_r / E_0)$, where *INT* means the integer of the argument.

Lord [11] further developed the CLL scattering model to include the case of *diffuse reflection with incomplete energy accommodation*. To construct a model including the case of diffuse elastic reflection, i.e. the speed of the reflected molecule retains the value of the incoming molecule but the direction of reflection follows the cosine law of the diffuse reflection, first we obtain the distribution probability that the resultant speed reflects from c_i into c_r according to CLL model when the tangential and the normal energy accommodation coefficients are the same : $\alpha_t = \alpha_n = \alpha$. In general, it is required to find the probability that the incident velocity with u_i and v_i not zero and $w_i = 0$ scatters into $c_r(u_r, v_r, w_r)$. This requires integration over all directions in four-dimensional polar coordinates [11]. This probability is the same as the probability that the veloc-

154   3. INTERACTION OF MOLECULES WITH SOLID SURFACE

ity $u_i = c_i$, $v_i = w_i = 0$ scatters into $c_r(u_r, v_r, w_r)$. The derivation of the latter is given here. In the spherical polar coordinates u, v, w are related with c, θ, ϕ

$$\begin{cases} u = c\cos\theta \\ v = c\sin\theta\cos\phi \\ w = c\sin\theta\sin\phi \end{cases} \qquad (3.66)$$

The searched probability $P(c_i \to c_r')dc_r$ is the product of $R(u_i \equiv c_i \to u_r)$ (see Eq. (3.44)), $R(0 \to v_r)$ and $R(0 \to w_r)$ (see Eq. (3.33)) integrated over all directions of θ, ϕ

$$P(c_i, c_r')dc_r = \int_0^{2\pi}\int_0^{\pi/2} R(u_i \equiv c_i, u_r) R(0, v_r) R(0, w_r) c_r'^2 \sin\theta\, d\theta\, d\phi\, dc_r =$$

$$\int_0^{2\pi}\int_0^{\pi/2} \frac{2c_r \cos\theta}{\alpha} I_0\left[\frac{2(1-\alpha)^{1/2}}{\alpha} c_r c_i \cos\theta\right] \exp\left[-\frac{c_r^2 \cos^2\theta + (1-\alpha)c_i^2}{\alpha}\right] \times$$

$$\frac{1}{(\pi\alpha)^{1/2}} \exp\left[-\frac{c_r^2 \sin^2\theta \cos^2\phi}{\alpha}\right] \cdot$$

$$\frac{1}{(\pi\alpha)^{1/2}} \exp\left[-\frac{c_r^2 \sin^2\theta \sin^2\phi}{\alpha}\right] c_r'^2 \sin\theta\, d\theta\, d\phi\, dc_r =$$

$$\frac{4c_r^3}{\alpha^2} \exp\left[-\frac{c_r^2 + (1-\alpha)c_i^2}{\alpha}\right] \int_0^{\pi/2} I_0\left[\frac{2(1-\alpha)^{1/2}}{\alpha} c_r c_i \cos\theta\right] \cos\theta \sin\theta\, d\theta\, dc_r =$$

$$\frac{4c_r^3}{\alpha^2} \exp\left[-\frac{c_r^2 + (1-\alpha)c_i^2}{\alpha}\right] \int_0^1 I_0\left[\frac{2(1-\alpha)^{1/2}}{\alpha} c_r c_i t\right] t\, dt\, dc_r \,. \qquad (3.67)$$

For the modified Bessel function one has

$$\frac{dI_n(x)}{dx} = I_{n-1}(x) - \frac{n}{x} I_n(x),$$

and

3.4 THE CLL GAS SURFACE INTERACTION MODEL

$$\frac{dxI_1(Ax)}{dx} = AxI_0(Ax).$$

Thus the above equation can be written

$$P(c_i, c_r)dc_r = \frac{4c_r^3}{\alpha^2}\exp\left[-\frac{c_r^2 + (1-\alpha)c_i^2}{\alpha}\right] \times$$

$$\frac{\alpha}{2(1-\alpha)^{1/2}c_rc_i} \cdot tI_1\left[\frac{2(1-\alpha)^{1/2}c_rc_i}{\alpha}t\right]_0^1 dc_r =$$

$$\frac{2c_r^2}{c_i\alpha(1-\alpha)^{1/2}}I_1\left[\frac{2(1-\alpha)^{1/2}c_rc_i}{\alpha}\right]\times\exp\left[-\frac{c_r^2+(1-\alpha)c_i^2}{\alpha}\right]dc_r. \quad (3.68)$$

The corresponding kinetic energy distribution is

$$P(E_r)dE_r = \frac{E_r^{1/2}}{\alpha\left[(1-\alpha)E_i\right]^{1/2}}I_1\left[\frac{2\{(1-\alpha)E_rE_i\}^{1/2}}{\alpha}\right]\times$$

$$\exp\left[-\frac{E_r+(1-\alpha)E_i}{\alpha}\right]dE_r. \quad (3.69)$$

The scattering kernel, in which the velocity is stipulated by Eq. (3.68) and the energy distribution is stipulated by Eq. (3.69), also should satisfy the space distribution of the complete diffuse reflection, i.e., the azimuth angle ϕ is uniformly distributed between 0 and 2π and the polar angle θ is distributed according to the cosine law between 0 and $\pi/2$. Such a scattering kernel is

$$R(c_i, c_r) = \frac{2\cos\theta}{\pi c_i\alpha(1-\alpha)^{1/2}}I_1\left[\frac{2(1-\alpha)^{1/2}c_rc_i}{\alpha}\right]\times\exp\left[-\frac{c_r^2+(1-\alpha)c_i^2}{\alpha}\right]. \quad (3.70)$$

It is easy to see that equation (3.70) satisfies the normalization condition, for

$$\int_0^\infty\int_0^{2\pi}\int_0^{\pi/2} R(c_i,c_r)c_r^2\sin\theta d\theta d\phi dc_r = \int_0^\infty P(c_i,c_r)dc_r.$$

The last integral is 1 because of the equation (3.67) and the normalization of $R(u_i \to u_r), R(0 \to v_r)$ and $R(0 \to w_r)$. The reciprocity condition of this scattering kernel can be written as (see Eq. (3.24))

$$c_r \cos\theta_r R(-\mathbf{c}_r, -\mathbf{c}_i) \exp(-c_r^2) = -c_i \cos\theta_i R(\mathbf{c}_i, \mathbf{c}_r) \exp(-c_i^2),$$

and is satisfied by Eq. (3.70).

The *implementation in DSMC simulation* of this diffuse reflection model with incomplete energy accommodation according to the derivation is as follows. First sample u_r according to the method of sampling the reflected normal velocity component (with incident value of u_i), then sample v_r and w_r according to the method of sampling the reflected tangential velocity component (with incident value of 0) with accommodation coefficient taken as $\alpha_n = \alpha_t = \alpha$, the value of c_r is taken as $\sqrt{u_r^2 + v_r^2 + w_r^2}$. The distribution of the direction of $\mathbf{c}_r$ is sampled according to that the azimuth angle ϕ is uniformly distributed between 0 and 2π

$$\phi = 2\pi ranf, \tag{3.71}$$

and the sampling of θ follows the cosine law of diffuse reflection. The probability that the reflected molecule is scattered by the surface into the solid angle $d\omega = \sin\theta d\theta d\phi$ is proportional to

$$\cos\theta d\omega. \tag{3.72}$$

which can be seen from the expression Eq. (2.14) of the flux of magnitude passing unit area, $\mathbf{c} \cdot \mathbf{l} = c\cos\theta$. The above probability can be written as

$$\cos\theta \sin\theta d\theta d\phi = \frac{1}{2} d(\sin^2\theta) d\phi. \tag{3.73}$$

Thus, $\sin^2\theta$ is a random fraction uniformly distributed between 0 and $\pi/2$. So the sampling rule is

$$\begin{cases} \sin\theta & is \quad sampled \quad as \quad \sqrt{ranf}, \\ \cos\theta & is \quad sampled \quad as \quad \sqrt{1-ranf}. \end{cases} \tag{3.74}$$

3.4 THE CLL GAS SURFACE INTERACTION MODEL

By using the above sampling method, provided that the values of E_i and α are determined, the distribution of E_r obtained by sampling is accurately agreed with the analytical expression Eq. (3.69) (see Fig. 3.3).

Take this opportunity let us get more acquainted with the nature of cosine law of the diffuse reflection. For this draw a sphere with an arbitrary R as the radius, with point O as the center which is tangent to the surface at point P of reflection of the molecules (see Fig. 3.4). The flux of diffusely reflected molecules at Q is proportional to $\cos\theta \sin\theta d\theta d\phi$ according to Eq. (3.73). If introduce the polar coordinates θ^*, ϕ at O, we have $\theta^* = 2\theta$, so the above mentioned flux is proportional to $\sin\theta^* d\theta^* d\phi$, which is just proportional to the elementary area of sphere at Q. Consequently, the number of molecules diffusely reflected onto the spherical surface near Q per unit time per unit spherical area is independent of θ^*, ϕ. If make a hollow glass sphere, evacuate it, and put a source at P emitting mercury vapor, then after a certain time on the whole spherical inner surface a layer of mercury with uniform thickness will be plated provided the sphere is small enough to neglect the effect of gravity. This result can be verified numerically by direct simulation, using the method of implementation of the diffuse reflection with help of sampling formulas for u, v, w, Eq. (3.18), Eq. (3.19), Eq. (3.15) and Eq. (3.20).

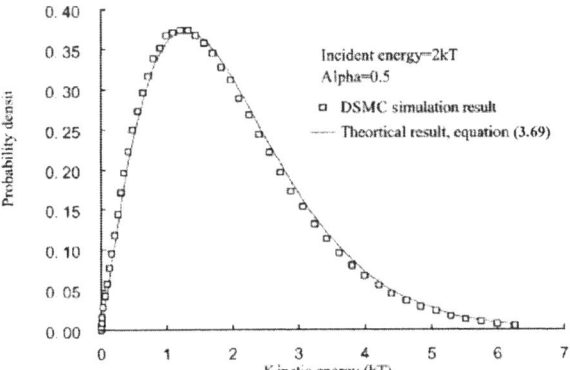

Fig. 3.3 Energy distribution of the diffuse reflection with incomplete energy accommodation. Comparison of the sampling result with the theoretical prediction Eq. (3.69)

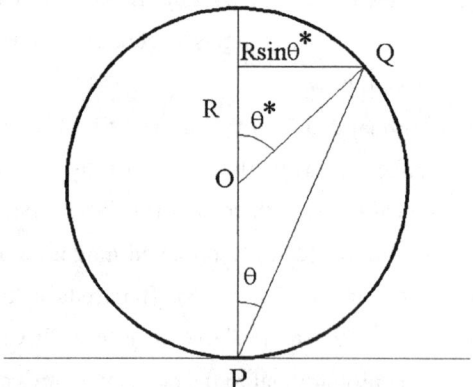

Fig. 3.4 For the derivation of uniform flux distribution of the diffusely reflected molecules on the sphere tangent to the surface

REFERENCES

1. Maxwell JC (1879) Phil Trans Roy Soc 1, Appendix
2. Maxwell JC (1890) Scientific Papers, 2: 704
3. Knudsen M (1934) The Kinetic Theory of Gases. London
4. Cercignani C (1969) Mathematical Methods in Kinetic Theory. Plenum Press
5. Wenaas EP (1971) Scattering at gas-surface interface. J Chem. Phys 54: 376-388
6. Kuscer J (1971) Reciprocity in scattering of gas molecules by surfaces. Surface Sci., 25: 225-237
7. Cercignani C, Lampis M (1971) Kinetic models for gas-suface interactions. Transport Theory and Statistical Physics, 1(2):101-114
8. Hinchen JJ and Foley WM (1966) Scattering of molecular beams by metallic surfaces. In: JH de Leeuw edited Rarefied Gas Dynamics. 2: 505
9. Cercignani C (1975) Theory and Applications of the Boltzmann Equation. Scottish Academy Press, Edinburgh and London
10. Lord RG (1991) Application of the C-L scattering kernel to DSMC calculation., In: AE Beylich edited Rarefied Gas Dynamics, VCH. 1427-1433
11. Lord RG (1991) Some extensions to the C-L gas-surface scattering kernel. Phys of Fluids, A3: 706-710

4. FREE MOLECULAR FLOW

The free molecular flow is the regime of rarefied gas flow with the highest degree of rarefaction. With the continuous increase of the degree of rarefaction, the mean free path of the gas molecules may surpass the characteristic size of the body by many times, and the molecules scattered from the body surface would move to a distance far away from the body before colliding with the oncoming incident molecules. When $Kn = \lambda/L \to \infty$, the change of the velocity distribution function of the incident gas caused by the collisions with the scattered molecules can be entirely neglected. The neglect of such collisions is the starting point of the theory of the free molecular flow. One can start from the basic equation of the free molecular flow, the collision-less Boltzmann equation, to solve the problems in free molecular flow regime. For the steady flows around bodies the velocity distribution function of the undisturbed oncoming flow is the equilibrium distribution, i.e., the Maxwell distribution, Eq. (2.184) or Eq. (2.196)

$$f_0 = n\left(\frac{m}{2\pi kT}\right)^{3/2} \exp\left[-\frac{m}{2kT}(u'^2 + v'^2 + w'^2)\right] = n\left(\frac{\beta}{\pi^{1/2}}\right)^3 \exp(-\beta^2 c'^2). \qquad (4.1)$$

Here n, T is the number density of molecules and the temperature of the oncoming gas flow, m is the mass of the molecule. The momentum transfer and the energy transfer of the incident molecules onto the body surface can be easily calculated through finding moments of the Maxwellian distribution function. If the interaction of the molecules with the surface is known (see Chapter 3), the momentum fluxes and the energy fluxes carried away from the surface by the reflected molecules are also easily calculated.

4.1 THE NUMBER AND MOMENTUM FLUXES OF MOLECULES IN GASES

The expression of the flux of the quantity Q carried by the gas molecules in unit time across unit area has been found in the beginning of section 2.2 (see Eq. (2.14)). Now consider the gas moving with velocity U relative to the surface element A. The external normal of the surface is l. The coordinate system is chosen in such a way that the x axis is directed to the opposite direction of l, U lies in the plane of xy and constitutes acute angle with the x axis and the $-y$ axis, and the surface element is lying in the plane of yz (see Fig.4.1, θ is the angle between U and the x axis). Eq (2.14) gives the flux of quantity Q relative to the surface in the direction of x (inward flux) as

$$n\overline{Qc \cdot l} = n\overline{Qu} = \int_{u>0} Qufdc = \int_{-\infty}^{\infty}\int_{-\infty}^{\infty}\int_{0}^{\infty} Qufdudvdw, \qquad (4.2)$$

where c is the velocity of the gas molecule relative to the surface framework, in the condition of free molecular flow f is the undisturbed Maxwellian distribution function Eq. (4.1). Equation (4.2) at the same time is the flux of Q carried by the gas molecules striking on the surface, when the body surface moves with velocity $-U$ in the quiescent gas, the components u, v, w of c along the axes x, y, z are related with the thermal velocity u', v', w' by the following formula

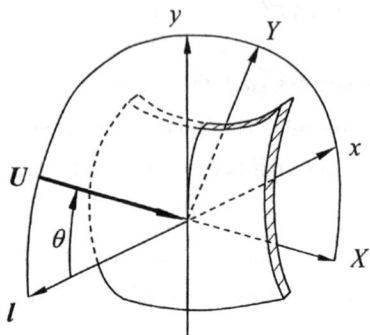

Fig. 4.1 The direction of the velocity U of the incident molecule, the direction of the external normal l, the direction X,Y of the drag and lift, the directions x,y in the expressions of the normal and tangent stresses

4.1 THE NUMBER AND MOMENTUM FLUXES OF MOLECULES IN GASES   161

$$\begin{cases} u = u' + U\cos\theta \\ v = v' - U\sin\theta \\ w = w' \end{cases} \quad (4.3)$$

First calculate the number flux N_i of molecules across surface A, the suffix i denotes incident flow. In Eq. (4.2) let $Q = 1$ and make use of Eq. (4.1), one obtains N_i

$$N_i = \int_{-\infty}^{\infty}\int_{-\infty}^{\infty}\int_{0}^{\infty} u f \, du\, dv\, dw =$$

$$\frac{n\beta^3}{\pi^{3/2}} \int_{-\infty}^{\infty}\int_{-\infty}^{\infty}\int_{0}^{\infty} u \exp\left\{-\beta^2\left(u'^2 + v'^2 + w'^2\right)\right\} du\, dv\, dw =$$

$$\frac{n\beta^3}{\pi^{3/2}} \int_{-\infty}^{\infty}\int_{-\infty}^{\infty}\int_{-U\cos\theta}^{\infty} \left(u' + U\cos\theta\right) \times$$

$$\exp\left\{-\beta^2\left(u'^2 + v'^2 + w'^2\right)\right\} du'\, dv'\, dw'. \quad (4.4)$$

With the help of the table of integrals in Appendix II (II.19), (II.20), one obtains

$$N_i = n\sqrt{\frac{kT}{2\pi m}} \left\{ e^{-(S\cos\theta)^2} + \sqrt{\pi}\left(S\cos\theta\right)\left[1 + erf\left(S\cos\theta\right)\right]\right\}. \quad (4.5)$$

Here *erfa* is the error function

$$erfa = \frac{2}{\sqrt{\pi}} \int_0^a e^{-y^2} dy, \quad (4.6)$$

and

$$S = U\beta = U/\sqrt{2RT} = \sqrt{\gamma/2} M_a, \quad (4.7)$$

is called the *molecular speed ratio*, it often appears in the calculation of free molecular flow instead of the Mach number Ma.

In the case of quiescent gas, i.e., when $U = 0$, $S = 0$, Eq. (4.5) yields the number flux of molecules as

$$\Gamma_n = N_i\Big|_{S=0} = n_i\sqrt{\frac{RT}{2\pi}} = \frac{p}{\sqrt{2\pi mkT}} = \frac{n\bar{c}}{4}. \tag{4.8}$$

This is an important and often cited result. It can be obtained with a some what different reasoning: In section 2.11 of Chapter 2 it is obtained that the average value of the component of the thermal speed in the positive direction is $(1/2)\bar{c}$ (see Eq. (2.204)), and the molecules moving in the positive direction constitute half of the molecules in the whole volume, so the number flow of molecules passing unit area in unit time is

$$\frac{1}{2}n\frac{1}{2}\bar{c} = \frac{1}{4}n\bar{c}.$$

The flux of the normal momentum of the gas across surface A, or the normal pressure produced by the oncoming molecules on the body surface, is obtained by letting $Q = mu = m(u' + u\cos\theta)$ in Eq.(4.2) (see Eqs. (II.19), (II.20) and (II.21) in Appendix II)

$$p_i = \int_{-\infty}^{\infty}\int_{-\infty}^{\infty}\int_{0}^{\infty} mu^2\, f du dv dw =$$

$$\frac{nmU^2}{2\sqrt{\pi}S^2}\Big\{(S\cos\theta)e^{-(S\cos\theta)^2} +$$

$$\sqrt{\pi}\Big[\frac{1}{2} + (S\cos\theta)^2\Big]\Big[1 + erf(S\cos\theta)\Big]\Big\}. \tag{4.9}$$

In the case of quiescent gas Eq. (4.9) yields

$$p_i\Big|_{S=0} = \frac{nm}{4\beta^2} = \frac{\rho}{2}RT = \frac{p}{2}, \tag{4.10}$$

i.e., in the quiescent gas the normal pressure on the body surface originated from the incident molecules is half the gas pressure.

The flux of the tangential (in the y direction) momentum of the gas molecules across the surface A, or the shear stress τ_i produced by the incident molecules on the body, is obtained by putting in Eq. (4.2) $Q = mv = m(v' + U\sin\theta)$ (note, the integration is the same as used to find N_i)

$$\tau_i = \int_{-\infty}^{\infty}\int_{-\infty}^{\infty}\int_0^{\infty} muvf\,du\,dv\,dw =$$

$$-\frac{nmU^2}{2\sqrt{\pi}S}\sin\theta\left\{e^{-(S\cos\theta)^2} + \sqrt{\pi}(S\cos\theta)\left[1+erf(S\cos\theta)\right]\right\}. \qquad (4.11)$$

4.2 THE AERODYNAMIC FORCES ACTED ON BODIES

Application of the results in the previous section for the incident molecules, finding the contributions of the reflected molecules to the force and adding them up yields the aerodynamic force acted on the body surface. As the theoretical and experimental research on the reflection of molecules is far for attaining perfection, the calculation of aerodynamic forces in free molecular flow is limited to the application of the Maxwellian type boundary conditions, i.e., σ portion of the molecules is reflected diffusely, and the $(1-\sigma)$ portion is reflected specularly. This corresponds to use a unique σ to replace the accommodation coefficient σ' of the normal momentum and σ of the tangential momentum in Eq. (3.22) and Eq. (3.23), and the σ here means the portion that reflects diffusely. Note that several authors use ε to denote the portion of specular reflection, then $\varepsilon = 1 - \sigma$.

The total pressure and the shear stress subjected by the body surface consist of the incident part p_i, τ_i and the reflection part p_r, τ_r,

$$p = p_i + p_r, \qquad (4.12)$$

$$\tau = \tau_i - \tau_r. \qquad (4.13)$$

Then making use of Eq.(3.22) and Eq. (3.23) and letting $\sigma' = \sigma$, one has

$$p = (2-\sigma)p_i + \sigma p_w, \qquad (4.14)$$

$$\tau = \sigma \tau_i. \qquad (4.15)$$

In Eq. (4.14) p_w is a quantity not determined yet, according to definition p_w is the flux of normal momentum of the molecules scattered with the Maxwellian distribution corresponding to the body surface temperature T_w, which can be obtained by letting in Eq. (4.10) the density be $\rho_w = mn_w$, and the temperature be T_w

$$p_w = \frac{mn_w}{2} RT_w, \qquad (4.16)$$

where n_w is the number density of the imaginary gas scattered from the internal of the body surface with the Maxwellian distribution. For its calculation the condition of no accumulation of molecules at the surface, or the condition of equality of the number fluxes of the incoming and the reflected flows, is used. The number flux N_i of the incident molecules is given by Eq. (4.5), the number flux N_w of the reflected molecules, or the number flux of molecules scattered with the Maxwellian distribution corresponding to T_w, can be obtained letting in Eq. (4.8) $n = n_w$, $T = T_w$

$$N_w = n_w \sqrt{\frac{RT_w}{2\pi}}. \qquad (4.17)$$

Equating Eq. (4.17) and Eq. (4.5), one obtains

$$n_w = N_i \sqrt{\frac{2\pi}{RT_w}}.$$

And according to Eq. (4.16)

$$p_w = mN_i \sqrt{\frac{\pi RT_w}{2}}, \qquad (4.18)$$

where N_i is given by Eq. (4.5).

Now it is easy to write the expressions of the normal pressure and the shear stress acting on the element of the body surface. Substituting Eq. (4.9), Eq. (4.18) and Eq. (4.11) into Eq. (4.14) and Eq. (4.15), one obtains

$$p = \frac{\rho_\infty U^2}{2S^2} \left[\left(\frac{2-\sigma}{\sqrt{\pi}} S\cos\theta + \frac{\sigma}{2}\sqrt{\frac{T_w}{T}} \right) e^{-(S\cos\theta)^2} + \right.$$

$$\left\{(2-\sigma)\left[(S\cos\theta)^2+\frac{1}{2}\right]+\right.$$

$$\left.\frac{\sigma}{2}\sqrt{\frac{\pi T_w}{T}}(S\cos\theta)\right\}[1+erf(S\cos\theta)]\Big],$$
(4.19)

$$\tau=-\frac{\sigma\rho_\infty U^2\sin\theta}{2\sqrt{\pi}S}\left\{e^{-(S\cos\theta)^2}+\sqrt{\pi}(S\cos\theta)[1+erf(S\cos\theta)]\right\}. \quad (4.20)$$

The expression Eq. (4.19) is obtained under the assumption that the portion of molecules conducting diffuse reflection is scattered with a Maxwellian distribution corresponding to the wall temperature T_w. If the portion of molecules conducting diffuse reflection has not accommodated completely with the surface and is scattered with a Maxwellian distribution corresponding to temperature T_r, then one can let $T_w = T_r$ in Eq. (4.19) and get the result for the normal pressure in this case

$$p=\frac{\rho_\infty U^2}{2S^2}\left[\left(\frac{2-\sigma}{\sqrt{\pi}}S\cos\theta+\frac{\sigma}{2}\sqrt{\frac{T_r}{T}}\right)e^{-(S\cos\theta)^2}+\right.$$

$$\left\{(2-\sigma)\left[(S\cos\theta^2)+\frac{1}{2}\right]+\right.$$

$$\left.\frac{\sigma}{2}\sqrt{\frac{\pi T_r}{T}}(S\cos\theta)\right\}[1+erf(S\cos\theta)]\Big].$$
(4.19)'

To find the force acting on the body in certain direction one can integrate the component forces acting in this direction over the body surface. From Eq. (4.19) and Eq. (4.20) it is seen, in general the resultant force is dependent on T_r and σ. In the following the results of the drag and lift coefficients for plane plate, cylinder and sphere are given.

The drag and lift coefficients for plane plate

The equations (4.19)' and (4.20) can be used to calculate directly the drag and lift coefficients for plane plate with incident angle α. If the area of one side of the plate is taken as the characteristic area the lift coefficient obviously can be written as

$$C_L = \frac{1}{\frac{1}{2}\rho U^2}\left\{\left[P\Big|_{\theta=\frac{\pi}{2}-\alpha} - P\Big|_{\theta=\frac{3\pi}{2}-\alpha}\right]\cos\alpha + \right.$$

$$\left[\tau\Big|_{\theta=\frac{\pi}{2}-\alpha} - \tau\Big|_{\theta=\frac{3\pi}{2}-\alpha}\right]\sin\alpha\right\} =$$

$$\frac{\cos\alpha}{S^2}\left\{\frac{2}{\sqrt{\pi}}(2-2\sigma)S\sin\alpha e^{-(S\sin\alpha)^2} + \right.$$

$$\left[2(2-2\sigma)(S\sin\alpha)^2 + (2-\sigma)\right]erf(S\sin\alpha) + \sigma\sqrt{\pi}\frac{S^2}{S_r}\cdot\sin\alpha\right\}, \qquad (4.21)$$

where

$$S_r = U/\sqrt{2RT_r} \ . \qquad (4.22)$$

Analogously, the drag coefficient can be written

$$C_D = \frac{1}{\frac{1}{2}\rho D^2}\left\{\left[P\Big|_{\theta=\frac{\pi}{2}-\alpha} - P\Big|_{\theta=\frac{3\pi}{2}-\alpha}\right]\sin\alpha + \right.$$

$$\left[\tau\Big|_{\theta=\frac{\pi}{2}-\alpha} - \tau\Big|_{\theta=\frac{3\pi}{2}-\alpha}\right]\cos\alpha\right\} =$$

$$\frac{1}{S^2}\left\{\frac{S}{\sqrt{\pi}}[4\sin^2\alpha + 2\sigma\cos(2\alpha)]\cdot e^{-(S\sin\alpha)^2} + \right.$$

$$\sin\left[1 + 2S^2 + (1-\sigma)(1 - 2S^2\cos(2\alpha))\right]\cdot$$

$$erf(S\sin\alpha) + \sigma\sqrt{\pi}\frac{S^2}{S_r}\sin^2\alpha\right\}. \qquad (4.23)$$

The drag coefficient for cylinder

4.2 THE AERODYNAMIC FORCES ACTED ON BODIES

For the straight cylinder with axis perpendicular to the direction of the incoming flow the projected area $2aL$ in the direction of the incoming flow is taken as the characteristic area, where L is the length of cylinder, a is the radius. Integrating Eq. (4.19)' and Eq. (4.20) (see Eqs. (II.30), (II.31) and (II.32) in Appendix II) one can obtain the following formula for the drag coefficient

$$C_D = \frac{1}{\frac{1}{2}\rho U^2 2aL} 2\int_0^\pi (p\cos\theta + \tau\sin\theta) aL d\theta =$$

$$\frac{4-\sigma}{3}\frac{\sqrt{\pi}e^{-S^2/2}}{S}\left[\left(S^2 + \frac{3}{2}\right)I_0\left(\frac{S^2}{2}\right) + \right.$$

$$\left.\left(S^2 + \frac{1}{2}\right)I_1\left(\frac{S^2}{2}\right)\right] + \frac{\sigma\pi^{3/2}}{4S_r}, \qquad (4.24)$$

where I_0, I_1 are the zero order and first order modified Bessel functions [1]

$$I_0(z) = \frac{1}{\pi}\int_0^\pi \exp(z\cos\theta)d\theta,$$

$$I_1(z) = \frac{1}{\pi}\int_0^\pi e^{z\cos\theta}\cos\theta\, d\theta.$$

The drag for sphere

For a sphere of radius a take πa^2 as the characteristic area, the drag coefficient is obtained as (see the integral formulae (II.25),(II.26) and (II.27))

$$C_D = \frac{\int_0^\pi (p\cos\theta + \tau\sin\theta)2\pi a^2 \sin\theta d\theta}{\frac{1}{2}\rho U^2\pi a^2} =$$

$$\frac{e^{-S^2}}{\sqrt{\pi}S^3}(1 + 2S^2) + \frac{4S^4 + 4S^2 - 1}{2S^4}erfS + \frac{2\sigma\sqrt{\pi}}{3S_r}. \qquad (4.25)$$

It is seen from Eq. (4.25), the difference between the drag coefficients under diffuse reflection and specular reflection in free molecular flow is that under diffuse reflection there is a term $2\sigma\sqrt{\pi}/(3S_r)$, but under specular reflection there is not such a term. This term is the contribution of the diffusely reflected molecules and is dependent on the temperature they are accommodated with. In the case of high speed incoming flow, cold wall and comparatively complete thermal accommodation $S_r \gg 1$, the sphere drag under diffuse reflection and specular reflection is the same and yields the result of 2 for the drag coefficient in the hypersonic limit. This result for the diffuse reflection is easily understandable: because of the cold wall the momentum contribution of the reflected molecules in comparison with the hypersonic incoming flow can be neglected, and the total momentum flux of the oncoming flow in the direction of the drag across the whole sphere is $\rho U^2 \cdot \pi a^2$, when normalized by $(1/2) \rho U^2 \cdot \pi a^2$, the drag coefficient is obtained as 2. For the specular reflection on the front windward side of the sphere the contribution of the reflected normal pressure p_r is almost the same as that of p_i, but when $\theta > 45°$, the contribution of the reflected molecules to the drag is negative, and on the whole spherical surface the shear stress is zero. The coupling of these factors makes the drag of the sphere under diffuse reflection and diffuse reflection the same.

The limit of the hypersonic flow

The case of $S \gg 1$ and $S_r \gg 1$ is a meaningful circumstance usually encountered in aerospace practice, and it is desired to have a formula giving fast estimate of the drag. In the related formulas Eq. (4.5), Eq. (4.19) anf Eq. (4.20) if assume $S\cos\theta \to \infty$, and consequently $e^{-(S\cos\theta)^2} \to 0$ and $erf(S\cos\theta) \to 1$, then

$$\begin{cases} N_i = nU\cos\theta, \\ p = (2-\sigma)\rho U^2 \cos\theta, \\ \tau = \sigma \rho U^2 \cos\theta \sin\theta. \end{cases} \tag{4.26}$$

In the case of complete diffuse reflection

$$\begin{cases} p = \rho U^2 \cos^2\theta, \\ \tau = \rho U^2 \cos\theta \sin\theta. \end{cases} \tag{4.27}$$

This corresponds to the complete neglect of the thermal speed of the molecules in the oncoming flow and the speed of the diffusely reflected molecules in

4.2 THE AERODYNAMIC FORCES ACTED ON BODIES   169

comparison with the ultra high speed of the oncoming flow. Such flow is called *superthermal flow*. Equations (4.27) can be obtained directly in the assumption that the oncoming molecules incidence onto the surface with parallel U and that the contribution of the reflected molecules is completely neglected. By using Fig. 4.1 it is easy to see that the force acting on the surface element dA along the direction of X axis (the drag) is

$$dD = (p\cos\theta + \tau\sin\theta)dA = \rho U^2 \cos\theta dA = \rho U^2 dA_p, \quad (4.28)$$

where $dA_p = \cos\theta dA$ is the projected area of dA in the direction of the oncoming flow. The total drag on the body is obtained through integrating over the whole body surface

$$D = \rho U^2 \int_{\cos\theta \geq 0} dA_p = \rho U^2 A_p. \quad (4.29)$$

A_p is the projected area of the entire body in the direction of the oncoming flow. Comparison with the exact formulas Eqs. (4.5), (4.19) and (4.20) shows, that provided $S\cos\theta > 2$, the approximation is quite accurate.

In the condition of superthermal flow the drag coefficient of both cylinder and sphere is 2. This is seen from Eq. (4.29), provided the projected area is used as the characteristic area. For the plate, when $S\sin\theta \to \infty$, from Eq. (4.23) one has

$$C_D = 2\sin\alpha - 2(1-\sigma)\sin\alpha\cos(2\alpha). \quad (4.30)$$

The first term in the above formula is the result for the diffuse reflection $\sigma = 1$, or the contribution of the oncoming flow, the latter term is the additional contribution of the specularly reflecting molecules. Under the condition of complete diffuse reflection

$$C_D = 2\sin\alpha. \quad (4.31)$$

This can be obtained from Eq. (4.29) directly, for a plate with incident angle α the projected area is $A\sin\alpha$.

Although the drag coefficient in the hypersonic limit is 2 for both the diffuse and specular reflection, but in fact different reflection model would yield completely different results for sphere drag. As an example consider the drag of sphere under the diffuse elastic reflection model in the hypersonic limit. In the so called diffuse elastic reflection molecules reflect from the surface elastically, i.e., there is no energy exchange and the speed preserves the value U of the oncoming flow but the direction is completely random, obeying the cosine law of the diffuse reflection. According to such reflection rule one can find after some manipulation that the contribution to the drag coefficient of sphere of the scattered molecules is 1. As indicated already, the contribution of the oncoming molecules is two, so under the diffuse elastic model in the hypersonic limit the sphere drag coefficient is 3.

4.3 HEAT TRANSFER TO SURFACE ELEMENT

As the calculation of the force acting on the surface element the calculation of the heat transfer to the body surface in free molecular flow is also conducted separately for the oncoming and reflected molecules. And, for the oncoming molecules the heat flux is divided into two portions, i.e., the flux q_{itr} of transitional kinetic energy and the flux $q_{i,int}$ of vibrational and rotational energy. The flux q_{itr} of transitional kinetic energy of molecules is obtained by setting $Q = (1/2)mc^2$ in Eq. (4.2)

$$q_{itr} = \int_{-\infty}^{\infty}\int_{-\infty}^{\infty}\int_{0}^{\infty} \frac{1}{2} m(u^2 + v^2 + w^2) u f du dv dw . \qquad (4.32)$$

Making use of the equations (II.11),(II.13),(II.19),(II.20) and (II.21) in Appendix II yields

$$q_{itr} = \rho RT \sqrt{\frac{RT}{2\pi}} \left\{ (S^2 + 2) e^{-(S\cos\theta)^2} + \sqrt{\pi} \left(S^2 + \frac{5}{2} \right) \times \right.$$

$$\left. (S\cos\theta) \left[1 + erf(S\cos\theta) \right] \right\} . \qquad (4.33)$$

In the case of quiescent gas Eq. (4.33) gives

$$q_{iJr}\big|_{S=0} = \rho RT \sqrt{\frac{2RT}{\pi}},\qquad (4.34)$$

if introduce the number flux Γ_n of molecules in quiescent gas (see Eq. (4.8)), one has

$$H = q_{iIr}\big|_{S=0} = 2kT\frac{nc}{4} = 2kT\cdot\Gamma_n \qquad (4.35)$$

Note that the average transitional kinetic energy of molecules is (see Eq. (2.31) or Eq. (2.201))

$$e_{tr} = \frac{1}{2}\overline{mc^2} = \frac{3}{2}kT \qquad (4.36)$$

Equation (4.35) shows, that the average transitional kinetic energy carried by a molecule crossing a surface in quiescent gas is not e_{tr}, but is $2kT$, i.e. is larger than the average transitional kinetic energy in a portion of 4:3. This happens because the probability of molecules with higher speed to cross a surface is higher than the slower molecules.

To obtain the internal energy flux carried by the incident molecules the number flux N_i (Eq. 4.5) of the oncoming flow is multiplied by the average internal energy carried by each molecule. The latter is

$$\frac{\varsigma}{2}kT = \frac{\varsigma}{2}mRT, \qquad (4.37)$$

where ς is the number of the internal degrees of freedom and is related with the total number of degrees of freedom

$$i = \varsigma + 3. \qquad (4.38)$$

The specific heat ratio γ is

$$\gamma = \frac{i+2}{i} = \frac{\varsigma+5}{\varsigma+3}, \qquad (4.39)$$

so

4. FREE MOLECULAR FLOW

$$\zeta = (5 - 3\gamma)/(\gamma - 1). \tag{4.40}$$

The internal energy flux carried by the incident molecules is

$$q_{i,\text{int}} = \frac{5-3\gamma}{\gamma-1} \frac{mRT}{2} N_i =$$

$$\frac{\rho RT}{2} \sqrt{\frac{RT}{2\pi}} \left(\frac{5-3\gamma}{\gamma-1} \right) \left\{ e^{-(S\cos\theta)^2} + \sqrt{\pi}(S\cos\theta) \cdot \left[1 + \text{erf}(S\cos\theta) \right] \right\}. \tag{4.41}$$

To obtain the transitional kinetic energy flux q_{wtr} of the diffusely reflected molecules entirely accommodated to the wall surface, one should let $T = T_w$, $n = n_w$ in Eq. (4.34), and according to Eq. (4.17) one has

$$q_{wtr} = 2mRT_w n_w \sqrt{\frac{RT_w}{2\pi}} = 2mRT_w N_w = 2mRT_w N_i. \tag{4.42}$$

Correspondingly, the internal energy flux of such reflected molecules is

$$q_{w,\text{int}} = \frac{\zeta}{2} mRT_w N_w = \frac{\zeta}{2} mRT_w N_i. \tag{4.43}$$

Consequently, the *total energy flux of the reflected molecules* is

$$q_w = q_{wtr} + q_{w,\text{int}} = (4+\zeta)\frac{mRT_w}{2} N_i = \frac{\gamma+1}{2(\gamma-1)} mRT_w N_i. \tag{4.44}$$

Now q_i is calculated using Eq. (4.33) and Eq. (4.41), then by using Eq. (3.21) and Eq. (4.44) the *total heat transfer to the body surface* of the incident and reflected molecules in free molecular flow with thermal accommodation coefficient α can be calculated as

$$q = q_i - q_r = \alpha(q_i - q_w) =$$

$$\alpha \rho RT \sqrt{\frac{RT}{2\pi}} \left(\left[S^2 + \frac{\gamma}{\gamma-1} - \frac{\gamma+1}{2(\gamma-1)} \frac{T_w}{T} \right] \times \right. \tag{4.45}$$

$$\left\{ e^{-(S\cos\theta)^2} + \sqrt{\pi}(S\cos\theta)[1+ erf(S\cos\theta)] \right\} - \frac{1}{2}e^{-(S\cos\theta)^2} \right).$$

If assume that molecules are reflected at the body surface with the Maxwellian boundary condition, i.e., σ portion reflects diffusely and $(1-\sigma)$ portion specularly, and the diffuse reflection corresponds to a temperature T_r different from the wall temperature T_w, then the total heat transfer to the body surface is

$$q = \sigma\left(q_i - q_w\big|_{\tau = \tau}\right),$$

i.e., one should replace α by σ and T_w by T_r in Eq. (4.45)

$$q = \sigma\rho RT\sqrt{\frac{RT}{2\pi}}\left(\left[S^2 + \frac{\gamma}{\gamma-1} - \frac{\gamma+1}{2(\gamma-1)}\frac{T_r}{T}\right]\right.$$

$$\left\{ e^{-(S\cos\theta)^2} + \sqrt{\pi}(S\cos\theta)[1+ erf(S\cos\theta)] \right\} - \frac{1}{2}e^{-(S\cos\theta)^2} \right). \tag{4.45)'}$$

If the wall temperature of the adiabatic body surface adopts such a value T_a, so that the surface heat transfer is zero, then this temperature T_a is called the *adiabatic temperature*. From Eq. (4.45) it is seen

$$\frac{T_a}{T} = \frac{2(\gamma-1)}{\gamma+1}\left(\frac{\gamma}{\gamma-1} + S^2 - \frac{\frac{1}{2}e^{-(S\cos\theta)^2}}{e^{-(S\cos\theta)^2} + \sqrt{\pi}S\cos\theta[1+erf(S\cos\theta)]}\right). \tag{4.46}$$

In the continuum flow regime the isentropic stagnation temperature T_0 is (Eq. (0.17))

$$\frac{T_0}{T} = 1 + \frac{\gamma-1}{2}Ma^2,$$

or, according to Eq. (4.7)

$$\frac{T_0}{T} = 1 + \frac{\gamma-1}{\gamma}S^2. \tag{4.47}$$

From Eq. (4.46) and Eq. (4.47) it is seen that under the condition of large S and frontal incidence the ratio of the adiabatic temperature T_{aFM} in free molecular flow to the stagnation temperature T_0 in continuum flow is

$$\frac{T_{aFM}}{T_0}\bigg|_{S\to\infty} \to \frac{2\gamma}{\gamma+1}. \qquad (4.48)$$

That is T_{aFM} is always higher than T_0. This situation is controversy with that in the viscous continuous flow: the adiabatic temperature in the boundary layer is always less than T_0. The *physical explanation that* T_{aFM} *is higher than* T_0 *when* $S\to\infty$ is as follows. In the limit case of large S the energy carried by unit mass of the incident flow in either continuum or free molecular flow is $(1/2)U^2$. For continuum flow, when the flow is retarded, out of this energy besides a portion going to increase the kinetic energy $(3/2)RT_0$ of the thermal motion (the consideration is limited to the case of monatomic molecules) there is still a portion RT_0 used for building up the pressure. Thus when the continuum flow is retarded the energy $(1/2)U^2$ transforms to a total enthalpy of $(5/2)RT_0$. But in free molecular flow in the equilibrium the energy of the oncoming flow is emitted from the surface by scattering an energy of $2RT_a$ per unit mass (see Eq. (4.37)). Consequently, for monatomic gas we have $T_{aFM}/T_0 = 5/4$. Surely this result is obtained also by letting $\gamma = 1.667$ in Eq. (4.48). For diatomic molecules $\gamma = 1.4$, and $T_{aFM}/T_0 = 7/6$.

4.4 FREE MOLECULAR EFFUSION AND THERMAL TRANSPIRATION

The phenomenon of the molecules in an equilibrium gas streaming through a small hole in a thin wall (or through thin plate made of porous material) into the vacuum is called *molecular effusion* or *transpiration*. Historically, the flow through porous thin plates has been first studied experimentally, later (in 1909) Knudsen observed and studied the effusion of hydrogen, oxygen and carbon dioxide through small holes. If the size of the hole is small in comparison with the molecular mean free path of the gas, then the effusion of a molecule through the hole would not be influenced by the other molecules flown through (in or out of)

4.4 FREE MOLECULAR EFFUSION AND THERMAL TRANSPIRATION

the hole, and the result of the free molecular flow obtained in section 4.1 can be used. The number flux of molecules is given by Eq. (4.8)

$$\Gamma_n = \frac{n\bar{c}}{4} = n\sqrt{\frac{RT}{2\pi}}.$$

Obviously the *mass flux of the free molecular effusion* is

$$\Gamma_{m,FM} = m\Gamma_n = \rho\left(\frac{RT}{2\pi}\right)^{1/2}. \tag{4.49}$$

It is interesting to compare this result with the mass flux of an isentropic outflow from a stagnation gas in continuum gas dynamics. The ρ and T in Eq. (4.49) should correspond to the stagnation density and temperature in the isentropic flow. From the isentropic flow theory it is known, the mass velocity reaches its maximum when the flow speed at the exit is sonic (see Eq. (4.17) of reference [2]), if denote the stagnation density and temperature in the isentropic flow by ρ and T, then the *maximum mass flux of the continuum flow* through a hole is

$$\Gamma_{m,C} = \left(\frac{2}{\gamma+1}\right)^{\frac{\gamma+1}{2(\gamma-1)}} \rho(\gamma RT)^{1/2}. \tag{4.50}$$

The ratio of the free molecular effusion to the continuum mass outflow is

$$\frac{\Gamma_{m,FM}}{\Gamma_{m,C}} = \frac{1}{(2\pi\gamma)^{1/2}}\left(\frac{\gamma+1}{2}\right)^{\frac{\gamma+1}{2(\gamma-1)}} =$$

$$\begin{array}{l} 0.5494, \text{ for monatomic molecule } \gamma = 1.667 \\ 0.5826, \text{ for diatomic molecule } \gamma = 1.4. \end{array} \tag{4.51}$$

In the free molecular flow regime the outflow is a kind of effusion of individual molecules in a quiescent equilibrium state by means of the molecular thermal motion, whereas in continuum regime it is a stream with macroscopic velocity reaching sonic speed at the exit. The latter is bigger than the former by a factor of 1.7~1.8.

At the end of Chapter 3 we discussed the cosine law of the special distribution of the diffusely reflected molecules. The spatial distribution of the gas molecules of the effusion is identical with that of the diffusely reflected molecules. They are both based on the assumption that there is a gas (for diffuse reflection, a hypothetic gas) with equilibrium Maxwellian distribution emitting molecules. From the discussion there it is seen, that the effusing molecules would strike with even probability on any point of a sphere with an arbitrary radius R, and which is tangent to the surface where the effusing hole is located (see Fig. 3.4)). The number density flux will decay inversely with the squire of R.

The above result related to the mass flux can be applied to the case in which on each side of the thin plate with small hole there is a container containing gas of the same nature but with different equilibrium temperatures T_A and T_B. If the size of the hole is small enough, the free molecular flow theory holds. The gas on each side would effuse according to Eq. (4.49) to the other side. At the beginning the fluxes are not the same, but the fluxes would regulate the densities in both containers and reach a state that the fluxes towards different directions equal each other

$$\rho_A(RT_A)^{1/2} = \rho_B(RT_B)^{1/2}.$$

As the gases on both sides are of the same nature, the R is the same, by using the state equation $p = \rho RT$, the ratio of the pressures on the two sides is obtained

$$(p_A/p_B) = (T_A/T_B)^{1/2}. \tag{4.52}$$

Thus the container with higher temperature retains higher pressure. This is the so called phenomenon of *thermal transpiration*. Reynolds conducted the experiment on thermal transpiration the earliest (in 1879), in which the two container-thermostats were connected not through small hole but through thin plug made of porous material. He observed that provided the size of the pores in the porous material is small enough and the pressure of the gas is sufficient low, the above theory holds. With the rising of the gas pressure deviation from the theory occurs.

Applying the above result to the free molecular effusion of a mixture of different gases (with molecular mass m_A, m_B) through a thin film of porous material into vacuum can explain the phenomenon of so called *differential effusion*. The number densities of the gases before effusion are supposed to be n_A, n_B. After a certain period of time the ratio of the number densities n_A', n_B' of the two gases collected in the vacuum chamber will be different from the original ratio. According to Eq. (4.8) one has

$$\frac{n'_A}{n'_B} = \frac{n_A}{n_B}\left(\frac{m_B}{m_A}\right)^{1/2}. \tag{4.53}$$

It can be seen that by using this formula the molecular weight can be determined, and also, by employing the phenomenon of differential effusion the gases of different molecular weights can be separated, of course, with poor accuracy. Ramsey was the first to use this phenomenon in the separation of helium from nitrogen employing unglazed clay as the porous material. From Eq. (4.53) it is seen, that the helium effuses through the clay at a rate in 2.6 times faster than the nitrogen. After multiple repetitions helium of fairly purity can be obtained. The gas separation caused by the differential effusion is the simplest one in gas separation based on the phenomena of rarefied gas dynamics. The separation of gases and isotopes employing the rarefied gas effects had been the subject of considerable attention of the scientific community in the 70s of the 20th century, readers interested in gaining comprehensive knowledge of this subject are referred to the reviews [3, 4].

4.5 COUETTE FLOW AND HEAT TRANSFER BETWEEN PLANE PLATES

In this section two simple one-dimensional steady flow problems, namely the Couette flow problem and the heat transfer problem between plane plates will be discussed.

We begin from the discussion of the *Couette flow problem in free molecular flow*. Consider the gas flow between two plates, let the lower one be at rest and the upper one move with velocity U in its own plane, the gas flow caused by the shear is the Couette flow. When the molecular mean free path of the gas between the plates is large in comparison with the distance between the plates, the free molecular flow takes place. In the case of complete diffuse reflection $\sigma = 1$, the shear stress subjected by the lower plate can be obtained from Eq. (4.15) and Eq. (4.11), and in the latter one should put $\theta = \pi/2$

$$\tau_L = \frac{\rho U^2}{2\sqrt{\pi S}} = \rho U (RT/2\pi)^{1/2}. \tag{4.54}$$

The discussion will be more complicated, when at the plate surface σ portion of molecules reflects diffusely, and $(1-\sigma)$ portion reflects specularly. This time the macroscopic velocity component along the direction of the plate of molecules leaving the lower plate is not zero, suppose it to be cU. Because of the symmetry, the macroscopic velocity component along the plate of molecules leaving the upper plate must be $(1-c)U$. Out of them there is σ portion reflecting diffusely and having the macroscopic velocity component U, the other $(1-\sigma)$ portion reflects specularly and has the same macroscopic velocity cU as the lower plate, consequently

$$\sigma U + (1-\sigma)cU = (1-c)U,$$

from where we obtain $c = (1-\sigma)/(2-\sigma)$. That is to say, the macroscopic velocity $(1-c)U$ leaving the upper plate is

$$U_U = \frac{1}{2-\sigma}U.$$

Then from Eq. (4.20) by setting $\theta = \pi/2$ and the macroscopic velocity to be $U/(2-\sigma)$, one obtains

$$\tau = \frac{\sigma}{2-\sigma}\rho U(RT/2\pi)^{1/2}. \tag{4.55}$$

Now consider the heat transfer between the plates when the distance between them is small in comparison with the molecular mean free path, i.e., the *one-dimensional heat conduction problem in free molecular flow.*

4.5 COUETTE FLOW AND HEAT TRANSFER BETWEEN PLANE PLATES

Suppose the temperatures of the upper and lower plates are T_1 and T_2, and the accommodation coefficients are α_1 and α_2. If let q_1 and q_2 to denote the energy fluxes carried by the molecules leaving the upper and lower plates, then the application of the definition Eq. (3.21) of the accommodation coefficient to the lower plate (for the latter q_1 is the energy flux of the incident molecules, q_2 is the energy flux of the reflected molecules) yields

$$q_1 - q_2 = \alpha_2 (q_1 - q_{w,2}), \tag{4.56}$$

where $q_{w,2}$ is the energy flux of the reflected molecules with the Maxwellian distribution at the temperature T_2 of the lower plate. Analogously, application of the equation (3.21) to the upper plate yields

$$q_2 - q_1 = \alpha_1 (q_2 - q_{w,1}), \tag{4.57}$$

where $q_{w,1}$ is the energy flux of the reflected molecules with the Maxwellian distribution at the temperature T_1 of the upper plate. From the above two equations it is easy to obtain

$$q_1 - q_2 = \frac{\alpha_1 \alpha_2}{\alpha_1 + \alpha_2 - \alpha_1 \alpha_2} (q_{w,1} - q_{w,2}). \tag{4.58}$$

$q_{w,1}$ and $q_{w,2}$ are given by Eq. (4.42)

$$q_{w,i} = 2mRT_i N_{w,i}, \quad i = 1,2, \tag{4.59}$$

$N_{w,i}$, $i=1,2$, are the actual number fluxes of molecules leaving the upper or lower plates (in unit time per unit area). Assume that the molecules leaving the upper and lower plates are approximately at the Maxwellian distribution but corresponding to $T_{r,1}$ and $T_{r,2}$, respectively, and the number densities of the molecular flows are n_1 and n_2, respectively. The actual number fluxes of molecules leaving the upper and lower plates, according to the derivation of Eq. (4.8), should be

$$N_{w,i} = 2n_i \sqrt{\frac{RT_{r,i}}{2\pi}}. \tag{4.60}$$

This formula in comparison with Eq. (4.8) has an extra factor 2, because that n_i represents only the density of molecules leaving one of the plates towards the other but not the density of the whole gas (see the some what different reasoning of the derivation of Eq. (4.8) below Eq. (4.8)), the density n of the gas between the plates should be the sum of n_1 and n_2

$$n = n_1 + n_2. \tag{4.61}$$

According to the condition that there is not accumulation and disappearance of the molecules at the wall surface $N_{w,1} = N_{w,2}$, from Eq. (4.60) one should have

$$n_1 \sqrt{T_{r,1}} = n_2 \sqrt{T_{r,2}}. \tag{4.62}$$

From the above two formulas it is easy to obtain

$$n_1 = \frac{n\sqrt{T_{r,2}}}{\sqrt{T_{r,1}} + \sqrt{T_{r,2}}}, \quad n_2 = \frac{n\sqrt{T_{r,1}}}{\sqrt{T_{r,1}} + \sqrt{T_{r,2}}}. \tag{4.63}$$

Now suppose the temperature of gas between the plates is T, the number flux N_w of this gas to the upper and lower plates equal to $N_{w,1}$ and $N_{w,2}$

$$N_w = N_{w1} = N_{w2}. \tag{4.64}$$

And

$$N_w = n\sqrt{\frac{RT}{2\pi}} = n_1\sqrt{\frac{RT_{r,1}}{2\pi}} + n_2\sqrt{\frac{RT_{r,2}}{2\pi}}. \tag{4.65}$$

Then according to Eq. (4.58), Eq. (4.59) and Eq. (4.64) we can write the *heat flux between the two plates (from the upper to the lower) in free molecular flow* as

$$q_{FM} = q_1 - q_2 = \frac{\alpha_1 \alpha_2}{\alpha_1 + \alpha_2 - \alpha_1 \alpha_2} nm\sqrt{\frac{RT}{2\pi}} \cdot 2R(T_1 - T_2). \tag{4.66}$$

To apply Eq. (4.66) the temperature T should be expressed through the known wall temperatures. Substitution of Eq. (4.63) into Eq. (4.65) yields

4.5 COUETTE FLOW AND HEAT TRANSFER BETWEEN PLANE PLATES

$$\sqrt{T} = \frac{2\sqrt{T_{r,1}T_{r,2}}}{\sqrt{T_{r,1}} + \sqrt{T_{r,2}}}. \tag{4.67}$$

Noting that the heat fluxes $q_1, q_2, q_{w,1}, q_{w,2}$ can all be expressed through temperatures $T_{r,1}, T_{r,2}, T_1, T_2$ by using formulas analogous with Eq. (4.59) and that the factor $N_{w,1}$ in four cases is the actual number flux of molecules leaving the upper or lower plates, one can obtain from Eq. (4.56), Eq. (4.57) and Eq. (4.58)

$$T_{r,1} = \frac{\alpha_1 T_1 + \alpha_2(1-\alpha_1)T_2}{\alpha_1 + \alpha_2 - \alpha_1\alpha_2}, T_{r,2} = \frac{\alpha_2 T_2 + \alpha_1(1-\alpha_2)T_1}{\alpha_1 + \alpha_2 - \alpha_1\alpha_2}. \tag{4.68}$$

Expressing the temperature T in Eq. (4.66) through the temperatures of the plates and accommodation coefficients with the help of Eq. (4.67) and Eq. (4.68) one obtains the searched solution of the problem.

When the accommodation coefficients of the upper and lower plates are both 1, it is evident, that $T_{r,1} = T_1$, $T_{r,2} = T_2$, and $q_1 = q_{w,1}$, $q_2 = q_{w,2}$, and the energy fluxes of the molecules leaving the upper and lower plates are

$$q_1 = \frac{mn_1}{\sqrt{\pi}}(2RT_1)^{3/2}, \tag{4.69}$$

$$q_2 = \frac{mn_2}{\sqrt{\pi}}(2RT_2)^{3/2}. \tag{4.70}$$

By using Eq. (4.63) the total energy flux between the plates (from the upper to the lower) in free molecular flow with $\alpha_1 = \alpha_2 = 1$ is obtained

$$q_{FM} = \frac{\rho}{\sqrt{\pi}}(2R)^{3/2}\sqrt{T_1 T_2}\left(\sqrt{T_1} - \sqrt{T_2}\right). \tag{4.71}$$

Obviously, letting in Eq. (4.66) $\alpha_1 = \alpha_2 = 1$ and using Eq. (4.67) and Eq. (4.68) can yield the same result.

In *the continuum flow* the heat transfer between the plates is

$$q_C = -K\frac{dT}{dy}.$$

If as usual assume $K = CT^\omega$, obviously

$$q_C = -\frac{C}{\omega+1}\frac{dT^{\omega+1}}{dy}.$$

In the quiescent gas q_C should be constant, so that

$$T^{\omega+1} = -\frac{(\omega+1)q_C}{C}y + D,$$

the constant D is determined to be $T = T_2^{\omega+1}$ from the condition $T = T_2$ at $y = 0$, and from the condition $T = T_1$ at $y = h$ the quantity q_C is obtained as

$$q_C = -\frac{C\left(T_1^{\omega+1} - T_2^{\omega+1}\right)}{(\omega+1)h}. \tag{4.72}$$

From Eq. (4.71) it is seen, for the heat transfer problem, in free molecular flow the heat flux is proportional to the density between the plates and is independent of the width of the gap between the plates; but from Eq. (4.72) it is seen the heat flux in continuum flow is inversely proportional to the width of the gap between the plates and is independent of the density. The ratio of the heat fluxes of the two regimes is

$$\frac{q_C}{q_{FM}} \propto \frac{1}{\rho h} \propto \frac{\lambda}{h} \sim Kn. \tag{4.73}$$

When $Kn \gg 1$, i.e., in the free molecular flow, the heat flux calculated by the continuum theory would exaggerate the correct heat transfer approximately by a factor of Kn. When $Kn \ll 1$, the continuum theory holds, the heat flux calculated by the free molecular flow theory would exaggerate the correct heat transfer approximately by a factor of Kn^{-1}.

4.6 THE GENERAL SOLUTIONS, UNSTEADY FLOW

For solution of the unsteady problems in free molecular flow regime the natural means is to solve the collisionless Boltzmann equation. In Eq. (2.152) abandon the right hand side term and consider the case of no external force $F = 0$, the *collision-less Boltzmann equation* holds

4.6 THE GENERAL SOLUTIONS, UNSTEADY FLOW

$$\frac{\partial f}{\partial t} + c \cdot \frac{\partial f}{\partial r} = 0. \tag{4.74}$$

The following initial condition is set for Eq. (4.74)

$$f(c,r,t=0) = f_0(c,r), \tag{4.75}$$

$f_0(c,r)$ is an arbitrary known function of c, r in certain domain. The characteristic equations are (s is the parameter along the characteristics)

$$\frac{dt}{ds} = 1, \quad \frac{dr}{ds} = c, \quad \frac{df}{ds} = 0. \tag{4.76}$$

The solution is

$$t = s, \quad r = c\,s + r_0, \quad f = f_0,$$

or in the form

$$f(c,r,t) = f_0(c, r - ct). \tag{4.77}$$

This is the *general solution of the collisionless Boltzmann equation*, indicating that the equation (4.74) stipulates that the distribution function remains constant along the characteristics, i.e., the molecular trajectories. This general solution was first pointed out by Yang and Lees [5] for the one-dimensional case when solving the Rayleigh problem. Narasimha [6] suggested use the moment of solution Eq. (4.77) (see Eq. (2.9))

$$n\overline{Q}(r,t) = \int_{-\infty}^{\infty} Q f_0(c, r - c\,t) dc \tag{4.78}$$

to obtain the macroscopic quantities of the flow field, and then employ the transformation

$$r' = r - ct \tag{4.79}$$

to ease the calculation of the right hand side integral. The Jacobian of the transformation $(r' \to c)$ is

$$\frac{\partial(x,y,z)}{\partial(u,v,w)} = -t^3,$$

so

$$n\bar{Q}(r,t) = \frac{1}{t^3}\int_{-\infty}^{\infty} Qf_0\left(\frac{r-r'}{t}, r'\right)dr', \tag{4.80}$$

the domain of integration is the extent of the gas in the moment $t=0$.

To illustrate the application of Eq. (4.80) first consider the problem of *one-dimensional free expansion*. A quiescent uniform monatomic gas (of temperature T_1 and density n_1) initially occupies the semi-space on the left of the plane $x=0$ and is separated by a thin wall from a vacuum on the right of the plane. At moment $t=0$ the separating plate is removed and the gas expands freely into the vacuum. When the coordinate x is small in comparison with the mean free path and the time t is small in comparison with the mean collision time, i.e., when the free molecular flow condition is satisfied, it is required to determine the number density $n(x,t)$ and the velocity $u_0(x,t)$ of the gas as functions of x and t.

For the one-dimensional problem under consideration Eq. (4.80) has the following form

$$n\bar{Q}(x,t) = \frac{1}{t}\int Qf_{u_0}\left(\frac{x-x'}{t}, x'\right)dx', \tag{4.81}$$

where f_{u_0} is the distribution function of the thermal velocity component in x direction in equilibrium gas, which was already obtained in Chapter 2 (see Eq. (2.203))

$$f_{u_0} = (n_1\beta_1/\pi^{1/2})\exp(-\beta_1^2 u^2), \qquad \beta_1 = (2RT_1)^{-\frac{1}{2}}, \tag{4.82}$$

and $x' = x - ut$.

Substituting Eq. (4.82) into Eq. (4.81) and letting $Q=1$ yields the number density

$$n(x,t) = \frac{1}{t}\int_{-\infty}^{0} n_1 \frac{\beta_1}{\pi^{1/2}} \exp\left(-\beta_1^2 u'^2\right) dx' =$$

$$\frac{n_1}{\pi^{1/2}} \int_{\beta_1 x/t}^{\infty} \exp\left(-\frac{\beta_1^2 (x-x')^2}{t^2}\right) d\left(\frac{\beta_1 (x-x')}{t}\right) = \frac{1}{2} nerfc\left(\frac{\beta_1 x}{t}\right). \tag{4.83}$$

Letting $Q = u$ in Eq. (4.81) yields the expression of the speed of the flow

$$nu_0 = \frac{1}{2}\frac{n_1}{\pi^{1/2}}\frac{1}{\beta_1}\exp\left[-\left(\frac{\beta_1 x}{t}\right)^2\right],$$

or

$$u_0 = \frac{1}{\pi^{1/2}\beta_1}\exp\left[-\left(\frac{\beta_1 x}{t}\right)^2\right]/erfc\left(\frac{\beta_1 x}{t}\right). \tag{4.84}$$

At the position $x = 0$ from Eq. (4.83) one obtains $n = (1/2)n_1$, from Eq. (4.84) one obtains $u_0 = 1/(\pi^{1/2}\beta_1)$. Both n and u_0 at $x = 0$ are not dependent on t, and generally they are functions of $(\beta_1 x/t)$. At $x = 0$ the number flux of molecules is $nu_0 = n_1/(2\pi^{1/2}\beta_1) = n_1\sqrt{RT/2\pi}$. The result is the same as the number flux of the steady effusion of molecules (see Eq. (4.8)).

Now discuss the problem of the *expansion of point cloud*. Consider the free expansion of the cloud of gas initially in the equilibrium state into the surrounding vacuum. At a distance far enough from the cloud the gas cloud can be considered as a point. Thus one can assume that at moment $t = 0$ there are N molecules concentrated at the origin $r = 0$ and having Maxwellain distribution (see Eq. (2.196))

$$f_0(c,r) = \delta(r)N\left(\frac{\beta}{\pi^{1/2}}\right)^3 \exp\left(-\beta^2 c^2\right), \tag{4.85}$$

where $\delta(r)$ is the Dirac function.

According to Eq. (4.80) let $Q = 1$, the expression of the number density is obtained

$$n(r,t) = \frac{1}{t^3}\int \delta(r') N\left(\frac{\beta}{\pi^{1/2}}\right)^3 \exp\left[-\beta^2\left(\frac{r-r'}{t}\right)^2\right]dr' =$$

$$\frac{N}{t^3}\left(\frac{\beta}{\pi^{1/2}}\right)^3 \exp(-\beta^2 x^2/t^2). \tag{4.86}$$

In Eq. (4.80) by letting $q = c$, one obtains

$$u_0 = r/t. \tag{4.87}$$

It is very natural to obtain the expression like Eq. (4.87) for the velocity, for only molecules having velocity $c = r/t$ in the gas cloud can reach r, then the macroscopic velocity u_0 is as well $c = r/t$. The thermal velocity of the molecules is $c' = c - u_0 = 0$, and the temperature is zero. The density at arbitrary point rises from zero at moment $t = 0$ to the maximum value at $t = x(2/3)^{1/2}\beta$

$$(3/2e\pi)^{3/2} N/x^3 \approx 0.0736 N/x^3, \tag{4.88}$$

and decreases at large time as t^{-3}.

Finally, consider the *Rayleigh problem in free molecular flow*. The upper half space above the plane $y = 0$ initially is filled with quiescent uniform gas with temperature T_1 and number density n_1. Suppose a plane plate at $y = 0$ acquires temperature T_w at $t = 0$ and starts to move in its own plane with a velocity U_w. Consider the flow of the gas when $t \ll t_c$ and $y \ll \lambda$ (t_c being the collision time, λ the molecular mean free path). The collision between molecules can be neglected, and the expression Eq. (4.80) of the moments of molecular quantities derived from the collisionless Boltzmann equation can be employed. Note, when the wall temperature is different from the initial temperature T_1, the macroscopic velocity except the component along the direction of the plate still has a component along the direction perpendicular to the plate, so this Rayleigh problem is not an one-dimensional flow problem. Write Eq. (4.80) in the following form

$$n\overline{Q}(y,t) = \frac{1}{t^3}\iiint Qf_0\left(\frac{x-x'}{t},\frac{y-y'}{t},\frac{z-z'}{t},y'\right)dx'dy'dz', \tag{4.89}$$

4.6 THE GENERAL SOLUTIONS, UNSTEADY FLOW

the integration domain is that occupied by the gas at $t=0$. Superficially, this domain is only the semi infinite space above the plane $y=0$, and the initial distribution function of the gas (see Eq. (2.196)) is

$$n_1\left(\frac{\beta_1}{\sqrt{\pi}}\right)^3 \exp\left[-\beta_1^2\left(u^2+v^2+w^2\right)\right] =$$

$$n_1\left(\frac{\beta_1}{\sqrt{\pi}}\right)^3 \exp\left\{-\beta_1^2\left[\left(\frac{x-x'}{t}\right)^2+\left(\frac{y-y'}{t}\right)^2+\left(\frac{z-z'}{t}\right)^2\right]\right\}.$$

But the gas in the flow partially is originated from the reflection at the surface (suppose the reflection is complete diffuse), one can imagine that this part of the gas was located initially below the plane $y=0$ and has the following initial distribution function

$$n_w\left(\frac{\beta_w}{\sqrt{\pi}}\right)^3 \exp\left\{-\beta_w^2\left[\left(u-U_w\right)^2+v^2+w^2\right]\right\} =$$

$$n_w\left(\frac{\beta_w}{\sqrt{\pi}}\right)^3 \exp\left\{-\beta_w^2\left[\left(\frac{x-x'}{t}-U_w\right)^2+\left(\frac{y-y'}{t}\right)^2+\left(\frac{z-z'}{t}\right)^2\right]\right\}.$$

Here the motion of the plate with velocity U_w has been taken into account. The number density n_w of the imaginary gas, as we has been doing for many times, is obtained from the condition of the equality of the number fluxes along the positive and negative y directions. From Eq. (4.8) we have

$$n_w = (\beta_w/\beta_1)n_1.$$

Thus, from Eq. (4.89) one can write

$$n\overline{Q} = n_1\left(\frac{\beta_1}{\sqrt{\pi}t}\right)^3 \int_{-\infty}^{\infty}\int_{0}^{\infty}\int_{-\infty}^{\infty} Q\exp\left\{-\beta_1^2\left[\left(\frac{x-x'}{t}\right)^2+\left(\frac{y-y'}{t}\right)^2+\right.\right.$$

$$\left.\left.\left(\frac{z-z'}{t}\right)^2\right]\right\}dx'dy'dz' + n_1\frac{\beta_w}{\beta_1}\left(\frac{\beta_w}{\sqrt{\pi}t}\right)^3 \tag{4.90}$$

$$\int_{-\infty}^{\infty}\int_{0}^{\infty}\int_{-\infty}^{\infty} Q \exp\left\{-\beta_w^2\left[\left(\frac{x-x'}{t}-U_w\right)^2+\left(\frac{y-y'}{t}\right)^2+\left(\frac{z-z'}{t}\right)^2\right]\right\}dx'dy'dz'.$$

Letting $Q=1$, u, v and $(1/3)c'^2/R$, the expressions of the number density, the velocity components along the x and y directions and the transitional temperature can be obtained

$$n=\frac{n_1}{2}\left\{\left[1+erf\left(\frac{\beta_1 y}{t}\right)\right]+\left(\frac{T_1}{T_w}\right)^{\frac{1}{2}} erfc\left(\frac{\beta_w y}{t}\right)\right\}, \qquad (4.91)$$

$$u_0=\frac{1}{2}U_w\left(\frac{T_1}{T_w}\right)^{\frac{1}{2}} erfc\left(\frac{\beta_w y}{t}\right)\bigg/(n/n_1), \qquad (4.92)$$

$$v_0=\frac{1}{2\sqrt{\pi}\beta_1}\left[\exp\left(-\frac{\beta_w^2 y^2}{t^2}\right)-\exp\left(-\frac{\beta_1^2 y^2}{t^2}\right)\right]\bigg/(n/n_1), \qquad (4.93)$$

$$T=T_1\left[1+\frac{1}{2}erfc\left(\frac{\beta_w y}{t}\right)\left(\frac{T_1}{T_w}\right)^{\frac{1}{2}}\left(\frac{T_w}{T_1}-1\right)\bigg/(n/n_1)\right]+$$

$$\frac{y\,v_0}{t\,3R}+\frac{U_w u_0}{3R}-\frac{1}{3R}\left(u_0^2+v_0^2\right). \qquad (4.94)$$

As the molecules reflected from the surface have not collided with the incident molecules, so the molecules coming into interaction with the surface are not disturbed. The pressure, the shear stress, the heat flux etc. subjected by the surface can be obtained according to the formulas of p, τ and q subjected by the gas moving steadily under free molecular conditions (see Eqs. (4.19), (4.20) and (4.45), where $\theta=\pi/2$, $\sigma=1, \alpha=1$)

$$p=\frac{mn_1}{4\beta_1^2}\left[1+\left(\frac{T_w}{T_1}\right)^{1/2}\right], \qquad (4.95)$$

$$\tau = \frac{mn_1 U_w}{2\sqrt{\pi}\beta_1}, \tag{4.96}$$

$$q = \frac{mn_1}{4\sqrt{\pi}\beta_1^3}\left[U_w^2\beta_1^2 + \frac{(\gamma+1)}{2(\gamma-1)}\left(1-\frac{T_w}{T_1}\right)\right]. \tag{4.97}$$

It is evident from Eqs. (4.91)~(4.94) that when the wall temperature of the plate is the same as the temperature T_1 of the gas, the impulse motion of the plate does not influence the gas density: $n = n_1$, and does not cause the gas motion perpendicular to the plate: $v_0 = 0$, and the transitional temperature has a disturbance related to the velocity component along the direction of the plate

$$T = T_1 + u_0(U_w - u_0)/3R. \tag{4.98}$$

The latter has a very simple expression

$$u_0 = \frac{1}{2}U_w \mathrm{erfc}(\beta_w y/t). \tag{4.99}$$

From Eqs. (4.95)~(4.97) it can be seen that when $T_w = T_1$

$$p = \rho_1 R T_1, \tag{4.100}$$

$$\tau = \frac{n\bar{c}_1}{4}mU_w = \Gamma \cdot (mU_w), \tag{4.101}$$

$$q = \frac{n\bar{c}_1}{4}\left(\frac{m}{2}U_w^2\right) = \Gamma \cdot \left(\frac{m}{2}U_w^2\right). \tag{4.102}$$

The expression $\Gamma_n = n\bar{c}_1/4$ (see Eq. (4.8)) is the number flux of molecules in the quiescent gas. It is seen that when $T_w = T_1$ the impulsive motion of the plate does not constitute disturbance to the pressure, and the shear stress and the heat transfer are originated from the momentum mU_w and $(1/2)mU_w^2$ carried by each molecule leaving the plate. The conclusion concerning the pressure $p = \rho_1 R T_1$ seems to be in contradiction with the expression of the temperature (except T_1 there is still disturbance term), but this is only superficial. In fact in the expression

Eq. (4.100) the part contributed by the incident molecules is not influenced by the reflected molecules but is provided by the gas with equilibrium distribution of temperature T_1. The disturbance term of T in Eq. (4.98) is the result of the calculation of the temperature considering two flows of the incident and the reflected molecular as one system.

REFERENCES

1. Reizhik YM and Gradshtein YS (1951) Tables of Integrals, Sums, Series and Products, Moscow (in Russian)
2. Shapiro AH. (1953) Compressible Fluid flow. Ronald Press, I
3. Muntz EP and Hamel BB (1974) Rarefaction phenomena in gas and isotope separations. In: M Becker and M Fiebig edited Rarefied Gas Dynamics, B 1-1, DFVLR, Germany
4. Muntz EP and Deglow TL (1979) Rarefaction phenomena in gas and isotope separation with emphasis on jets and beams. In: R Campargue' edited Rarefied Gas Dynamics, 1: 573-586
5. Yang HT and Lees L (1960) Rayleigh's problem at low Reynolds number according to the kinetic theory of gases. In: FM Devienne edited Rarefied Gas Dynamics, Pergamon Press, 201-238
6. Narasimha R (1962) Collisionless expansion of gases into vacuum. J Fluid Mech., 12: 294-308

5 CONTINUUM MODELS

5.1 INTRODUCTION

When the gas is slightly rarefied or, to be more exact, when the appropriately defined Kn number is in the scope less than 0.1 and larger than 0.01, the discrete molecular effects do not manifest themselves sharply (see the discussion in 0.1 of the Chapter Introduction), the gas flow can be investigated from the point of view of the continuum model. An established practice has been and remains to start from the Navier-Stokes equations with the employment of the slip boundary conditions to obtain the solutions of certain problems. The asymptotic theory starts from the Boltzmann equation by using the asymptotic analysis to obtain the equations of fluid mechanics more exact than the Navier-Stokes equations and corresponding slip boundary conditions to solve the problems in this regime. Qian Xuesen (H. S. Tsien) [1] pointed out early in the forties of 20th century, that the Burnett equations obtained as the second order terms of the Chapman-Eskog expansion should give results butter than those of the Navier-Stokes equations (first order terms of the Chapman-Enskog expansion) when Kn is not so small and Ma is large, and for the Burnett equations, as equations of order higher than the Navier-Stoke equations, more boundary conditions should be proposed. Own to the complexity of the Burnett equations and the instability problem related to the high frequency disturbances there have been suspicions concerning reliability of the Burnett equations. But recent investigations and the comparison with the DSMC method and the experimental results show that the Burnett equations in the slip flow regime are indeed superior than the Navier-Stokes equations. Grad [2] employed the Hermite polynomials expansion up to the third order terms to solve the Boltzmann equation yielding the thirteen moment equations, or the Grad equations. To the Grad equations had been also attached great importance, unfortunately the experimental and theoretical investigation did not verify their validity. In section 5.2 the Navier-Stokes equations, the Burnett equations and the Grad thirteen moment equations are presented, and the reliability of them as the basic equations of the continuum model is discussed. In section 5.3 the derivations of the slip velocity and temperature jump boundary conditions are given and the problem of the formulation of the boundary conditions in slip flow regime is discussed. In section 5.4 solutions of some simple problems are discussed starting from the Navier-Stokes equations and usual slip conditions. In section 5.5 the problem of thermal creep is discussed.

5.2 BASIC EQUATIONS

5.2.1 EQUATIONS OF MASS, MOMENTUM AND ENERGY CONSERVATION

In section 2.7 of Chapter 2 the moment equation or the Maxwellian transport equation (2.183) of the Boltzmann equation (2.152) is obtained

$$\frac{\partial}{\partial t}(n\bar{Q}) + \nabla \cdot \overline{ncQ} - nF \cdot \overline{\frac{\partial Q}{\partial c}} = \Delta[Q], \tag{5.1}$$

where the collision integral $\Delta[Q]$ can be expressed in the form of Eq. (2.160). From there it is seen that when Q is taken as the mass m, the momentum mc and kinetic energy $(1/2)mc^2$, the right hand side of Eq. (5.1) is zero. And the moments of Q and cQ are various macroscopic quantities in gas dynamics discussed in detail in section 2.2 of Chapter 2 (see formulae collected at the end of section 2.2). Thus, the following equations of mass, momentum and energy conservation are obtained (cf. reference [3], section 2 and section 3 of Chapter 3)

$$\begin{cases} \dfrac{d\rho}{dt} + \rho \dfrac{\partial u_i}{\partial x_i} = 0 \\[6pt] \rho \dfrac{du_i}{dt} + \dfrac{\partial p}{\partial x_i} + \dfrac{\partial \tau_{ij}}{\partial x_j} = 0 \\[6pt] \rho \dfrac{de}{dt} + p \dfrac{\partial u_i}{\partial x_i} + \tau_{ij} \dfrac{\partial u_i}{\partial x_j} + \dfrac{\partial q_i}{\partial x_i} = 0. \end{cases} \tag{5.2}$$

Equations (5.2) are the basic equations controlling gas flows, they can be derived as well from the mass, momentum and energy conservation equations starting from the continuum model assumption. These constitute an incomplete system of equations. The unknown functions are ρ, p (or e, related with ρ, p through Eq. (2.31)), $u_i, q_i \tau_{ij}$, among which 13 are independent (u_i, q_i contain 6 scalars, τ_{ij} according to the definition of Eq. (2.24) is a non-divergent symmetric tensor and contains 5 independent scalars). But the number of the equations (scalar equations) is five. Consequently the equations (5.2) must be added by some relations to form a closed system. The continuum model handles this problem by adopting certain relations between the stresses and the strain rates, and the heat transfer and temperature gradient (the temperature T is related with p, ρ through the equation of state Eq. (3.31)), thus make the system close. If let the relation between the stress and the strain rate be linear (the Newtonian fluid hypothesis) and the heat transfer be proportional to the temperature gradient (Fourier law), then the Navier-Stokes equations are obtained. If adopt the inviscid assumption by letting $\tau_{ij} = 0$, $q_i = 0$, then the equations (5.2) become the Euler equations. The kinetic theory of gases, starting from the solution of the Bolotzmann equation, obtains the expression of τ_{ij} and q_i through u_i and T and calculates the proportional coefficients (the viscosity and the conductivity). This is the method developed by Chapman and Enskog, expounded in detail in reference [3].

5.2.2 CHAPMAN-ENSKOG EXPANSION

To solve the Boltzmann equation Chapman and Enskog expanded the distribution function f into a series of the terms proportional to the powers of Knudsen number (in the scope considered Kn is less than 1)

$$f = \sum_{r=0}^{\infty} f^{(r)}, \tag{5.3}$$

where $f^{(0)}$ is the approximation of zero order and is taken as the equilibrium Maxwellian distribution Eq. (2.184), $f^{(r)}$ is the rth correction term of f. The corresponding stress tensor P_{ij} and the heat transfer vector q_i can be written

$$P_{ij} = \sum_{r=0}^{\infty} P_{ij}^{(r)}, \tag{5.4}$$

$$q_i = \sum_{r=0}^{\infty} q_i^{(r)}, \tag{5.5}$$

where

$$P_{ij}^{(r)} = m \int c'_i \, c'_j \, f^{(r)} dc, \tag{5.6}$$

$$q_i^{(r)} = \frac{m}{2} \int c'_i c'^2 f^{(r)} dc. \tag{5.7}$$

5.2.3 EULER EQUATIONS

If Take the zeroth order approximation as $f = f^{(0)}$, then

$$\tau_{ij}^{(0)} = 0, \quad q_i^{(0)} = 0.$$

Substituting into Eq. (5.2) yields the *Euler equations*.

5.2.4 NAVIER-STOKES EQUATIONS

The first order approximation $f^{(1)}$ of f is solved to be (see reference [3])

$$f^{(1)} = -f^{(0)} \left(\frac{4K\beta^2}{5nk} \left(\beta^2 c'^2 - \frac{5}{2} \right) c'_i \frac{\partial}{\partial x_i} \ln T + \frac{4\mu\beta^4}{\rho} \overset{\circ}{c'_i c'_j} \frac{\partial c_{0i}}{\partial x_j} \right), \tag{5.8}$$

where $\overset{\circ}{c'_i c'_j} = c'_i c'_j - c'^2 \delta_{ij}/3$ denotes the non-divergent tensor constituted from $c'_i c'_j$.

In the first approximation

$$\tau_{ij}^{(1)} = -2\mu \overline{\frac{\partial u_i}{\partial x_j}}, \qquad (5.9)$$

$$q_i^{(1)} = -K \frac{\partial T}{\partial x_i}, \qquad (5.10)$$

where μ is the *viscosity coefficient*, K is the *conductivity coefficient*, the Chapman Enskog theory gives

$$\mu = \frac{(5/8)(\pi mkT)^{1/2}}{(m/4kT)^4 \int_0^\infty c_r^7 \sigma_\mu \exp(-mc_r^2/4kT) dc_r}, \qquad (5.11)$$

$$K = \frac{15}{4} \frac{k}{m} \mu. \qquad (5.12)$$

See reference [4], Eq. (7.2), Chapter 10. The expression $\overline{\partial u_i/\partial x_j}$ in Eq. (5.9) is the non-divergent symmetric tensor constituted from $\partial u_i/\partial x_j$. In general the non-divergent symmetric tensor is formed as:

$$\overline{A}_{ij} = \frac{1}{2}(A_{ij} + A_{ji}) - \frac{1}{3}\delta_{ij} A_{kk}. \qquad (5.13)$$

Substituting Eq. (5.9) and Eq. (5.10) into the equations of momentum and energy conservation in Eq. (5.2), and introducing the enthalpy $h = e + p/\rho$, one obtains

$$\rho \frac{du_i}{dt} + \frac{\partial p}{\partial x_i} - \frac{\partial}{\partial x_j}\left[\mu\left(\frac{\partial u_i}{\partial x_j} + \frac{\partial u_j}{\partial x_i}\right) - \frac{2}{3}\mu \frac{\partial u_k}{\partial k_k}\right] = 0, \qquad (5.14)$$

$$\frac{dh}{dt} = \frac{1}{\rho}\frac{dp}{dt} + \frac{1}{\rho}\frac{\partial}{\partial x_i}\left(K \frac{\partial T}{\partial x_i}\right) + \frac{\phi}{\rho}, \qquad (5.15)$$

where ϕ is the dissipation function

$$\phi \equiv -\tau_{ij}^{(1)} \frac{\partial u_i}{\partial x_j} = 2\mu \overline{\frac{\partial u_i}{\partial x_j} \frac{\partial u_i}{\partial x_j}} = \frac{1}{2}\mu\left(\frac{\partial u_i}{\partial x_j}+\frac{\partial u_j}{\partial x_i}\right)^2 - \frac{2}{3}\mu\left(\frac{\partial u_k}{\partial x_k}\right)^2. \qquad (5.16)$$

Equations (5.14) and (5.15) are the *Navier-Stokes momentum equations* and the corresponding *energy equation* in the usual fluid mechanics.

5.2.5 BURNETT EQUATIONS

1. *The form of the equations*

Burnett [5] was the first to investigate the *second approximation* of f and gave the corresponding stress tensor and the heat transfer vector (see Chapter 15 of [3])

$$\tau_{ij}^{(2)} = K_1 \frac{\mu^2}{p}\frac{\partial u_k}{\partial x_k}\frac{\partial u_i}{\partial x_j} + K_2 \frac{\mu^2}{p}\left[-\overline{\frac{\partial}{\partial x_i}\frac{1}{\rho}\frac{\partial p}{\partial x_j}} - \overline{\frac{\partial u_k}{\partial x_i}\frac{\partial u_j}{\partial x_k}} - 2\overline{\frac{\partial u_i}{\partial x_k}\frac{\partial u_k}{\partial x_j}}\right] +$$

$$K_3\frac{\mu^2}{\rho T}\overline{\frac{\partial^2 T}{\partial x_i \partial x_j}} + K_4 \frac{\mu^2}{\rho p T}\overline{\frac{\partial p}{\partial x_i}\frac{\partial T}{\partial x_j}} + K_5 \frac{\mu^2}{\rho T^2}\overline{\frac{\partial T}{\partial x_i}\frac{\partial T}{\partial x_j}} + K_6 \frac{\mu^2}{p}\overline{\frac{\partial u_i}{\partial x_k}\frac{\partial u_k}{\partial x_j}}, \qquad (5.17)$$

$$q_i^{(2)} = \theta_1 \frac{\mu^2}{\rho T}\frac{\partial u_j}{\partial x_j}\frac{\partial T}{\partial x_i} + \theta_2 \frac{\mu^2}{\rho T}\left[\frac{2}{3}\frac{\partial}{\partial x_i}\left(T\frac{\partial u_j}{\partial x_j}\right) + 2\frac{\partial u_j}{\partial x_i}\frac{\partial T}{\partial x_j}\right] +$$

$$\left[\theta_3 \frac{\mu^2}{\rho p}\frac{\partial p}{\partial x_j} + \theta_4 \frac{\mu^2}{\rho}\frac{\partial}{\partial x_j} + \theta_5 \frac{\mu^2}{\rho T}\frac{\partial T}{\partial x_j}\right]\overline{\frac{\partial u_j}{\partial x_i}}. \qquad (5.18)$$

Here the notation of Eq. (5.13) is used, and K_i and θ_i [1] are both constants. For Maxwellain molecules their values are

---

[1] In the first edition of reference [3] (1939) θ_2 was mistaken as +45/8, θ_5 is mistaken as $3/2(5-\frac{T}{\mu}\frac{d\mu}{dT})=6$, Wang Chengshu and Uhlenbeck pointed out this error [6], and found out that the conclusion of Schamberg et al. about the impossibility to observe the sound dispersion in air was originated from the erroneous θ_2 and θ_5. In the second edition of reference [3] the errors in θ_2 and θ_5 are corrected.

$$\begin{cases} K_1 = 3.333, K_2 = 2, K_3 = 3, \\ K_4 = 0, K_5 = 3, K_6 = 8, \\ \theta_1 = 9.375, \theta_2 = -45/8, \theta_3 = -3, \theta_4 = 3, \\ \theta_5 = 3\left(35/4 + \dfrac{T}{\mu}\dfrac{\mu}{dT}\right) = 29.25. \end{cases} \qquad (5.19)$$

For hard sphere model their values are

$$\begin{cases} K_1 = 4.056, K_2 = 2.028, K_3 = 2.418, \\ K_4 = 0.681, K_5 = 0.219, K_6 = 7.424, \\ \theta_1 = 11.644, \theta_2 = -5.822, \theta_3 = -3.090, \\ \theta_4 = 2.418, \theta_5 = 25.157. \end{cases} \qquad (5.20)$$

Substituting

$$\tau_{ij} = \tau_{ij}^{(B)} = \tau_{ij}^{(1)} + \tau_{ij}^{(2)},$$

$$q_i = q_i^{(B)} = q_i^{(1)} + q_i^{(2)} \qquad (5.21)$$

into Eq. (5.2), with $\tau_{ij}^{(1)}, q_i^{(1)}$ substituted by Eq. (5.9) and Eq. (5.10) and $\tau_{ij}^{(2)}, q_i^{(2)}$ substituted by Eq. (5.17) and Eq. (5.18), the *Burnett equations* are obtained.

2. *The solution of the equations at hypersonic speed*

Qian Xuesen (H.S. Hsien) analyzed the magnitudes of the ratios of the second order terms to the first order terms and pointed out that in the slip flow regime at high Mach number the Burnett equations must be adopted. At the same time he pointed out, that as the additional heat transfer and the stress terms of the Burnett equations contain derivatives of order higher than the first, the order of the system of the partial differential equations is enhanced and more boundary conditions are needed than usually required in the gas dynamics.

The progress of the later time application of the Burnett equations in the hypersonic flows was not so successful. The Burnett equations are very complex and they are very difficult to be treated. There were not generally adopted opinions concerning the formulation of the correct boundary conditions suitable for the Burnett equations. And theoretical and experimental results sometimes showed evidence that Navier-Stokes equations were superior to the Burnett equations. This made people think that in the slip flow regime the basic equations still should rely on the Navier-Stokes equations [7]. This situation started to change from the late 80s of the 20th century. Fisco and Chapman [8] in 1988 investigated the structure of the one-dimensional shock wave to check the reliability of the Burnett equations. The merit of choosing such a problem is the possibility to get rid of the uncertainty in the formulation of the boundary conditions. The result showed that the Burnett equations agreed with the simulation result of DSMC method better that the Navier-Stokes equations. This and similar investigations changed the opinion that the Burnett equations were seemingly useless. In fact, before that there had been works showing that Burnett equations were superior in comparison with the Navier-Stokes equations. The measurement of the density distribution in the cylindrical Couette flow by the electronic beam fluorescence method published in 1971 by Alofs and Springer [9] showed that the result at $Kn = 0.25$ of using Burnett equations and higher order slip conditions obtained by T.C. Lin and Street [10] agreed better with the experimental result than that obtained by the Navier-Stokes equations and first order slip condition.

Zhong, McCormack and Chapman [11] proved that Burnett equations Eq. (5.2), Eq. (5.21) are not stable for short wave length disturbances and considered this is the reason why the numerical solution of the Burnett equations met difficulties. They suggested select some of the third order (the augmented) terms higher than the original Burnett terms and add to the Burnett stress and heat transfer terms:

$$\begin{cases} \tau_{ij} = \tau_{ij}^{(B)} + \tau_{ij}^{(a)} \\ q_i = q_i^{(B)} + q_i^{(a)} \end{cases}.$$ (5.22)

In the above formulas $\tau_{ij}^{(B)}, q_i^{(B)}$ are given by Eqs. (5.21), (5.9), (5.10), (5.17), and (5.18), the augmented terms $\tau_{ij}^{(a)}, q_i^{(a)}$ are

$$\begin{cases} \tau_{ij}^{(a)} = \dfrac{\mu^3}{p^2} \left[\dfrac{3}{2} K_7 RT \dfrac{\partial}{\partial x_j} \left(\dfrac{\partial^2 u_i}{\partial x_k \partial x_k} \right) \right] \\ q_i^{(a)} = \dfrac{\mu^3}{p\rho} \left[\theta_7 R \dfrac{\partial}{\partial x_i} \left(\dfrac{\partial^2 T}{\partial x_k \partial x_k} \right) + \theta_6 \dfrac{RT}{\rho} \dfrac{\partial}{\partial x_i} \left(\dfrac{\partial^2 \rho}{\partial x_k \partial x_k} \right) \right] \end{cases},$$ (5.23)

where

$$K_7 = 2\beta, \quad \theta_6 = -5\beta, \quad \theta_7 = 11/16.$$

Substitution of Eq. (5.22) and Eq. (5.23) into Eq. (5.2) yields the *augmented Burnett equations*, which refrain from the problem of un-stability encountered in the numerical calculations of the usual Burnett equations when the cell sizes are small than the mean free path. Zhong et al. [12, 13] solved the problems of hypersonic flows around two-dimensional blunted cylinder and axi-symmetric spherical head using the augmented Burnett equations with first order slip boundary conditions, and obtained better agreement with the calculation results of DSMC method than the Navier-Stokes equations. The Burnet equations and the first order slip boundary conditions were also used to solve the low density nozzle flow field and the hypersonic flow around the three-dimensional ellipsoid in transitional flow [14, 15].

If take the term containing the velocity gradient in Eq. (5.17) as the representative term of the Burnett equations, the ratio of the Burnett stress term to the ordinary stress term (see Eq. (5.9)) can be written as

$$\tau_{ij}^{(B)} \Big/ \tau_{ij}^{(1)} \approx \frac{\mu}{p}\frac{\partial u_k}{\partial x_k} \approx \frac{a\rho\lambda}{\rho a^2}\frac{U}{L} \approx MaKn, \tag{5.24}$$

where a is the speed of sound and $\mu \sim a\rho\lambda$ (see Eq. (2.222)). Similarly, from the analysis of the order of magnitudes of Eq. (5.18) and Eq. (5.10) one obtains that the ratio of the Burnett heat transfer to the ordinary term of heat transfer is also $MaKn$. This is the analysis of the order of magnitudes given by Qian Xuesen (H.S. Hsien) in reference [1], from where he drew the conclusion that under high Ma numbers when Kn is not small the Burnett equations should be adopted. The many years discussion and the scientific practice concerning the suitability of the Burnett equations verify the correctness of this conclusion.

3. *Low speed non isothermal flows*

For a class of low speed non-isothermal flows the following comparison of the orders of magnitudes between the ordinary stress and the Burnett stress can be obtained. Assume in such kind of flows $\text{Re} \leq O(1)$, and $Ma \approx \text{Re} Kn \ll 1$, but the variation of the temperature is rather large: $\Delta T = \theta T_0$, T_0 is the characteristic temperature, $\theta \leq O(1)$. Then the order of magnitude of the velocity is:

$u = \text{Re}\,\mu/\rho L \approx \text{Re}\,a\rho\lambda/\rho L \approx a Kn$, and the order of magnitude of the ordinary viscous stress is

$$\tau_{ij}^{(1)} \approx \mu \frac{\partial u_i}{\partial x_j} \approx a\rho\lambda u/L \approx a^2 \rho Kn^2 \ . \tag{5.25}$$

If separate specially the terms containing the first order and second order derivatives of temperature in the second order approximation Eq. (5.17) of the stress and call them the Burnett thermal stress

$$\tau_{ij}^{(T)} = K_3 \frac{\mu^2}{\rho T} \overline{\frac{\partial^2 T}{\partial x_i \partial x_j}} + K_5 \frac{\mu^2}{\rho T^2} \overline{\frac{\partial T}{\partial x_i} \frac{\partial T}{\partial x_j}} \ , \tag{5.26}$$

it is readily seen that the order of magnitude of $\tau_{ij}^{(T)}$ in the above defined slow non-isothermal flows is

$$\tau_{ij}^{(T)} \approx \frac{\mu^2 \theta}{\rho L^2} \approx a^2 \rho \lambda^2/L^2 \approx \theta a^2 \rho Kn^2 \ . \tag{5.27}$$

Comparison of Eq. (5.25) and Eq. (5.27) shows that if $\theta = O(1)$, then $\tau_{ij}^{(T)}$ and $\tau_{ij}^{(1)}$ have the same order of magnitude, i.e., in the *slow non-isothermal flows* the Burnett thermal stress and the usual (Navier-Stokes) stress have the same order of magnitude. The consequence of taking into account of the thermal stress is the occurrence of a new type of rarefied flows. Under usual pressure and ordinary sizes the steady temperature field can not produce steady flow of gases. But in gas of small *Kn* number when the temperature field is originated from the external heat transfer but not by the transformation of the kinetic energy into the thermal energy the slow non-isothermal flow convection takes place [16, 17]. This specific phenomenon in the continuum Burnett approximation is called *temperature stress convection*.

In rarefied gases the most remarkable flow phenomenon caused by the non-uniformity of the temperature is the thermal creep originated from the slip boundary condition, see the discussion in section 5.5. The asymptotic theory of

Sone (see section 5.2.7) shows that at small Kn number there is the second order thermal creep phenomenon called *thermal stress slip flow* taking place [18, 19].

4. *Experimental verification*

The most convincing experimental proof of the Burnett approximation should be the direct experimental verification of some one of the Burnett stress or heat flux terms (see Eq. (5.17), Eq. (5.18)), for example, the actual measurement of the so called "viscous heat flux" $q_i^{(2)V} = \theta_2(2\mu^2/3\rho T)\partial(T\partial u_j/\partial x_j)/\partial x_i$ (see the second term of Eq. (5.18)). Unfortunately, this heat flux virtually is not possible to be measured by experiment, for it takes place in the direction of the mass flux (the flow direction). But in the gas flow of polyatomic molecules the external magnetic field can produce a transverse heat flux perpendicular both with the velocity gradient and the magnetic field, i.e., the so called "viscous-magnetic heat flux". This heat flux can be measured experimentally. Vestner [20] developed a theory that can distinguish the Burnett bulk and the boundary layer contributions and obtained the heat flux expression in dependence of the magnetic field orientation. Hermans et al. [21] carried out the experiment for carbon monoxide and nitrogen and obtained results in excellent agreement with the theory. Thus, the Burnett approximation or the second order Chapman- Enskog approximation was verified quantitatively by the experiment.

5.2.6 GRAD'S THIRTEEN MOMENT EQUATIONS

Grad [2] suggested a different method to solve the Boltzmann equation considering $\rho, p, u_i, \tau_{ij}, q_i$ as 13 equal unknown functions and obtained 13 moment equations of the Boltzmann equation. For this he expanded the velocity distribution function into Hermite polynomials. If take the first four terms of this expansion and represent the fourth term by the contraction of the Hermite polynomials, then the function f can be written as

$$f = f^{(0)}\left[1 + \frac{\tau_{ij}}{2pRT}c_i'c_j' - \frac{q_i}{pRT}c_i'\left(1 - \frac{c'^2}{5RT}\right)\right]. \qquad (5.28)$$

Substituting this expression into the Boltzmann equation (2.152) and multiplying the equation successively by $m, c_i', (1/2)mc'^2, c_i'c_j', (1/2)c_i'c'^2$ and integrating over all possible velocities yields the equation (5.2) and the following equations

$$\frac{\partial \tau_{ij}}{\partial t} + \frac{\partial}{\partial x_k}(u_k \tau_{ij}) + \frac{2}{5}\overline{\frac{\partial q_i}{\partial x_j}} + \tau_{ik}\overline{\frac{\partial u_j}{\partial x_k}} + p\overline{\frac{\partial u_i}{\partial x_j}} = -\frac{q}{\mu}\tau_{ij}, \qquad (5.29)$$

$$\frac{\partial q_i}{\partial t} + \frac{\partial u_k q_i}{\partial x_k} + \frac{7}{5}q_k\frac{\partial u_i}{\partial x_k} + \frac{2}{5}q_k\frac{\partial u_k}{\partial x_i} +$$

$$\frac{2}{5}q_i\frac{\partial u_k}{\partial x_k} + RT\frac{\partial \tau_{ik}}{\partial x_k} + \frac{7}{2}\tau_{ik}R\frac{\partial T}{\partial x_k} - \frac{\tau_{ij}}{\rho}\frac{\partial P_{jk}}{\partial x_k} +$$

$$\frac{5}{2}pR\frac{\partial T}{\partial x_i} = -\frac{2}{3}\frac{p}{\mu}q_i. \qquad (5.30)$$

Equation (5.29) is an equation for the non-divergent symmetric tensor involving five independent scalar equations, Eq. (5.30) involves three equations. Together with the five equations of (5.2) there are thirteen independent equations. The temperature T in Eq. (5.29) can be expressed through the equation of state as $p/\rho R$. Equations (5.2), (5.29) and (5.30) contain altogether thirteen unknowns: $\rho, p, u_i, q_i, \tau_{ij}$. They constitute a complete system of equations, i.e., the *Grad thirteen moment equations*. Similar great importance had been attached to the Grad equations as to the Burnett equations. However, the expansion expression Eq. (5.28) assumes that the distribution function is continuous in the velocity space, but in fact, near the solid boundary the distribution function is always discrete in the normal to the boundary direction for the velocity. Moreover, the polynomial expansion can be proved to be divergent in some cases, for example in the shock wave problem when the oncoming *Ma* number is greater than 1.85. In contrast with the reattachment of great importance to the Burnett equations starting from the late eighties of the 20th century, the Grad equations have not yield results supported by the experiments and direct statistical simulations.

5.2.7 THE ASYMPTOTIC THEORY FOR SMALL KNUDSEN NUMBERS

Sony [18] established fluid-dynamic type system of equations and corresponding slip boundary conditions to treat rarefied gas dynamics problems under small Kn numbers by the systematic asymptotic analysis for the Boltzmann equation assuming small deviation (of the order of the Knudsen number) from the local Maxwellian distribution function. In the external region where the length scale is the characteristic length L of the flow the expansion of the distribution function in a power series of Kn is used and the fluid mechanics type equations thus obtained are very like the Navier-Stokes equations except a thermal stress term. Near the wall in the Knudsen layer (a layer of thickness of the mean free path) the Knudsen layer corrections are introduced to describe the rather rapid variation of the distribution function in the normal direction to the wall. These corrections are the slip boundary conditions. The main conclusions which could be drawn from the Sony's asymptotic theory are:

1. Up to the second order of the Knudsen number the behavior of the rarefied gas can be treated by the Navier-Stokes equations under the slip boundary conditions. The effect of rarefaction reveals itself only trough the boundary conditions of the slip type.

2. In the fluid-dynamic type equations except the usual Navier-Stokes terms there is a term called thermal stress which can cause another type of flow called *thermal stress slip flow* [19]. Under small Knudsen number the thermal stress at the surface acts on the gas and results in the thermal stress slip which is a second order thermal creep flow, the order of the velocity being aKn^2, the order of the thermal stress being $a^2 Kn^3 \rho$. This flow is caused by the gas rarefaction effect and is not expected in the classic gas dynamics.

3. Sony investigated the behavior in the continuum limit of gases by the asymptotic analysis of the Boltzmann equation and showed that only by using the analysis of the kinetic theory the correct description of the behavior in the

continuum limit can be obtained, i.e., the classic continuum gas dynamics is not complete. He showed by the asymptotic theory that something which only exists for $Kn \neq 0$ but does not exist in the limit of $Kn = 0_+$ would affect the temperature field in the limit $Kn = 0_+$ and called it the "ghost effect". In particular, the solution of the classic heat conduction equation is different from the temperature field of the quiescent gas obtained by the asymptotic analysis in the limit of positive zero of the mean free path (see the review paper [22] and references given there).

All these have important theoretical meaning and also appeal the experimental verification. Meanwhile, the usual slip boundary conditions obtained from the conditions of mass, momentum and energy conservation across the Knudsen layer (see the next section) are applicable for both low speed and hypersonic speed conditions and can involve situations of chemically non-equilibrium and catalytic wall for multi-component gases that the asymptotic theory has not treated yet because of the physical complexity. Consequently, despite the existence of the theoretically more strict asymptotic theory, in practice the solution of the Navier-Stokes equations with the usual slip boundary conditions is still widely used and has practical merits.

5.3 SLIP BOUNDARY CONDITIOINS

5.3.1 THE SIMPLE DERIVATION

The condition of equality of the tangential velocity components of the gas and of the solid surface (the no slip condition) proposed in the usual gas dynamics is not exact. Starting from the kinetic theory of gases a slip velocity near the wall can be derived, the magnitude of which is proportional to the mean free path λ, and under ordinary conditions and for usual sizes of the subject it can be neglected. In fact, the following simple reasoning can provide a roughly correct estimate of the slip velocity. Near the body surface half of the gas molecules come from the external flow and the other half are reflected from the body surface, the macroscopic velocity should be the average of the velocities of these two parts. Denote the velocity component of the gas along the direction of body surface as

u, and suppose the variation of u along the normal direction y is known, $u=u(y)$. The average velocity of the molecules migrated from the external flow arriving at the body surface should be the mean magnitude of the velocities attained by the molecules in the last collision, i.e., the velocity at a distance $\varsigma\lambda$ from the surface, ς being a numerical coefficient near one. Denote the unknown velocity of the gas near the wall by u_s, then the average velocity of the molecules from the external flow is

$$u_s + \varsigma\lambda\left(\frac{\partial u}{\partial y}\right)_0.$$

Assume that the σ portion of the reflected molecules is diffusely reflecting and the other $(1-\sigma)$ portion is specularly reflecting, then the average velocity of the half of molecules that are reflected from the surface is (the velocity of the surface is supposed to be zero)

$$\sigma\cdot 0 + (1-\sigma)\left[u_s + \varsigma\lambda\left(\frac{\partial u}{\partial y}\right)_0\right].$$

u_s should be the mean of the above two velocities

$$u_s = \frac{1}{2}\left\{u_s + \varsigma\lambda\left(\frac{\partial u}{\partial y}\right)_0 + (1-\sigma)\left[u_s + \varsigma\lambda\left(\frac{\partial u}{\partial y}\right)_0\right]\right\},$$

from where the expression of the slip velocity is obtained

$$u_s = \frac{2-\sigma}{\sigma}\varsigma\lambda\left(\frac{\partial u}{\partial y}\right)_0. \qquad (5.31)$$

The derivation of Eq. (5.31) given here is not strict, but the result concerning the dependence of u_s on σ and $(\partial u/\partial y)_0$ is correct.

5.3.2 THE CONSERVATION OF MOMENTUM AND ENERGY FLUXES IN THE KNUDSEN LAYER

Now we proceed to the derivation of the slip velocity and temperature jump boundary conditions for the mono-component gas. This derivation is based on the conservation of momentum and energy fluxes in the Knudsen layer. The so called *Knudsen layer* is a region of the gas near the body surface with the thickness of the order of the mean free path, where the behavior of the gas is governed by the kinetic theory of gases. At the outer edge of the Knudsen layer consider the balance of the momentum and energy fluxes being transferred. Introduce the total flux of the quantity being transferred $\phi(c')$ (the momentum or energy) in the direction (the y,v direction) of the normal to the body surface

$$F = \int_{-\infty}^{\infty}\int_{-\infty}^{\infty}\int_{-\infty}^{\infty} v'\phi(c')f_s(c')dc', \tag{5.32}$$

where f_s is the velocity distribution function at the outer edge of Knudsen layer. It is taken as the first Chapman Enskog approximation

$$f_s = f^{(0)} + f^{(1)}, \tag{5.33}$$

where $f^{(0)}, f^{(1)}$ are given by Eq. (2.184) and Eq. (5.8), respectively. Similarly the incident flux F_i, the specularly reflected flux F_{sp} and the diffusely reflected flux F_w of the quantity being transferred are introduced (at the outer edge of the Knudsen layer)

$$F_i = \int_{-\infty}^{\infty}\int_{-\infty}^{0}\int_{-\infty}^{\infty} v'\phi(c')f_s(c')dc', \tag{5.34}$$

$$F_{sp} = \int_{-\infty}^{\infty}\int_{0}^{\infty}\int_{-\infty}^{\infty} v'\phi(c')f_s(u',-v',w')dc', \tag{5.35}$$

$$F_w = \int_{-\infty}^{\infty}\int_{0}^{\infty}\int_{-\infty}^{\infty} v'\phi(c')f_w(c')dc', \tag{5.36}$$

where f_w is the Maxwellian velocity distribution function corresponding to the wall condition. Obviously, the total flux at the outer edge of Knudsen layer consists of the incident flux, σ portion of the diffusely reflected flux and $(1-\sigma)$ portion of the specularly reflected flux, that is

$$F = F_i + (1-\sigma)F_{sp} + \sigma F_w. \tag{5.37}$$

Substituting the quantity being transferred by the momentum and energy yields the equations of momentum and energy conservation.

5.3.3 THE DERIVATION OF THE SLIP VELOCITY FORMULA

Now apply Eq. (5.37) to the derivation of the slip velocity along the wall surface (in the x direction). The quantity being transferred in this case is the momentum in the x direction, $\phi = m(u_s + u')$, where u_s is the velocity of the gas at the outer edge of the Knudsen layer relative to the body surface. Suppose the macroscopic velocity varies along the normal to the wall surface, $u_0 = u_0(y)$, and the temperature varies along the wall surface, $T = T(x)$. Thus the distribution function f_s can be written (according to Eq. (5.33) and Eq. (5.8))

$$f_s = f^{(0)}\left[1 - \frac{4K\beta^2}{5nk}\left(\beta^2 c'^2 - \frac{5}{2}\right)u'\frac{1}{T}\frac{\partial T}{\partial x} - \frac{4\mu\beta^4}{\rho}u'v'\frac{\partial u_0}{\partial y}\right]. \tag{5.38}$$

If use the momentum flux P to replace F, then obviously

$$P_{sp} = -P_i, \quad P_w = 0. \tag{5.39}$$

Thus Eq. (5.37) has the form

$$P = \sigma P_i. \tag{5.40}$$

And P_i is

$$P_i = \int_{-\infty}^{\infty}\int_{-\infty}^{0}\int_{-\infty}^{\infty} v'(u_s + u')mn\left(\frac{\beta^3}{\pi^{3/2}}\right)\exp(-\beta^2 c'^2)\times$$

$$\left[1 - \frac{4K\beta^2}{5nk}\left(\beta^2 c'^2 - \frac{5}{2}\right)\frac{1}{T}\frac{\partial T}{\partial x}u' - \frac{4\mu\beta^4}{\rho}\frac{\partial u_0}{\partial y}u'v'\right]dc' =$$

$$\int_{-\infty}^{\infty}\int_{-\infty}^{0}\int_{-\infty}^{\infty} \rho\left(\frac{\beta^3}{\pi^{3/2}}\right)\exp(-\beta^2 c'^2)\left[\underset{①}{u_s v'} + \underset{②}{u'v'} - \right.$$

$$\frac{4K\beta^2}{5nk}\left(\beta^2 c'^2 - \frac{5}{2}\right)\frac{1}{T}\frac{\partial T}{\partial x}\left(u_s \underset{③}{u'v'} + \underset{④}{u'^2 v'}\right) -$$

$$\frac{4\mu\beta^4}{\rho}\frac{\partial u_0}{\partial y}\left(u_s \underset{⑤}{u'v'^2} + \underset{⑥}{u'^2 v'^2}\right)\bigg] du'dv'dw', \tag{5.41}$$

where the terms ②③⑤ after integration yield zero, for they are odd functions relative to u'. The terms ①④⑥ can be calculated readily according to Eqs. (II.12), (II.13), (II.14) and (II.15) in Appendix II

$$P_i = \frac{n\bar{c}}{4} mu_s - \frac{Km}{10kT}\frac{1}{\beta\pi^{1/2}}\frac{\partial T}{\partial x} + \frac{1}{2}\mu\frac{\partial u_0}{\partial y}. \tag{5.42}$$

The calculation of P has the same under-integral expression, but the limits of the integration for v' are from $-\infty$ to $+\infty$, so ①,④ after integration also yield zero for the under-integral expression being odd functions of v'. Thus

$$P = \mu\frac{\partial u_0}{\partial y}. \tag{5.43}$$

From Eq. (5.40) one has

$$\mu\frac{\partial u_0}{\partial y} = \sigma\left(\frac{n\bar{c}}{4}mu_s - \frac{Km}{10kT}\frac{1}{\beta\pi^{1/2}}\frac{\partial T}{\partial x} + \frac{1}{2}\mu\frac{\partial u_0}{\partial y}\right). \tag{5.44}$$

Solving Eq. (5.44) relative to u_s and taking into account that the framework connected with the body surface was chosen for the calculation, we have

$$u_{y=0} - u_w = \frac{2-\sigma}{\sigma}\frac{\mu\beta\pi^{1/2}}{\rho}\frac{\partial u_0}{\partial y} + \frac{Km}{5kT\rho}\frac{\partial T}{\partial x},$$

or (using the expressions (2.222) and (5.12) of μ and K for the hard sphere model)

$$u_{y=0} - u_w = 0.998\left(\frac{2-\sigma}{\sigma}\right)\lambda\left(\frac{\partial u_0}{\partial y}\right)_{y=0} + \frac{3}{4}\frac{\mu}{\rho T}\left(\frac{\partial T}{\partial x}\right)_{y=0}. \tag{5.45}$$

This is the expression for the *Maxwell slip velocity*.

Equation (5.45) can be written in a more general form as

$$u_{y=0} - u_w = C_m \lambda \frac{\partial u_0}{\partial y} + C_s \frac{\mu}{\rho T} \frac{\partial T}{\partial x}, \qquad (5.45)'$$

where C_m is the *velocity slip coefficient* and C_s is the *thermal creep coefficient*, there are many works devoted to determine the values of C_m and C_s more accurately under different models.

In the isothermal case Eq. (5.44) can be written as

$$\mu \frac{\partial u_0}{\partial y} = \sigma \left(\frac{n\bar{c}}{4} m u_s + \frac{1}{2} \mu \frac{\partial u_0}{\partial y} \right).$$

This equality can be assigned the following physical interpretation : at the body surface (the outer edge of the Knudsen layer) the viscous tangential stress $(\mu \partial u_0 / \partial y)$ partially consists of the viscous shear $(1/2)(\mu \partial u_0 / \partial y)$ of the molecules moving towards the surface (constituting half the total molecules), partially consists of the tangential momentum flux $(1/4) n \bar{c} (m u_s)$ originated from the slip velocity u_s, $(1/4) n \bar{c}$ is the number flux of the molecules (see Eq. (4.8)), and when the momentum reflection coefficient is σ, the action of the tangential momentum subjected by the surface is only σ portion of the that carried by the incident flow (see Eq. (4.15)). The theoretical analysis of the slip given by Maxwell [23] is much more complex, but the line of reasoning is roughly like this (see [24]).

5.3.4 THE DERIVATION OF THE TEMPERATURE JUMP EXPRESSION

Now apply again Eq. (5.37) but this time to the derivation of the temperature jump formula. The quantity concerned and being transferred in this case is the energy of molecules. Suppose the temperature varies along the normal to the wall surface, $T = T(y)$, and the influence of the velocity is not considered, thus the relevant distribution function f_s can be written as

$$f_s = f^{(0)} \left[1 - \frac{4K\beta^2}{5nk} \left(\beta^2 c'^2 - \frac{5}{2} \right) v' \frac{1}{T} \frac{\partial T}{\partial y} \right]. \qquad (5.46)$$

If use E denoting the energy flux to replace F in Eq. (5.37), then obviously

$$E_{sp} = -E_i. \qquad (5.47)$$

From Eq. (5.37) one has

$$E = \alpha E_i + \alpha E_w. \qquad (5.48)$$

Let us calculate the incident energy flux E_i, for monatomic molecules having only translational energy $\phi = (1/2)mc'^2$,

$$E_{itr} = \frac{1}{2} \int_{-\infty}^{\infty} \int_{-\infty}^{0} \int_{-\infty}^{\infty} v'mc'^2 nf^{(0)} \left[1 - \frac{4K\beta^2}{5nk} \left(\beta^2 c'^2 - \frac{5}{2} \right) v' \frac{1}{T} \frac{\partial T}{\partial y} \right] dc'$$

$$= \frac{1}{2} \int_{-\infty}^{\infty} \int_{-\infty}^{0} \int_{-\infty}^{\infty} \rho \left(\frac{\beta^3}{\pi^{3/2}} \right) \exp(-\beta^2 c'^2) \times$$

$$\left[v' - \frac{4K\beta^2}{5nk} \left(\beta^2 c'^2 - \frac{5}{2} \right) v'^2 \frac{1}{T} \frac{\partial T}{\partial y} \right] du'dv'dw'. \qquad (5.49)$$

According to Eqs. (II.12), (II.13), (II.14) and (II.15) in Appendix II one obtains

$$E_{itr} = 2kT \frac{n\bar{c}}{4} + \frac{1}{2} K \frac{\partial T}{\partial y}. \qquad (5.50)$$

The diffusely reflecting heat flux from the wall is (see Eq. (4.42))

$$E_{wtr} = -2kT_w \frac{n\bar{c}}{4}. \qquad (5.51)$$

If take into account that the molecules carry also the internal energy, then instead of Eq. (5.50) and Eq. (5.51) one has (see Eq. (4.40))

$$E_i = \frac{n\bar{c}}{4} \left[2kT + \frac{5-3\gamma}{2(\gamma-1)} kT \right] + \frac{1}{2} K \frac{\partial T}{\partial y}, \qquad (5.52)$$

$$E_w = -\frac{n\bar{c}}{4} \left[2kT_w + \frac{5-3\gamma}{2(\gamma-1)} kT_w \right]. \qquad (5.53)$$

E has the same under-integral expression, but the limits of the integration for v are from $-\infty$ to $+\infty$, obviously the contribution of the first term is zero, consequently

$$E = K\frac{\partial T}{\partial y}. \tag{5.54}$$

Substituting E, E_i, E_w into Eq. (5.48), one has

$$K\frac{\partial T}{\partial y} = \alpha\left[\frac{1}{2}K\frac{\partial T}{\partial y} + \frac{n\overline{c}}{4}k\frac{\gamma+1}{2(\gamma-1)}(T-T_w)\right],$$

from where

$$T_s - T_w = 2\frac{2-\alpha}{\alpha}\frac{2(\gamma-1)}{k(\gamma+1)}\frac{K}{n\overline{c}}\frac{\partial T}{\partial y}.$$

Making use of $\mu = 0.491\rho\overline{c}\lambda$, $k/m = c_p(\gamma-1)/\gamma$ and introducing $\Pr = \mu c_p/K$, one can write the above formula as

$$T\Big|_{y=0} - T_w = 0.982\frac{2\gamma}{\gamma+1}\frac{\lambda}{\Pr}\frac{2-\alpha}{\alpha}\left(\frac{\partial T}{\partial y}\right)_{y=0}. \tag{5.55}$$

This is the *Maxwell-Smoluchovski expression of the temperature jump.*

Analogous with Eq. (5.45)', Eq. (5.55) can be written as

$$T_{y=0} - T_w = C_t\lambda\frac{\partial T}{\partial y}, \tag{5.55)'}$$

C_t is the *temperature jump coefficient*, from more careful analysis more accurate result for different molecular models can be obtained. Smoluchovski first found experimentally that near the wall exists the phenomenon of temperature discontinuity predicted long ago by Poisson, and Maxwell by the reasoning analogous with that he used to obtain the velocity slip came to the formula for temperature jump. Some times Eqs. (5.45) and (5.55) are called Maxwell-Smoluchovski slip conditions.

5.3.5 THE EXTANSION TO CASES OF MULTI-COMPONENT GASES AND NON-EQUILIBRIUM FLOWS

The method of derivation of the slip boundary conditions adopted here, i.e., the method of employment of the momentum and energy flux conservation in the Knudsen layer, is put forward by Patterson [25] (he called it the compatibility condition) and further developed by Shidlowsky [26], but not alike their method here the Chapman-Eskog first approximation Eq. (5.38) and Eq. (5.46) have been used directly. Gupta, Scott and Moss [27] by the similar method and the employment of the Chapmen-Eskog distribution function for the multi-component gas mixture derived the slip boundary conditions for the multi-component non-equilibrium flow, giving the slip (jump) conditions of component concentrations, pressure, velocities and temperature for high flight altitudes and low Reynolds numbers taking into account the influence of the finite surface catalysis and recombination reactions. Tang Jinrong and Tao Bo [28] did the similar derivation, making some improvement in the definition of the component number density at the wall, the equation of balance of the normal momentum and the calculation of the catalytic reaction rate, providing the slip boundary conditions of general form. Tan Jinrong and Tao Bo also presented the thermo-chemically non-equilibrium slip boundary conditions corresponding to the basic equations based on the multi-temperature assumption. Zhao Jingye, Shen Ching and Tang Jinrong [29] discussed the formulation of the slip boundary conditions with injections from the wall and obtained the slip boundary conditions with catalytic reactions on the wall for multi-component gas mixture and with wall injections. Shen Ching [30] obtained the concentration jump coefficients for gas mixture by using the Boley Yip model equation for gas mixture.

5.4 THE SOLUTION OF SOME SIMPLE PROBLEMS

In the present section the slip boundary condition as the correction to the ordinary fluid mechanics is adopted to solve several classic unidirectional flow problems, i.e., the Couette flow, the Poiseuille flow and the Rayleigh problem under the low speed condition (so the incompressible assumption holds). These simple examples can demonstrate the method of the treatment of the slip problems and the effect of the slip boundary conditions. At the same time, as the interest towards the low speed problems in the micro-electric-mechanical systems (MEMS) is increasing, the solutions of these problems under low speed conditions can serve the bench mark merits to test some new computational methods.

5.4.1 COUETTE FLOW

Consider the plane Couette flow, i.e., a hypothetic shear flow between two plates of infinite length. Two parallel plates at a distance d apart each other with infinite length move to opposite directions along their own planes with velocity $U/2$ [2]. The center of the gap between plates is chosen as the origin of the coordinate system, the direction of motion is taken as the x axis, with y axis perpendicular to it (see Fig.5.1). It is evident that the flow is independent of x, the velocity has only component u in the x direction, and u and the shear stress are only dependent on the y coordinate. The equation of balance of forces acting on a gas element is (see Eqs. (5.2) and (5.14))

$$\frac{d\tau_{xy}}{dy} = 0. \tag{5.56}$$

That is

$$\tau_{xy} = const = \tau_w,$$

τ_w is the shear stress on the plate surface. In the scope of the Navier-Stokes approximation $\tau_{xy} = \mu(du/dy)$, one obtains

$$u = \tau_w \int_0^y \frac{dy}{\mu}. \tag{5.57}$$

For the case of slow motion the variation in temperature is small and μ is a constant. Thus the incompressible assumption yields the linear velocity distribution

---

[2] This formulation is equivalent to that in section 4.5 where the lower plate is quiescent and the upper moves with a velocity U.

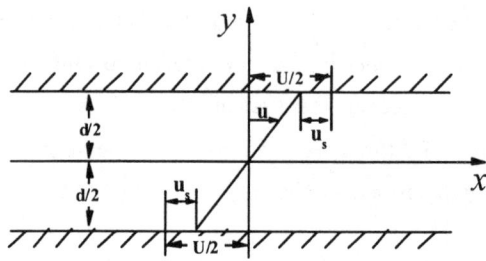

Fig. 5.1 The Couette flow with slip

$$u = \tau_w \frac{y}{\mu}. \tag{5.58}$$

For the *flow without slip boundary condition*, $u = U/2$ at $y = d/2$, the shear stress is obtained

$$\tau_w = \mu \frac{U}{d}. \tag{5.59}$$

When the mean free path λ of the gas is not negligible in comparison with d, the influence of the *slip boundary condition* must be taken into account. Then at $y = d/2$

$$u\big|_{y=d/2} = U/2 - u_s, \tag{5.60}$$

where u_s is the slip velocity. According to Eq. (5.45)

$$u_s = \frac{2-\sigma}{\sigma}\lambda\frac{du}{dy} \equiv \zeta\frac{du}{dy}, \quad \zeta = \frac{2-\sigma}{\sigma}\lambda, \tag{5.61}$$

ζ is called the *slip coefficient*. Equation (5.58) remains the solution of the problem, but with the boundary condition Eq. (5.60) the solution of the slip boundary condition problem is

$$\tau_w = \frac{\mu U/d}{1 + 2\zeta/d}, \tag{5.62}$$

5.4 THE SOLUTION OF SOME SIMPLE PROBLEMS

$$u = \frac{yU/d}{1+2\zeta/d}. \tag{5.63}$$

Under the model of hard sphere by using Eq. (2.221) the viscosity μ can be connected with the mean free path λ. Introducing the Knudsen number $Kn = \lambda/d$ and using the expression of the shear τ_{FM} in free molecular flow (see Eq. (4.55)) to normalize τ_w, it is readily to obtain

$$\frac{\tau_{NSslip}}{\tau_{FM}} = \frac{5\pi Kn}{8\left[\dfrac{\sigma}{2-\sigma} + 2Kn\right]}. \tag{5.64}$$

And the velocity normalized by the plate velocity can be written

$$\frac{u_{NSslip}}{U/2} = \frac{2}{1+2(2-\sigma)Kn/\sigma}\frac{y}{d}. \tag{5.65}$$

The solution Eq. (5.64) in principle is suitable only for slip regime, it involves the case of non-slip case (compare Eq. (5.62) for $Kn=0$ with Eq. (5.59)) and is approximately correct in the free molecular flow limit: from Eq. (5.64) one has $\tau_{NSslipKn\to\infty} \approx \tau_{FM}$. Remarkably, it has also a fairly good agreement with the numerical solutions of the Boltzmann equation in the transitional regime, but this of course is only accidental coincidence.

5.4.2 THE POISEUILLE FLOW

Now consider the *plane Poiseuille flow*, i.e., a hypothetic flow of gas between two parallel quiescent plates at a distance d apart under a constant pressure gradient (see Fig.5.2). In the case of low speed the temperature variation is neglected, μ is a constant, from the Navier-Stokes equation (5.14) one obtains

$$\frac{d^2u}{dy^2} = \frac{dp/dx}{\mu}, \tag{5.66}$$

where u is the velocity component along the direction of the plate (x direction), y is the coordinate perpendicular to x, the center of the gap between

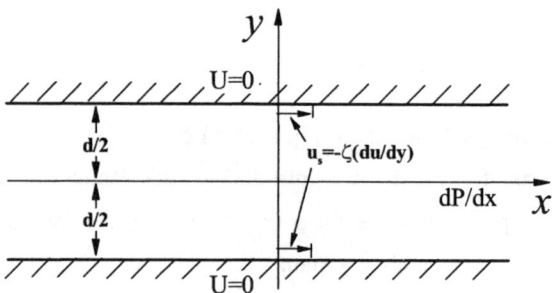

Fig. 5.2 The schematic of the Poiseuille flow with slip

plates is again taken as the origin of the coordinate system, the pressure gradient dp/dx is a constant. The equation (5.66) is readily integrated

$$\frac{du}{dy} = \frac{y}{\mu}\frac{dp}{dx}, \quad u = \frac{y^2}{2\mu}\frac{dp}{dx} + C. \tag{5.67}$$

Owing to the symmetry about the axis of the flow the constant of integration of the first order is zero. When Kn number is small but is not negligible, the *slip boundary condition* must be taken into account

$$u\big|_{y=d/2} = -\zeta \frac{du}{dy}, \quad \zeta = \frac{2-\sigma}{\sigma}\lambda. \tag{5.68}$$

The constant C can be determined from Eq. (5.68)

$$C = -\left(\frac{\zeta d}{2\mu} + \frac{d^2}{8\mu}\right)\frac{dp}{dx},$$

from where the solution is obtained

$$u = -\frac{1}{8\mu}\left(d^2 - 4y^2 + 4\zeta d\right)\frac{dp}{dx}. \tag{5.69}$$

Define the mean velocity as $U_A = \int_0^{d/2} u\,dy/(d/2)$, its value can be found

$$U_A = -\frac{1}{12\mu}\left[1 + 6\frac{2-\sigma}{\sigma}Kn\right]\frac{dp}{dx}.$$

5.4 THE SOLUTION OF SOME SIMPLE PROBLEMS

Now let us search the mass flux Q_m flown in unit time through the gap between the two plates (unit length in the z direction is taken). Obviously

$$\frac{1}{2}Q_m = \rho \int_0^{\frac{d}{2}} u\,dy = \frac{-p\,dp}{RT\,dx}\frac{1}{8\mu}\left(\frac{1}{3}d^3 + 2\zeta d^2\right), \qquad (5.70)$$

where the gas density ρ is written as p/RT to show explicitly that the formula is dependent of the location. Similarly, as $\varsigma = (2-\sigma)\lambda/\sigma$ and $\rho\lambda = const$, so writing ς as ς_1/p reveals the dependence of ς on the position through p. Multiplying Eq. (5.70) by dx and integrating it along the x axis from $-l/2$ to $+l/2$, l being the length of plate under consideration, keeping in mind that Q_m is constant along x and making the isothermal assumption, the value of Q_m is obtained

$$Q_m = \frac{d^3}{12l\mu}\frac{1}{RT}\left[\frac{1}{2}(p_1^2 - p_2^2) + 6\frac{\varsigma_1}{d}(p_1 - p_2)\right], \qquad (5.71)$$

where p_1, p_2 are the pressure values at $x = -l/2$ and $x = l/2$. If write the pressure distribution along x in the linear form

$$p = p_0(1 - \alpha x/d), \qquad (5.72)$$

obviously, $p_0 = (p_1 + p_2)/2, (p_1 - p_2)/l = -dp/dx = \alpha p_0/d$. Writing ς_1 approximately as $\varsigma_1 = \varsigma/p_0$, the mass flux can be expressed as

$$Q_m = \frac{\alpha\rho_0 d^2}{12\mu}\left[1 + 6\frac{2-\sigma}{\sigma}Kn\right]\rho_0 RT . \qquad (5.73)$$

Using the relation Eq. (2.221) between λ and μ, and introducing $c_m = \sqrt{2RT}$ as the normalization velocity, the dimensionless Q_m can be written

$$\frac{Q_m}{\alpha\rho_0 c_m d} = \frac{2}{15\sqrt{\pi}Kn}\left[1 + 6\frac{2-\sigma}{\sigma}Kn\right]. \qquad (5.74)$$

Analogously, the dimensionless velocity can be expressed as

$$\frac{u}{\alpha c_m} = \frac{1}{5\sqrt{\pi}K_n}\left[1 - \frac{4y^2}{d^2} + 4\frac{(2-\sigma)}{\sigma}Kn\right]. \qquad (5.75)$$

5.4.3 THE RAYLEIGH PROBLEM

Now consider the *Rayleigh problem*. The infinite plane plate acquires momentarily velocity U along its own plane and causes the motion of gas above it. This is an unsteady flow problem including two independent arguments, i.e., the time t counted from the start of the motion and the ordinate y (see Fig. 5.3). We consider the case when the ratio of t to the collision time τ is much greater than 1: $t/\tau \gg 1$, i.e., the continuum flow regime, and obtain the solutions both without slip and with the slip boundary condition taken into account basing on Navier-Stokes equation. Under the low speed assumption, $\rho = const$, and $\mu = const$, from the momentum Navier-Stokes equation (5.14) one obtains

$$\frac{\partial u}{\partial t} = \nu \frac{\partial^2 u}{\partial y^2}, \tag{5.76}$$

where $\nu = \mu/\rho = const$ is the kinematic viscosity.

For the *case of no slip boundary condition*, the initial and boundary conditions can be written

$$u(y,0) = 0, \tag{5.77}$$

$$u(0,t) = U. \tag{5.78}$$

The employment of the Laplace transformation

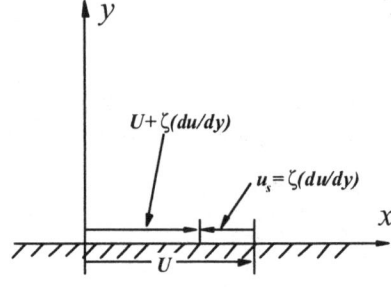

Fig. 53 The schematic of boundary condition for the Rayleigh problem with slip

$$\bar{u} = \int_0^\infty u(y,t)e^{-pt}dt,\qquad(5.79)$$

can transform the equation (5.76) and the boundary condition (5.78) into

$$p\bar{u} = v\frac{d^2\bar{u}}{dy^2},\qquad(5.80)$$

$$\bar{u}\big|_{y=0} = \frac{U}{p}.\qquad(5.81)$$

The solution of the equation (5.80) that is finite at $y=0$ is

$$\bar{u} = Ce^{-\sqrt{\frac{p}{v}}y}.\qquad(5.82)$$

Letting it satisfy the boundary condition (5.81), one obtains

$$\bar{u} = \frac{U}{p}e^{-\sqrt{\frac{p}{v}}y}.\qquad(5.83)$$

The original function whose image is Eq. (5.83) is

$$u = U\left(1 - \mathrm{erf}\frac{y}{2\sqrt{vt}}\right).\qquad(5.84)$$

This is the solution under the no slip boundary condition.

Now discuss the situation *with slip boundary condition taken into account* [31], where the following formula should replace Eq. (5.78)

$$u(0,t) = U + \zeta\frac{\partial u}{\partial y}\bigg|_{y=0}.\qquad(5.85)$$

The solution of the Laplace transformed equation is still Eq. (5.82), the boundary condition is transformed into

$$\bar{u}\big|_{y=0} = \frac{U}{p} + \zeta\frac{d\bar{u}}{dy}\bigg|_0.\qquad(5.86)$$

From Eq. (5.82) and this boundary condition the value of C is readily determined

$$C = \frac{1}{\left(1+\frac{\zeta}{\sqrt{\nu}}\sqrt{p}\right)}\frac{U}{p}.$$

Thus the solution of the transformed equation is

$$\bar{u} = \frac{U}{p\left(1+\frac{\zeta}{\sqrt{\nu}}\sqrt{p}\right)}e^{-\sqrt{p}\frac{y}{\sqrt{\nu}}},$$

or

$$\bar{u} = U\frac{\alpha}{p(\alpha+\sqrt{p})}e^{-\sqrt{p}k}, \quad \alpha = \frac{\sqrt{\nu}}{\zeta}, \quad k = \frac{y}{\sqrt{\nu}}. \tag{5.87}$$

The original function of this image (see the Mathematics Hand Book, p.564, Eq. (6)) is

$$u = erfc\left(\frac{k}{2\sqrt{t}}\right) - e^{\alpha k}e^{\alpha^2 t}erfc\left(\alpha\sqrt{t}+\frac{k}{2\sqrt{t}}\right).$$

Substituting the magnitudes of α and k yields the solution of the Rayleigh problem in the slip flow regime

$$u = erfc\frac{y}{2\sqrt{\nu t}} - \exp\left(\frac{y}{\zeta}+\frac{\nu t}{\zeta^2}\right)erfc\left(\frac{\sqrt{\nu t}}{\zeta}+\frac{y}{2\sqrt{\nu t}}\right). \tag{5.88}$$

5.5 THERMAL CREEP AND THERMOPHORESIS

The phenomena caused by, and consequences on the flow originated from the three terms of velocity and temperature gradients in the expressions of slip velocity Eq. (5.45) and temperature jump Eq. (5.55) are different, despite that all they are proportional to the mean free path. The velocity and temperature gradients $\partial u/\partial y$ and $\partial T/\partial y$ along the normal to the surface cause the velocity slip and temperature jump of the gas near the wall. As the flows discussed in section 5.4 reveal, they only cause some corrections to the main flow and are a kind of second order effects in the framework of viscous fluid flow. In this sense their effects are always weak. The temperature gradient $\partial T/\partial x$ along the body surface also causes the velocity slip of the gas, which is called the *thermal creep*. Under certain circumstances it could be the dominant factor, and some phenomena happen that would not occur starting from the classic fluid mechanics.

Basing on the classic fluid mechanics a body set in quiescent gas with temperature gradient does not subject to the force action. But when the size of the body is sufficient small that the Knudsen number with the characteristic length equal to the body size is not negligible, the temperature gradient along the body surface pushes the gas from the region of lower temperature to the region of higher temperature (thermal creep). If the body is not moving, the gas produces the so called radiometric force on the body, if the body is not restrained, it moves from the region of higher temperature to the region of lower temperature. The transport of particles in the quiescent gas with temperature gradient is called the *thermophoresis* (phoresis - being carried in Greek). The phenomenon of thermophoresis has wide applications in aerosol research and industry. For example, the thermal dust remover is very effective for the removal of particles under $1\mu m$ in diameter, the sample collector based on the thermaophoresis principle is extensively used for collecting small particles in the aerosol, the thermophoresis causes the contamination on the heat exchanger in oil refining industry, etc. Here the thermophoretic force on a spherical particle in the gas with temperature gradient is given. Epstein [32] was the first to calculate this thermophoretic force, but for a defect in the formulation of the boundary condition in the slip regime the result is erroneous. Brock [33] gave the correct result. The derivation given here follows the analysis in reference [34], this is a simple and detailed derivation.

Consider a spherical particle of radius a in the quiescent gas with temperature gradient (see Fig. 5.4). Suppose the gas temperature is $T_g = T_0(1+\tau_g)$ and the particle temperature is $T_p = T_0(1+\tau_p)$. τ_g and τ_p satisfy the Laplace equation, the condition of heat flux continuity and the temperature jump condition (see Eq. (5.55)'), τ_g satisfies also the condition of constancy of the temperature gradient ∇T at infinity

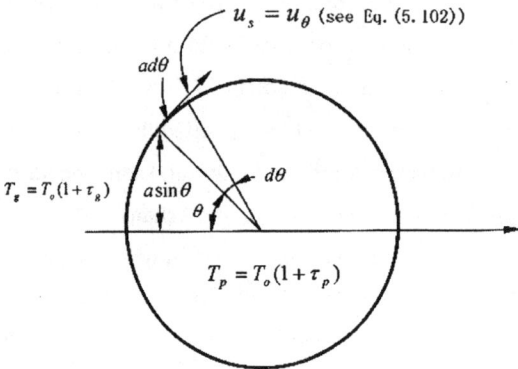

Fig. 5.4 A spherical particle in the field of temperature gradient

$$\nabla^2 \tau_g = 0, \tag{5.89}$$

$$\nabla^2 \tau_p = 0, \tag{5.90}$$

$$K_g \left(\partial \tau_g / \partial r \right)_{r=a} = K_p \left(\partial \tau_p / \partial r \right)_{r=a}, \tag{5.91}$$

$$\tau_g - \tau_p = C_t \lambda \left(\partial \tau_g / \partial r \right), \quad \text{at} \quad r = a, \tag{5.92}$$

$$\tau_g \to (G/a) r \cos\theta, \quad \text{at} \quad r \to \infty. \tag{5.93}$$

where $G = |\nabla T| a / T_0$ is the dimensionless temperature gradient; C_t is the temperature jump coefficient; K_g and K_p is the conductivities of the gas and the particle, respectively.

Write Eq. (5.89) in the spherical coordinate system

$$\frac{1}{r^2} \frac{\partial}{\partial r} \left(r^2 \frac{\partial \tau_g}{\partial r} \right) + \frac{1}{r^2 \sin\theta} \frac{\partial}{\partial \theta} \left(\sin\theta \frac{\partial \tau_g}{\partial \theta} \right) = 0. \tag{5.94}$$

By employing the method of variable separation, letting $\tau_g = R(r)\cos\theta$, from Eq. (5.94) an ordinary differential equation for $R(r)$ is obtained

$$r^2 R'' + 2rR' - 2R = 0.$$

Substituting the solution of this Euler equation into $\tau_g = R(r)\cos\theta$, one obtains

$$\tau_g = \left(a_1 r + a_2 r^{-2}\right)\cos\theta. \tag{5.95}$$

The solution of τ_p has the similar form but as it should be finite at $r=0$, so one has

$$\tau_p = b_1 r \cos\theta. \tag{5.96}$$

The three undetermined constants a_1, a_2, b_1 are readily determined from the boundary conditions Eq. (5.91), Eq. (5.92) and Eq. (5.93), and the temperature τ_g of the gas and the temperature τ_p of the particle are obtained

$$\tau_g = \left(1 + \frac{k + C_t\lambda/a - 1}{1 + 2k + 2C_t\lambda/a}\frac{a^3}{r^3}\right)\frac{r}{a} G\cos\theta, \tag{5.97}$$

$$\tau_p = \frac{3k(r/a)G\cos\theta}{1 + 2k + 2C_t\lambda/a}, \tag{5.98}$$

where $k = K_g/K_p$.

The velocity field satisfies the Navier-Stokes equation (5.14) simplified according to the steady and low speed conditions under consideration, i.e., the Stokes equation and the continuity equation

$$\nabla^2 \boldsymbol{u} = \frac{1}{\mu}\nabla p, \tag{5.99}$$

$$\nabla \cdot \boldsymbol{u} = 0. \tag{5.100}$$

The conditions that the normal component of the velocity at the wall is zero and the velocity at infinity is zero are

224   5 CONTINUUM MODELS

$$u_r\big|_{r=a} = 0; u_r \to 0, u_\theta \to 0, \quad \text{at} \quad r \to \infty \tag{5.101}$$

And the velocity slip and the thermal creep condition at the wall (see Eq. (5.45)') is

$$u_\theta = C_m \lambda \left[r \frac{\partial}{\partial r}\left(\frac{u_\theta}{r}\right) + \frac{1}{r}\frac{\partial u_r}{\partial \theta} \right] + \frac{C_s v}{a T_\infty}\left(\frac{\partial T}{\partial \theta}\right)_{r=a}. \tag{5.102}$$

The first term of Eq. (5.102), the velocity slip, is written in such a form, because the calculation of the momentum conservation Eq. (5.40) should be carried out in the spherical coordinates. Now in the last term of the under-integral expression in place of $(\partial u/\partial y)\mu v'$ there should stand the product $2e_{r\theta} u_r' u_\theta'$ of the strain tensor $\partial c_{0i}/\partial x_j$ and the dyad of the thermal velocities $u_i' u_j'$. The expression $e_{r\theta}$ is the r, θ component of the strain tensor in the spherical coordinates, that is

$$\frac{r}{2}\frac{\partial}{\partial r}\left(\frac{u_\theta}{r}\right) + \frac{1}{2r}\frac{\partial u_r}{\partial \theta}.$$

After the integration and calculation, the result gives that in Eq. (5.45) in place of $\partial u_0/\partial y$ there is $r\partial(u_\theta/r) + (\partial u_r/\partial \theta)/r$, i.e., the Eq. (5.102) holds.

Write Eq. (5.99) and Eq. (5.100) in the spherical coordinates

$$\begin{cases} \dfrac{\partial^2 u_r}{\partial r^2} + \dfrac{1}{r^2}\dfrac{\partial^2 u_r}{\partial \theta^2} + \dfrac{2}{r}\dfrac{\partial u_r}{\partial r} + \dfrac{\operatorname{ctg}\theta}{r^2}\dfrac{\partial u_r}{\partial \theta} - \dfrac{2}{r^2}\dfrac{\partial u_\theta}{\partial \theta} - \dfrac{2u_r}{r^2} - \dfrac{2\operatorname{ctg}\theta}{r^2}u_\theta = \dfrac{1}{\mu}\dfrac{\partial p}{\partial r}, \\[4pt] \dfrac{\partial^2 u_\theta}{\partial r^2} + \dfrac{1}{r^2}\dfrac{\partial^2 u_\theta}{\partial \theta^2} + \dfrac{2}{r}\dfrac{\partial u_\theta}{\partial r} + \dfrac{\operatorname{ctg}\theta}{r^2}\dfrac{\partial u_\theta}{\partial \theta} + \dfrac{2}{r^2}\dfrac{\partial u_r}{\partial \theta} - \dfrac{u_\theta}{r^2 \sin^2 \theta} = \dfrac{1}{\mu}\dfrac{\partial p}{\partial \theta}, \\[4pt] \dfrac{\partial u_r}{\partial r} + \dfrac{1}{r}\dfrac{\partial u_\theta}{\partial \theta} + \dfrac{2u_r}{r} + \dfrac{u_\theta \operatorname{ctg}\theta}{r} = 0. \end{cases} \tag{5.103}$$

Write the solution of the momentum equation in the following form

$$\begin{cases} u_r(r,\theta) = f(r)\cos\theta, \\ u_\theta(r,\theta) = -g(r)\sin\theta, \\ p(r,\theta) = \mu h(r)\cos\theta. \end{cases} \tag{5.104}$$

Substituting Eq. (5.104) into Eq. (5.103), the equations can be written as

5.5 THERMAL CREEP AND THERMOPHORESIS

$$g = \frac{1}{2}f'r + f, \tag{5.105}$$

$$h = \frac{1}{2}f''r^2 + 3rf' + 2f, \tag{5.106}$$

$$r^3 f^{iv} + 8r^2 f''' + 8rf'' - 8f' = 0. \tag{5.107}$$

The general solution of the Euler equation (5.107) is

$$f = \frac{A}{r^3} + \frac{B}{r} + C + Dr^2.$$

The boundary conditions Eq. (5.101) give $C = D = 0$ and $B = -A/a^2$, consequently

$$f = A'\left(\frac{a^3}{r^3} - \frac{a}{r}\right),$$

where $A' = A/a^3$. The velocity components thus obtained have the following expressions (through Eq. (5.104) and Eq. (5.105))

$$u_r = A'\left(\frac{a^3}{r^3} - \frac{a}{r}\right)\cos\theta, \tag{5.108}$$

$$u_\theta = \frac{A'}{2}\left(\frac{a^3}{r^3} + \frac{a}{r}\right)\sin\theta. \tag{5.109}$$

The solution of the pressure p is obtained with the help of Eq. (5.106)

$$p = -A'\mu \frac{a}{r^2}\cos\theta. \tag{5.110}$$

The value of the constant A' is determined from the velocity slip and thermal creep condition Eq. (5.102), in particular when searching the temperature gradient the solution of the temperature field Eq. (5.97) is employed

$$A' = -(3C_s \mu\, G/\rho a)\frac{k + C_t \lambda/a}{(1 + 3C_m \lambda/a)(1 + 2k + 2C_t \lambda/a)}. \tag{5.111}$$

To find the total force on the particle, substituting Eq. (5.108), Eq. (5.109) and Eq. (5.110) into the expressions of the normal stress and tangential stress on the sphere surface

$$P_{rr} = -p + 2\mu \frac{\partial u_r}{\partial r},$$

$$P_{r\theta} = \mu \left(\frac{1}{r}\frac{\partial u_r}{\partial \theta} + \frac{\partial u_\theta}{\partial r} - \frac{u_\theta}{r} \right),$$

on the spherical surface $r = a$ one has

$$P_{rr}\big|_{r=a} = -3\frac{\mu A'}{a}\cos\theta,$$

$$P_{r\theta}\big|_{r=a} = -3\frac{\mu A'}{a}\sin\theta.$$

Projecting the stress acting on the ring (with an area $2\pi a^2 \sin\theta d\theta$), whose normal has an angle θ with the incident flow (the axis) and a width of $ad\theta$, on the axial direction and integrating over the whole spherical surface, one has

$$W = \int_0^\pi (P_{rr}\cos\theta - P_{r\theta}\sin\theta) 2\pi a^2 \sin\theta d\theta =$$

$$2\pi a^2 \int_0^\pi \left(-3\frac{\mu A'}{a}\cos^2\theta + \frac{3\mu A'}{a}\sin^2\theta \right) \sin\theta d\theta = 4\pi a\mu A'.$$

According to Eq. (5.111) the thermophoretic force acting on the sphere is finally found

$$W = -12\pi\mu\nu GC_s \frac{k + C_t\lambda/a}{(1 + C_m\lambda/a)(1 + 2k + 2C_t\lambda/a)}. \tag{5.112}$$

This expression of the thermophoretic force acting on the spherical particle owing to its derivation is only suitable for slip flow regime. But Talbot et al. [35] found, when appropriate velocity slip coefficient C_m, thermal creep coefficient C_s and temperature jump coefficient C_t are chosen, it agrees quite well with the experimental data obtained early and in [35] in the whole transitional regime and

5.6 SECOND ORDER SLIP-JUMP CONDITIONS

can serve as the semi-empirical formula bridging the continuum and free molecular flow.

When the thermophoretic force and the thermophoretic velocity of the particles in the temperature field are known, the problem of deposition of particles carried by the gas flow around cold body onto the surface can be solved. In the papers [36] and [37] Shen discussed the general method of calculation of the *thermophoretic deposition* of particles in the external flow around cold surfaces. The effect of density variation (despite the small velocity) is explicitly shown in solving the external flows, some solutions of typical two-dimensional and axi-symmetric thermophoretic deposition problems of particles are given.

5.6 SECOND ORDER SLIP-JUMP CONDITIONS

In section 5.3.1 the first order slip velocity expression Eq. (5.31) was obtained basing on a simple physical derivation. Near the body surface half of the gas molecules was assumed to come from the external flow and the other half reflected from the body surface. The average velocity from the external flow arriving at the body surface was assumed to be the mean magnitude of the velocities at a distance $\varsigma\lambda$ from the surface

$$u_s + \varsigma\lambda(\frac{\partial u}{\partial y})_0.$$

This expression would be exact if the velocity were linearly distributed along the normal direction to the body surface. In general case of the velocity variation, more terms of the Taylor series expansion of the velocity about u_s can be involved. If retain only terms up to second order, the average velocity at a distance $\varsigma\lambda$ from the surface would be

$$u_s + \varsigma\lambda(\frac{\partial u}{\partial y})_0 + \frac{\varsigma^2\lambda^2}{2}(\frac{\partial^2 u}{\partial y^2})_0. \tag{5.113}$$

Repeating the derivation in section 5.3.1 instead of Eq. (5.31) we arrive at

$$u_s = \frac{2-\sigma}{\sigma}[\varsigma\lambda(\frac{\partial u}{\partial y})_0 + \frac{\varsigma^2\lambda^2}{2}(\frac{\partial^2 u}{\partial y^2})_0] \tag{5.114}$$

Referring to Eq. (4.45)', Eq. (5.114) can be written in a more general form (with $\sigma = 1$)

$$u_{y=0} - u_w = C_{m1}\lambda(\frac{\partial u}{\partial y})_0 + C_{m2}\lambda^2(\frac{\partial^2 u}{\partial y^2})_0, \tag{5.115}$$

where C_{m1} and C_{m2} are the first and second order velocity slip coefficients. The second order slip conditions are obtained by Schamberg [38] ($C_{m1} = 1.0$, $C_{m2} = 1.309$) Cercignani [39] ($C_{m1} = 1.1466$, $C_{m2} = 0.8247$) Deissler [40] ($C_{m1} = 1.0, C_{m2} = 1.125$) and Hsia and Domoto [41] ($C_{m1} = 1.0, C_{m2} = 0.5$) basing on different derivations and determined different values of C_{m1} and C_{m2}. With appropriate slip coefficients the second order slip conditions together with Navier-Stokes equations can extend the applicability region of slip flow method into the range of higher Kn numbers (e.g. see Sreekanth [42] in solving the Poiseulle problem).

REFERENCES

1. Tsien HS (1946) Superaerodynamics, mechanics of rarefied gases. J Aeronautical Sci, 13: 653-664
2. Grad H (1949) On the kinetic theory of rarefied gases. Communication On Pure and Applied Math, 2: 331-407
3. Chapman S and Cowling TG (1970) The Mathematical Theory of Non-uniform Gases. 3rd Edition, Cambridge Univ. Press
4. Vincenti WG and Kruger GH. (1965) Introduction to Physical Gas Dynamics. John Wiley and Sons, New York
5. Burnett D (1935) The distribution of molecular velocities and the mean motion in a non-uniform gas. Proc London Math Soc, 40: 382
6. Wang Chang CS and Uhlenbeck GE (1970) On the transport phenomena in rarefied gases. Univ. of Michigan; (1948) Studies in Statistical Mechanics, vol 5, North Holland Publ Co,

7. Schaaf S and Chambre PL (1958) Flow of Rarefied Gas, Part H of Fundamentals of Gas Dynamics. Princeton Univ. Press
8. Fisco KA and Chapman DR (1989) Comparison of Burnett, super-Burnett and Monte Carlo solutions for hypersonic shock structure. Progress in Astro and Aeronautics, 118: 374-395
9. Alofs DJ and Springer GS (1971) Cylindrical Couette flow experiments in transition regime. Phys Of Fluids, 14: 298-305
10. Lin TC and Street RE (1953) Effect of variable viscosity and thermal conductivity on high-speed slip flow between concentric cylinders. NACA TN 2895
11. Zhong X, MacCormack RW and Chapman DR. (1991) Stabilization of the Burnett equations and application to high-altitude hypersonic flows. AIAA paper, 91-0770
12. Zhong X (1993) On numerical solutions of Burnett equations for hypersonic flow past 2-D circular blunt leading edges in continuum transition regime. AIAA paper, 93-3092
13. Zhong X and Furumoto G (1994) Solutions of Burnett equations for axi-symmetric hypersonic flow past spherical blunt bodies. AIAA paper, 94-1959
14. Deng ZT and Liaw GS (1994) Numerical investigations of low-density nozzle flow fields by solving N-S and Burnett equations. AIAA paper 94-2055
15. Deng ZT and Liaw GS (1998) Computation of hypersonic transitional flows by solving 3-D Burnett equations. AIAA paper 98-1607
16. Galkin VS, Kogan MN and Fridlender OG (1971) Some kinetic effects in continuum gas flows. Izv AN SSSR, Mekh Zhidk i gaza, No 3, 13
17. Alexandrov V, (1997) et al. Thermal stress effects and its experimental detection. In: C Shen edited Rarefied Gas Dynamic, Peking Univ. Press, 79-84
18. Sone Y. (1969, 1971) Asymptotic theory of flow of rarefied gas over a smooth boundary I, II, In: L Trilling and HY Wachman edited Rarefied Gas Dynamics, AP, 1: 243-253, and D Dini edited Rarefied Gas Dynamics, ETS, 737-49
19. Sone Y (1972) Flow induced by thermal stress in rarefied gas. Phys of Fluids, 15: 1418-1424
20. Vestner H. (1976) Theory of the viscomagnetic heat flux. Z. Naturforsch. 31a, 540-552
21. Hermans L J F et al. (1979) The use of a magnetic field in an experimental verification of transport theory for rarefied gas. In R Campargue edited Rarefied Gas Dynamics, 2:799-806

22. Sone Y (1997) Continuum gas dynamics in the light of kinetic theory and new features of rarefied gas flows. In: C Shen edited Rarefied Gas Dynamics-20, Peking Univ. Press, 2-24
23. Maxwell J C (1878) Phil Trans, R Soc 1, Appendix
24. Kennard EH (1938) Kinetic Theory of Gases, McGraw-hill
25. Patterson GN (1956) Molecular Flow of Gases. John Wiley and Sons
26. Shidlovsky VP (1967) Introduction of Dynamics of Rarefied Gases, Elsevier
27. Gupta RN, Scott CD, Moss JN (1985) Slip-boundary equations for multi-component non-equilibrium airflow. NASA Technical Report 2452
28. Tang JR and Tao B (1995) Wall slip-boundary equations with catalysis for multi-component chemical non-equilibrium gas flows. In: J Harvey and G Lord edited Rarefied Gas Dynamics-19, 2:1096-1099
29. Zhao JY, Shen C, Tang JR (1999) Effect of the wall injection on the slip boundary conditions. Acta Mechanica Sinica 31: 268 (in Chinese)
30. Shen C, (1983) The concentration-jump coefficient in a rarefied binary gas mixture. J Fluid Mechanics, 137: 221
31. Shen C (1962) Rarefied Gas Dynamics, Dept of Modern Mechanics. Chinese Univ. of Science and Technology (in Chinese)
32. Epstein PS (1929) Zeits. Physik, 54: 537
33. Brock JR (1962) On the theory of thermal forces acting on aerosol particles. J Colloid Sci, 17: 768-780
34. Shen C (1983) Thermophoresis of spherical particle and the dependence of the jump coefficients on the accommodation coefficient. Acta Mechanica Sinica, 15: 7 (in Chinese)
35. Tallbot L, Cheng RK, Schefer RW and Willis DR (1980) Thermophoresis of particles in a heated boundary layer. J Fluid Mech., 101, part 4: 737-758
36. Shen C (1989) Thermophoretic deposition of particles onto cold surface of bodies in 2-D and axisymmetric flows. Journal of Colloid and Interface Science, 127 (1): 104-105
37. Shen C (1989) Thermophoretic deposition of small particles suspended in gas flow onto cold surfaces. In High-Temperature Dust Laden Jets, ed. by Solonenko and Fedorcheko, VSP, 349-369

6 TRANSITIONAL REGIME

6.1 GENERAL OVERVIEW

In section 0.4 the transitional regime is defined as the scope of Kn number (λ/L) between 0.1 and 10 (see Eq. (0.8)), i.e., the molecular mean free path λ is not too large, nor too small in comparison with the typical flow size L. In such case the collisions between molecules and the collisions of molecules with the surface must be taken into account simultaneously, both the comparatively simple free molecular flow theory and the mature continuum method are not appropriate, one has to solve the Boltzmann equation or to evoke equivalent to it physic-mathematical handling. To solve the integral-differential Boltzmann equation with the collision term and an unknown function having as many as seven arguments is so difficult that many researchers envisaged various methods to solve the problems in transitional regime, some of these methods seemed to carry the implication of roundabout tactics, but could lessen the difficulty and make the problems readily tackled.

The methods of solution of the transitional regime problems can be distinguished into two categories: *the analytic and numerical methods*. The analytic methods without exception start from the Boltzmann equation, but owing to the complexity of this equation they often adopt the small disturbance or linearization assumption (*the linearized Boltzmann equation method*), or make some assumption relative to the form of the distribution function (for example *the moment method*), or make some simplification even of the collision term itself (*the model equation method*), these constitute the contents of the sections 6.2~6.4. These methods of solution despite their belonging to the analytical category seldom attain the analytical solutions, to obtain the final results certain numerical calculations are required. It is noted that except the solutions of the linearized Boltzman equation in the scope of small disturbances these solutions can hardly be called exact solutions. This is especially true for the model equation method, this is a modification of the Boltzmann equation proper and replacement of it by a simplified equation, showing how difficult is the direct solution of the Boltzmann equation. The moment method owing to the arbitrariness and un-uniqueness of the form of the distribution function does not possess either the merit of exact solution.

Two kinds of the numerical methods can be distinguished: the direct numerical solution of the Boltzmann equation and the direct simulation of the flow physics itself, even though the line of demarcation between them becomes not so absolute because the direct simulation method can be proved to be consistent with the Boltzmann equation. Among the methods of direct numerical solution of the Boltzmann equation there is the method (*finite difference method*) which uses specific algorithm (the Monte Carlo method of quadrature) to calculate the collision integral and uses the mature finite different method of the computational fluid dynamics (CFD) to solve the Boltzmann equation, the method (*discrete ordinate* or *discrete velocity method*) which assumes that the velocity space only has finite number of discrete values thus significantly simplifying the calculation of the collision integral, and the method (*the integral method*) that writes the Boltzman equation in the integral from and solves it, they are introduced in sections 6.5~6.7. These methods attempt to base the formulation on the exact Boltzmann equation and have the property of exact solution in the limits of errors (of course, provided the calculation method is correct).

With the development of the electronic computers a kind of methods staring directly from the simulation of physics of the flows appears, including the method of deterministic simulation —*the molecular dynamics (MC) method* and the probability simulation method which can be distinguished further into the *test particle Monte-Carlo method* and the *direct simulation Monte-Carlo (DSMC) method*. The direct simulation Monte-Carlo method deserves separate discussion owing to its success in solving rarefied gas flow problems, especially the non-equilibrium flow problems in the transitional regime. The DSMC method possesses the same physical basis and assumptions (the molecular chaos and dilute gas assumption). But DSMC method is different in some aspects from the Boltzmann equation. The latter depends on the assumption of the reverse collisions so could not treat the three body collision problem, but the DSMC method can be applied to complex problems such as with recombination chemical reactions where three-body collisions are involved. In treating the problems of molecular models and molecule-surface interaction when introducing physical models the DSMC method because of its nature of physical simulation can naturally and easily introduce more complicated and more close to reality models. But it is quite difficult to introduce the real and complex models into the mathematical formulation of the Boltzmann equation. For the gas flow cases accompanied by the chemical reactions and radiation the mathematical formulation in the framework of the Boltzmann equation is difficult to accomplish for practical application, but it is easy for the DSMC method to implement the simulation of these flows. Even for flow cases without such processes as chemical reactions and radiation to solve directly the Boltzmann equation is rather difficult and sometimes is limited to simple geometry and sometimes to low speed. Thus, the deep rooted view point, i.e., the view point that a simulation process can be accepted only if it is strictly derived from the Boltzmann equation, seems to be antiquated and non-practical. The DSMC method has been verified experimentally both in the aspect of global flow characteristics and in the aspect of micro-level characteristics such as the velocity distribution function. It is safe to say that the DSMC method is a method that solves successfully various practical flow problems in the transitional flow regime, its meaning and role has been recognized by the scientific community. Owing to the important meaning of the DSMC method in the transitional regime Chapter 7 will be devoted specially to the discussion of this method. And this Chapter will give an overall outline of various methods in the transitional regime including a section 6.8 giving a parallel overview of the direct simulation method.

6.2 LINEARIZED BOLTZMANN EQUATION

It has been stated above that due to the complicated non-linearity of the collision integral of the Boltzmann equation the direct solution of it is very difficult. The existence of the solution of the Boltzmann equation under equilibrium of the gas, the Maxwellian distribution (see section 2.10), allows us to suppose to find the solution of the linearized problem near this solution, this limits the solution to low speed flow problems, i.e., the small disturbance solution about the quiescent gas. For the slowly moving gas flow slightly deviated from the stationary state the difference of its distribution function $f(r,c,t)$ from the Maxwellian distribution is small, and it can be written in the following form

$$f(r,c,t) = f_0\left[1 + \varphi(r,c,t)\right], \tag{6.1}$$

where

$$f_0 = n_0\left(\beta/\sqrt{\pi}\right)^3 \exp(-C^2), \tag{6.2}$$

$$C = c\beta \equiv \frac{c}{\sqrt{2\dfrac{k}{m}T_0}}. \tag{6.3}$$

Substituting Eq. (6.1) into the Boltzmann equation (Eq. (2.152) in section 5.2) and retaining only the first order of φ yields

$$\frac{\partial \varphi}{\partial t} + c_i \frac{\partial \varphi}{\partial x_i} = J(\varphi), \tag{6.4}$$

where $J(\varphi)$ is a collision operator linear relative to φ which can be written as

$$J(\varphi) = \int f_{01}\left(\varphi^* + \varphi_1^* - \varphi - \varphi_1\right) c b \, db \, d\varepsilon \, dc_1. \tag{6.5}$$

In obtaining Eq. (6.4) the fact, that $f_0^* f_{01}^* = f_0 f_{01}$ and $f(c')$ can be taken out of the integration, has been taken into account.

The linearized Boltzmann equation (6.4) is still a complicated integral-differential equation, but as it is linear, in comparison with the Boltzmann equation it is much simpler. Many authors discussed in detail this equation, see Cercignani [1, 2]. These small disturbance solutions provide with important data having value of reference when they are obtained without the assistance of arbitrary assumptions about the relevant parameters or the modification of the equation, they can serve as merits for other numerical methods and have practical application value in the aspects of gas flow problems in MEMS. Even though the equation is linearized, the available solutions are limited mainly to certain class of relative simple boundary conditions. Sometimes some works are discussing not the linearized Boltzmann equation but the linearized model equation, one should be careful when applying such results.

In the following the equation (6.4) will be written for the case when the space variation occurs only in the y direction and the gas molecules are hard spheres, and the formulation of the boundary conditions for the linearized Bolzmann equation will be discussed for the plane boundary case [3, 4]. When y is the only space coordinate by using the dimensionless molecular velocity C (see Eq. (6.3)) Eq. (6.4) can be written

$$\beta\frac{\partial\varphi}{\partial t}+Cy\frac{\partial\varphi}{\partial y}=J(\varphi). \qquad (6.6)$$

For hard sphere molecule it is easy to obtain

$$J(\varphi)=\frac{1}{\sqrt{2\pi}\,\mathcal{T}^2\lambda}\int_0^{2\pi}d\varepsilon\int_0^{\pi}\sin\theta d\theta\int_{-\infty}^{\infty}Ge^{-C_1^2}\{\varphi^*+\varphi_1^*-\varphi-\varphi_1\}dC_1, \qquad (6.7)$$

where $G=\beta c_r \equiv c_r/\sqrt{2kT/m}$, the definitions of ε, θ, c_r see Chapter 2, $\varphi^*=\varphi(C^*), \varphi_1^*=\varphi(C_1^*)$; C^*, C_1^* are the velocities after collision of molecules whose velocities before collision are C, C_1. The relations between C^*, C_1^* and C, C_1 can be found in Chapter 2 Eq. (2.59)*, Eq. (2.59)**, for example in the direction of x

$$\begin{cases} C_x^* = C_x + (C_{1x}-C_x)\cos^2\frac{\chi}{2} + \\ \qquad \frac{1}{2}\Big[G^2-(C_{1x}-C_x)^2\Big]^{1/2}\sin\chi\cos\varepsilon, \\ C_{1x}^* = C_{1x} - (C_{1x}-C_x)\cos^2\frac{\chi}{2} - \\ \qquad \frac{1}{2}\Big[G^2-(C_{1x}-C_x)^2\Big]^{1/2}\sin\chi\cos\varepsilon. \end{cases} \qquad (6.8)$$

We discuss the case of the Maxwellian type boundary conditions, i.e., after incidence onto the surface σ portion of the molecules reflects diffusely and the other $(1-\sigma)$ portion reflects specularly. Obviously, when $\sigma \neq 0$, the incident molecules and the reflected molecules have different distribution functions. For the case of a plane surface perpendicular to the y axis the distribution function is non-continuous with respect to the normal to the surface velocity component C_y. Then two distribution functions which are continuous by themselves are to be introduced, they are defined each only in the half velocity space [3, 4]

$$\begin{cases} f = f^+(y, \mathbf{C}, t), C_y > 0, \\ f = f^-(y, \mathbf{C}, t), C_y < 0. \end{cases} \quad (6.9)$$

Correspondingly, different disturbance distribution functions φ^+, φ^- are introduced

$$\begin{cases} f^+ = f_0\left[1 + \varphi^+(y, \mathbf{C}, t)\right], C_y > 0, \\ f^- = f_0\left[1 + \varphi^-(y, \mathbf{C}, t)\right], C_y < 0. \end{cases} \quad (6.10)$$

For the case when there is a velocity u_0 of the plane boundary, at plane $y = y_0$ the boundary condition is (denote $S_0 = \beta u_0 \equiv u_0/\sqrt{2kT/m}$)

$$f^\pm(y_0, \mathbf{C}, t) = (1-\sigma) f^\mp(y_0, C_x, -C_y, C_z, t) +$$

$$\sigma n_0 \left(\beta/\sqrt{\pi}\right)^3 \exp\left\{-\left[(C_x - S_0)^2 + C_y^2 + C_z^2\right]\right\}$$

Neglecting the second order of S_0, it can be written as

$$f^\pm(y_0, \mathbf{C}, t) = (1-\sigma) f^\mp(y_0, C_x, -C_y, C_z, t) +$$

$$\sigma n_0 \left(\beta/\sqrt{\pi}\right)^3 \exp(-C^2)[1 + 2C_x S_0]. \quad (6.11)$$

Substituting Eq. (6.10) into Eq. (6.11) and taking into account Eq. (6.2), one obtains

$$\varphi^\pm(y_0, C_x, C_y, C_z, t) = (1-\sigma)\varphi^\mp(y_0, C_x, -C_y, C_z, t) + 2\sigma C_x S_0. \quad (6.12)$$

6.2 LINEARIZED BOLTZMANN EQUATION

The basic equation (6.6) and boundary condition Eq. (6.12) are applicable for cases when the flow varies only in the direction normal to the plane boundary and the boundary conditions are of the Maxwellian type.

After finding φ the physical macroscopic quantities interesting us are easily obtained, e.g., the macroscopic velocity c_0 and the stress P_{ij}

$$c_{0i} = \frac{1}{n_0}\int c_i f_0 (1+\varphi) dc = \beta^{-1}/\pi^{3/2} \int C_i \varphi e^{-C^2} dC \qquad (6.13)$$

$$P_{ij} = m\int (c_i - c_{0i})(c_j - c_{0j}) f_0 (1+\varphi) dC = \frac{2n_0 kT}{\pi^{3/2}} \int C_i C_j \varphi e^{-C^2} dC \qquad (6.14)$$

With the help of the linearized Boltzmann equation Gross and Ziering [3] solved the problem of stress flow between two plates, i.e., the Couette flow problem, Gross and Jackson [4] solved the problem of unsteady flow caused by the instant motion of the plane plate, i.e., the Rayleigh problem. They suggested to express $\varphi^\pm$ as

$$\varphi^\pm = a_0^\pm (y,t) C_x + a_1^\pm (y,t) C_x C_y . \qquad (6.15)$$

Such form of the dependence on the velocity originated from Chapman-Enskog expansion, but as the discontinuous φ relative to C_y is introduced, so $a_0^+ \neq a_0^-, a_1^+ \neq a_1^-$. Substituting the function φ defined in the whole velocity space

$$\varphi = \varphi^+ \left(\frac{1+ signC_y}{2}\right) + \varphi^- \left(\frac{1- signC_y}{2}\right), \qquad (6.16)$$

where

$$signC_y = 1, C_y > 0, signC_y = -1, C_y < 0, \qquad (6.17)$$

into the basic equation (6.6), and multiplying both sides of the equation on C_x and $C_x C_y$ and searching the moments separately in the upper half velocity space and the lower half velocity space, the simultaneous equations relative to $a_0^+, a_0^-, a_1^+, a_1^-$ are obtained, the right hand sides are the integral terms including the collision term Eq. (6.7). For the hard sphere molecule the integral terms can be evaluated. The boundary conditions after substitution of Eq. (6.15) can be transformed into the boundary conditions relative to $a_0^+, a_0^-, a_1^+, a_1^-$. So the whole problem can be solved. In searching the solution of the linearized Boltzmann equation the moment method (see section 6.3) has been employed. These were the early attempts of the solution of the transitional regime.

Grad [5] and Cercignani [6] investigated the problem of further transformation of the linearized collision operator $J(\varphi)$, the integration for ε is expressed through zero order Bessel function of the first kind I_0 (Eq. (3.43)), and for the hard sphere model the integration relative to θ can also be accomplished. Thus the linearized Boltzmann equation in the one-dimensional steady case is written as

$$C_y \frac{\partial \varphi}{\partial y} = \frac{1}{(\sqrt{\pi}/2)\lambda}\left[L_1(\varphi) - L_2(\varphi) - v(C)\varphi\right], \tag{6.18}$$

where

$$L_1(\varphi) = \frac{1}{\sqrt{2\pi}} \int \frac{1}{|C-\xi|} \exp(-\xi^2) + \frac{|C\times\xi|^2}{|C-\xi|^2} \varphi(y,\xi) d\xi, \tag{6.19}$$

$$L_2(\varphi) = \frac{1}{2\sqrt{2\pi}} \int |C-\xi| \exp(-\xi^2) \varphi(y,\xi) d\xi, \tag{6.20}$$

$$v(C) = \frac{1}{2\sqrt{2}}\left[\exp(-C^2) + \left(2C + \frac{1}{C}\right)\int_0^C \exp(-t^2) dt\right]. \tag{6.21}$$

In the above expressions ξ is the integration variable corresponding to C, $d\xi = d\xi_1 d\xi_2 d\xi_3$, $C\times\xi$ is the vector production of C and ξ, the integration domain of L_1 and L_2 is the whole space of the molecular velocity C (or ξ).

Sone, Owwada and Aoki et al. developed efficient numerical methods of employing the above linearized Boltzmann equations (6.18)~(6.21) in solving some half space boundary problems – the temperature jump and Knudsen layer problem [7], the problem of evaporation and condensation [8], the problem of shear and thermal creep [9] and some problems of flow between two plates – the Poiseuille flow and thermal transpiration problem [10] and the Couette flow problem [11]. The solutions of these steady boundary value problems are obtained as the stabilized solutions of the initial and boundary-value problems of the unsteady equation obtained from Eq. (6.18) by adding the term $\partial \varphi / \partial t$ to its left hand side. The unsteady problems are solved by the standard implicit finite difference scheme. The key question is the calculation of the collision term. The distribution function (or the reduced distribution function) is expanded by a set of the basis functions in analogue with the basis functions in the finite element method. Thus, the collision integral can be written as the matrix product of the collision integral kernel and the values of distribution function on lattice points. The collision integral kernel is the collision integral of the basis functions at the lattice points and thus can be evaluated universally. As the computation of the collision integral is most time-consuming, the method of computation of the collision integrals of a current problem by using the prepared universal collision integral kernels demonstrates the high efficiency advantage of the method.

6.3 THE MOMENT METHOD

The practice of the so called moment method is not to solve the Boltzmann equation itself but to solve its moment equation. The latter is obtained by multiplying the Boltzmann equation by some molecular quantity Q and integrating over the entire velocity space (i.e., the so called finding moment). For arbitrary form of Q we have obtained the general form of the moment equation (section 2.9, Eq. (2.183))

$$\frac{\partial}{\partial t}\left(n\bar{Q}\right) + \nabla \cdot \left(n\overline{cQ}\right) - nF \cdot \overline{\frac{\partial Q}{\partial c}} = \Delta[Q]. \qquad (6.22)$$

The macroscopic characteristics that interest us are the average values of some microscopic molecular quantities, i.e., the moments of the distribution functions (see the expressions of the macroscopic properties in section 2.2). For example, the density is the zero order moment, the velocity is the first order moment, P_{ij} is the second order moment, p is the contraction of the second moment, Eq. (2.25), q_i is the contraction of the third moment, Eq. (2.27)), etc. Substitution of different quantities Q into Eq. (6.22) yields the system of equations for the macroscopic quantities. In particular, when Q are the collision invariants m, mc and $(1/2)mc^2$, we have $\Delta[Q]=0$, and obtain the conservation equations of the macroscopic quantities. When Q is certain power of c, as there is the term $\overline{cQ}$ on the left hand side of Eq. (6.22), this causes the appearance of a moment of order one order higher than before. When turning to the moment equation in expectation to obtain equation for higher order moment, this term of higher order leads to even higher order moment, and eventually to infinite equations and moments. The moment method assumes that the distribution function can be expressed as function of certain moments (and the molecular velocity c, the space coordinate x and time t), with specific forms of expression through the moments and c, but the variation with the space coordinates is to be determined. The moment equation thus obtained would have higher order of c, but for some flow cases and with some specific treatment, it can be expressed through moments of lower orders, and one attains closed system of equations characterizing the variation in physical space.

In discussing the basic equations of the continuum media (section 5.2 in Chapter 5) we considered the Chapman-Enskog solution of the Boltzmann equation, this is the expansion of the distribution function into the series of terms proportional to the powers of Kn number

$$f = \sum_{r=0}^{\infty} f^{(r)}.$$

$f^{(0)}$ is the Maxwellian distribution, the corresponding zero order moment equation is the Euler equation (5.2) (with $\tau_{ij}=0, q_i=0$), the unknown functions are ρ, u_0 and T. The first order distribution function of this expansion is the Chapman-Enskog distribution function Eq. (5.8), yielding the expression of stress and heat flux in the ordinary fluid mechanics. Substituting them into the corresponding conservation equations or the moment equations, one obtains the continuum Navier-Stokes equations (5.14), (5.15). The second order distribution function of this expansion yields the expressions of the stress τ_{ij} and heat flux q_i, different from the ordinary continuum media mechanics, and gives the Burnett equations. They have been listed in Chapter 2 as the basic equations of the continuum media, but they are applicable to larger scope of Kn, where the Navier-Stokes equations become invalid. For their nature of the expansion into the positive powers of the Kn number, they could not yield good results when Kn is much larger than 0.1, but they can provide results more close to the exact solutions than the ordinary slip flow solutions. How deep they can allow the continuum equations method to penetrate into the transitional regime and to what extent they can provide more exact solution depends on the method of solution and the comparison with experiments and solutions having the merits of verifying standard. The development of effective solution methods for Burnett equations is still an issue needing continuous efforts. The thirteen moment equations of Grad Eq. (5.2), Eq. (5.29) and Eq. (5.30) have ρ, u_0, T, τ_{ij} and q_i as the basic arguments and possess the character of general basic equations, but application of them to the solution of peculiar problems didn't lead to credible results, the possible reason for this has been discussed in section 5.2 of Chapter 5.

Except those above mentioned methods leading to basic equations possessing general meaning, individual moment methods have been developed for different problems and boundary geometries. Aiming at the physical feathers of different problems the expressions of the distribution functions are proposed and useful solutions have been obtained.

Among this kind of the moment methods the early and the most famous one is certainly the method of *binomial distribution function* developed by Mott-Smith for dealing with the *problem of shock wave structure* [12]. The distribution function f in the shock wave transition zone is expressed as the combination of the upstream and downstream uniform distribution functions

$$f = a_1(x)f_1 + a_2(x)f_2 \equiv N_1(x)F_1(c) + N_2(x)F_2(c), \tag{6.23}$$

where

$$F_1(c) = \frac{f_1}{n_1} = (\beta_1/\sqrt{\pi})^3 \exp\left\{-\beta_1^2\left[(u-u_{01})^2 + v^2 + w^2\right]\right\}, \tag{6.24}$$

$$F_2(c) = \frac{f_2}{n_2} = (\beta_2/\sqrt{\pi})^3 \exp\left\{-\beta_2^2\left[(u-u_{02})^2 + v^2 + w^2\right]\right\}, \tag{6.25}$$

in which f_1, f_2 are the upstream and downstream equilibrium distribution functions (see Eq. (2.196)), n_1, n_2 are the upstream and downstream uniform number densities, u_{01}, u_{02} are the upstream and downstream uniform velocities, $a_1(x) \equiv N_1(x)/n_1$ and $a_2(x) \equiv N_2(x)/n_2$ represent the proportions the two known upstream and downstream distribution functions occupy at different positions of x. Obviously

$$N_1 = n_1, \quad N_2 = 0, \text{ when } x \to -\infty$$
$$N_1 = 0, \quad N_2 = n_2, \text{ when } x \to \infty.$$

The mass, momentum and energy across the shock wave are conserved

$$m\int uf_1 dc = m\int uf_2 dc, \quad \text{i.e. } \rho_1 u_{01} = \rho_2 u_{02} = j, \tag{6.26}$$

$$m\int u^2 f_1 dc = m\int u^2 f_2 dc, \quad \text{i.e. } ju_{01} + kn_1 T_1 = ju_{02} + kn_2 T_2, \tag{6.27}$$

and

$$\frac{m}{2}\int uc^2 f_1 dc = \frac{m}{2}\int uc^2 f_2 dc,$$

i.e.

$$j\left(\frac{u_{01}^2}{2} + \frac{5}{2}\frac{k}{m}T_1\right) = j\left(\frac{u_{02}^2}{2} + \frac{5}{2}\frac{k}{m}T_2\right). \tag{6.28}$$

These are the well known Rankine-Hugonio relations. To find the distribution function, the mass conservation is applied to some upstream point and some point inside the shock wave to yield

$$m\int uf_1 dc = m\int uf dc,$$

i.e.,

$$n_1 u_{01} = n u_0. \tag{6.29}$$

From Eq. (6.23) and the definition of density (see Eq. (2.5)) one has

$$\int f dc = n(x) = N_1(x) + N_2(x). \tag{6.30}$$

And according to Eq. (6.23) and the definition of the macroscopic velocity (see Eq. (2.8))

$$n u_0 \int uf dc = N_1(x) u_{01} + N_2(x) u_{02}. \tag{6.31}$$

Thus from Eq. (6.29) one has

$$u_{01} N_1(x) + u_{02} N_2(x) = n_1 u_{01}. \tag{6.32}$$

To determine three unknown functions $n(x), N_1(x), N_2(x)$, except Eq. (6.30) and Eq. (6.32) one extra moment equation must be employed. Take $Q = u^2$, for the one-dimensional steady problem under discussion, this equation is

$$\frac{d}{dx}\left(\overline{nu^3}\right) = \Delta\left[u^2\right]. \tag{6.33}$$

Discuss at first the right hand side of this equation with the purpose to express it through the unknown functions n, N_1, N_2. The expression $\Delta[u^2]$ is a collision term, this integral has been evaluated for Maxwellian molecules and VHS molecules with $\xi = 1/2$ (see Eq. (2.175) and Eq. (2.175)')

$$\Delta\left[u^2\right] = \frac{p}{m}\frac{\tau_{xx}}{\mu}.$$

For both these molecular models one has $\mu = CT$ (see Eq. (2.93) and Eq. (2.96)'), and with the employment of the expression of their mean free path Eq. (2.217)', the viscosity μ can be written

6 TRANSITIONAL REGIME

$$\mu = \frac{\rho_1 \lambda_1}{2}\left(\frac{2\pi k}{mT_1}\right)^{1/2} T,$$

thus, one has

$$\Delta[u^2] = \frac{(2kT_1/\pi m)^{1/2}}{n_1 \lambda_1} \frac{n}{m} \tau_{xx}. \tag{6.34}$$

According to the definition of τ_{xx} (Eq. (2.24) and Eq. (2.25)), one has

$$\tau_{xx} = mn\left(\frac{\overline{c^2}}{3} - \overline{u'^2}\right) = mn\left(\frac{\overline{u'^2}+\overline{v'^2}+\overline{w'^2}}{3} - \overline{u'^2}\right) = \frac{2}{3}mn\left(\overline{v'^2}-\overline{u'^2}\right). \tag{6.35}$$

When writing the last expression, the equality $\overline{v'^2} = \overline{w'^2}$ has been used, for the problem is symmetric with the x axis. Obviously, $\overline{u'^2}$ can be written

$$\overline{u'^2} = \overline{(u-u_0)^2} = \overline{u'^2} - 2\overline{uu_0} + u_0^2 = \overline{u^2} - u_0^2.$$

According to the definition

$$n\overline{u^2} = \int u^2 f dc = N_1 \int u^2 F_1 dc + N_2 \int u^2 F_2 dc,$$

where the first term is

$$N_1 \int u^2 F_1 dc = N_1 \frac{\beta_1^3}{\pi^{3/2}} \int_{-\infty}^{\infty}\int_{-\infty}^{\infty}\int_{-\infty}^{\infty} [u_{01} + (u-u_{01})]^2 \times$$

$$\exp\left\{-\beta_1^2\left[(u-u_{01})^2 + v^2 + w^2\right]\right\} dudvdw =$$

$$N_1 \frac{\beta_1^3}{\pi^{3/2}}\left[u_{01}^2\left(\frac{\pi^{1/2}}{\beta_1}\right)^3 + 2u_{01}\cdot 0 + \frac{1}{2\beta_1^2}\left(\frac{\pi^{1/2}}{\beta_1}\right)^3\right] =$$

$$N_1\left(u_{01}^2 + \frac{1}{2\beta_1^2}\right) = N_1\left(u_{01}^2 + RT_1\right).$$

Analogously, the second term is

$$N_2 \int u^2 F_2 dc = N_2 \left(u_{02}^2 + RT_2 \right).$$

And $\overline{nv'^2}$ can be written as

$$\overline{nv'^2} = N_1 \int v'^2 F_1 dc + N_2 \int v'^2 F_2 dc = N_1 R T_1 + N_2 R T_2.$$

Substituting the expressions of $\overline{u'^2}$ and $\overline{v'}$ into Eq. (6.35), one obtains

$$\tau_{xx} = \frac{2}{3} m \left(nu_0^2 - N_1 u_{01}^2 - N_2 u_{02}^2 \right).$$

Substitution of Eq. (6.30) and Eq. (6.31) enable the above equation to be written as

$$\frac{n}{m}\tau_{xx} = \frac{2}{3}\left[(N_1 u_{01} + N_2 u_{02})^2 - N_1(N_1 + N_2) u_{01}^2 - N_2(N_1 + N_2) u_{02}^2 \right] =$$

$$-\frac{2}{3} N_1 N_2 (u_{01} - u_{02})^2 = -\frac{2}{3} N_1 (n_1 - N_1) \frac{u_{01}}{u_{02}} (u_{01} - u_{02})^2. \quad (6.36)$$

In writing the last equality Eq.(6.32) has been employed.

Now discuss the left hand side term of Eq. (6.33). Similarly with the evaluation of $\overline{nu^2}$, according to the definition

$$\overline{nu^3} = N_1 \int_{-\infty}^{\infty} u^3 F_1 dc + N_2 \int_{-\infty}^{\infty} u^3 F_2 dc =$$

$$N_1 u_{01} \left(u_{01}^2 + 3RT_1 \right) + N_2 u_{02} \left(u_{02}^2 + 3RT_2 \right) =$$

$$N_1 u_{01} \left(u_{01}^2 + 3RT_1 - u_{02}^2 - 3RT_2 \right) + n_1 u_{01} \left(u_{02}^2 + 3RT_2 \right).$$

With the help of Eq. (6.28) the expression ($\left(u_{01}^2 + 3RT_1 - u_{02}^2 - 3RT_2 \right)$) can be written as $(2/5)(u_{01}^2 - u_{02}^2)$, and the latter is independent of x, so one has

$$\frac{d\left(\overline{nu^3}\right)}{dx} = \frac{2}{5} u_{01} \left(u_{01}^2 - u_{02}^2 \right) \frac{dN_1}{dx}. \quad (6.37)$$

Substituting Eq. (6.36) (through Eq. (6.34)) and Eq. (6.37) into the moment equation (6.33) yields the equation relative to $N_1(x)/n_1$

$$\frac{d(N_1/n_1)}{d(x/\lambda_1)} = -\alpha(N_1/n_1)\left(1 - \frac{N_1}{n_1}\right), \qquad (6.38)$$

where

$$\alpha = \frac{5}{3\pi^{1/2}} \frac{(2kT_1/m)^{1/2}}{u_{02}} \frac{u_{01} - u_{02}}{u_{01} + u_{02}}. \qquad (6.39)$$

The solution of Eq. (6.38) is

$$\frac{N_1}{n_1} = \frac{1}{1 + \exp\left[\alpha(x/\lambda_1)\right]}. \qquad (6.40)$$

Substituting this expression of $N_1(x)$ and that of $N_2(x)$ (obtained from Eq. (6.32)) into Eq. (6.23), one obtains the distribution function of the entire shock wave transition zone. The density across the shock transition zone (with the employment of Eq. (6.30) and Eq. (6.32)) is

$$n(x) = N_1(x) + N_2(x) = N_1 + (n_1 - N_1)\frac{u_{01}}{u_{02}} = N_1\left(1 - \frac{n_2}{n_1}\right) + n_2,$$

i.e.,

$$\frac{n}{n_1} = \frac{n_2}{n_1}\left(1 - \frac{N_1}{n_1}\right) + \frac{N_1}{n_1}.$$

Substituting Eq. (6.40) into it yields

$$\frac{n}{n_1} = \frac{1 + \frac{n_2}{n_1}\exp(\alpha x/\lambda_1)}{1 + \exp(\alpha x/\lambda_1)}. \qquad (6.41)$$

The post and pre shock density ratio n_2/n_1 can be expressed through the Ma number in front of the shock (see for example [13]), e.g., for monatomic molecules the specific heat ratio $\gamma = 5/3$, one has

$$\frac{n}{n_1} = \frac{1 + \frac{4Ma^2}{3 + Ma^2}\exp(\alpha x/\lambda_1)}{1 + \exp(\alpha x/\lambda_1)}. \qquad (6.42)$$

The solution of Mott-Smith is extended by Muckenfus [14] to other inverse power law molecular models. When the power is chosen by fitting the viscosity law, the Mott-Smith method under relatively strong *Ma* number can predict fairly well the shock structure. The result of the method is dependent on the choice of u^2 as the object of averaging in the moment method [15]. Despande and Narasimha [16] pointed out that the employment of u^3 should be better than the employment of u^2. However, the main problem of the Mott-Smith method is it could not provide the correct description of the distribution function (see section 7.1).

The moment methods are mostly applied to steady one-dimensional problems, for example the heat transfer problem between two plates [17] and the problem of evaporation of the plane condensed phase [18], and others. Gross and others employed also the moment method in solving the Couette flow [3] and Rayleigh problem [4] staring from the linearised Boltzmann equation. For the two-dimensional and axi-symmetric flows it is difficult to construct distribution function suitable for the flow field.

6.4 MODEL EQUATIOINS

The Boltzmann equation (Eq. (2.152))

$$\frac{\partial f}{\partial t} + c \cdot \frac{\partial f}{\partial r} + F \cdot \frac{\partial f}{\partial c} = \left(\frac{\partial f}{\partial t}\right)_c \equiv \int_{-\infty}^{\infty}\int_0^{4\pi} \left(f^* f_1^* - ff_1\right) c_r \sigma d\Omega dc_1 \qquad (6.43)$$

causes so tremendous difficulties to the mathematical solution for the complexity of the right hand side collision term, that many researchers proposed the employment of simplified collision term or collision model to replace it. The most famous model equation was put forward by Bhatnagar, Gross and Krook [19] and is called the *BGK equation* (also was called Krook equation). This equation has the following form

$$\frac{\partial f}{\partial t} + c \cdot \frac{\partial f}{\partial r} + F \cdot \frac{\partial f}{\partial c} = v(f_e - f), \qquad (6.44)$$

where $f_e = f_e(c_0, T)$ is the local equilibrium distribution or Maxwellian distribution and is the function of the density ρ, the flow velocity c_0 and the temperature T, and they are obtained by integration of f (see Eq. (2.5), Eq. (2.8) and Eq. (2.31))

$$f_e = \frac{n}{\left(2\pi \frac{k}{m} T\right)^{3/2}} \exp\left[-\frac{(c_i - c_{0i})^2}{2\frac{k}{m}T}\right], \qquad (6.45)$$

$$n = \int f dc, \qquad (6.46)$$

$$c_{0i} = \frac{1}{n} \int c_i f dc, \qquad (6.47)$$

$$T = \frac{1}{3\frac{k}{m}n} \int (c_i - c_{0i})^2 f dc. \qquad (6.48)$$

So the equation (6.44) is still a non-linear integral-differential equation, v is the collision frequency, it is proportional to the density and is related to the temperature but is not dependent on the molecular velocity. Welander [20] almost simultaneously independently proposed the same model equation, so this equation sometimes is also called Boltzmann-Krook-Welander equation (for short *BKW equation*).

It is evident that BGK equation gives the correct solution $f = f_0$ at equilibrium. It also provides the correct collision-less and free molecular solution, for here the collision term is irrelevant. Employing the Chapman-Enskog method to the BGK equation yields the conservation equations having the form of Navier-Stoke equations (for detail see Vincenti and Kruger [21]), unfortunately, the transport coefficients thus obtained, i.e. the viscosity and the heat conductivity, do not possess the correct values.

The collision integral of the Boltzmann equation is the sum of two terms : one term involves $-f(c)f(c_1)$, meaning the depletion of the molecules of class c

caused by collisions, the other involves $f(c^*)f(c_1^*)$, meaning the increase of the number of molecules of this class caused by collisions.

In the model equation (6.44) the term $-vf$ is used to replace the collision term in the Boltzmann equation causing the depletion of the c molecules. This can have some explanation for a peculiar but not realistic molecular model, i.e., the Maxwallian molecules. The function $f(c)$ in the second term at the right hand side of Eq. (6.43) is independent of c_1, and can be taken out of the symbol of integration, and the remaining expression

$$\int_{-\infty}^{\infty}\int_0^{4\pi} f_1 c_r \sigma d\Omega dc_1 \tag{6.49}$$

in general is dependent on c, for $c_r = c - c_1$, but in the specific case of Maxwell molecules it is independent of c, and is the collision frequency v [1]. So this term of the BGK equation gained some kind of proof for the Maxwell molecules.

Replacement of the first term on the right hand side of the Boltzmann equation (6.43) by vf_e does not have such a kind of proof. We only can understand it as an assumption, i.e., the number of molecules scattered out of the c class in collisions is assumed to be equal to the number of molecules scattered out of the c class by the molecules in local equilibrium with a collision frequency independent of the molecular velocity.

Employing the Chapman-Enskog method to the BGK equation yields the conservation equations in the form of Navier-Stoke equations with the transport coefficients ([21], p.384, Eq. (3.13))

$$\mu_{BGK} = \frac{nkT}{v}, \tag{6.50}$$

$$K_{BGK} = \frac{5}{2}\left(\frac{k}{m}\right)\frac{nkT}{v}. \tag{6.51}$$

---

[1] The expression of the collision frequency v is Eq. (2.206), when $\sigma_T c_r$ is independent of c_r, it can be written as the product of Eq. (6.49) and $\dfrac{1}{n}\int_{-\infty}^{\infty} fdc = 1$.

At the same time for the Boltzmann equation Chapman-Enskog method yields the following values of μ and K ([22], p.226, p.247)

$$\mu = 0.499 \rho \bar{c} \lambda, \tag{6.52}$$

$$K = \frac{15}{4} \frac{k}{m} (0.499 \rho \bar{c} \lambda). \tag{6.53}$$

For ensuring a reasonable expression for μ one can adopt an expedient measure, i.e., instead of $v = \bar{c}/\lambda$ the following expression is used for the collision frequency

$$v = (\pi/3.992)(\bar{c}/\lambda). \tag{6.54}$$

Thus the viscosity coefficient obtained from the BGK equation would have the correct expression equal to Eq. (6.52). This is beneficial for solving the flow problem with momentum exchange as the dominant effect, for it can lead to correct reference parameters. But for flow problems with dominant energy exchanges to ensure a reasonable expression for K the following expression for the collision frequency should be adopted

$$v = (\pi/5.988)(\bar{c}/\lambda). \tag{6.55}$$

Thus the conductivity coefficient K_{BGK} (Eq. (6.51)) obtained from the BGK equation would have the correct expression Eq. (6.53). Unfortunately, it is impossible to ensure μ and K to have the correct expressions simultaneously. From Eq. (6.50) and Eq. (6.51) it is seen, that for the BGK model the Prandtl number $Pr_{BGK} = c_p \mu_{BGK} / K_{BGK} = 1$ does not have the correct value of 2/3. This is already an indication of the limitation of this expedient measure, it is powerless in solving problems involving momentum and energy exchanges of equal importance.

The BGK model equation is widely used in the transitional regime for its simplicity. For problems of small disturbance deviated not far from the equilibrium the shortcoming of the approximation of the BGK equation becomes not so remarkable. For the Maxwellian molecules it has been proved that the linear form of the BGK equation is the first term of a model-series approximating the Boltzmann equation with arbitrary accuracy. There is a number of small disturbance problems with practical meaning that are solved using the BGK model but not the Boltzmann equation. Some methods of statistical models for constructing the collision term has been put forward which can provide with correct Prandtl number (see e.g. [24]). For the system of Boltzmann equations of multi-component gas the right hand sides involve the self-collision operator for molecules of one component and cross-collision operators for molecules from various components. For the self-collision operator the BGK model is applicable, for the cross-collision operators Boley and Yip [25] put forward the theory of obtaining the model equation basing on the eigenfunction theory and obtained the system of model equations for multi-component gas. There have been many works devoted to the research and the application of the model equations on solving small disturbance problems of simple geometry.

The BGK equation after all is an approximation using a simplified term to replace the exact term basing on solid physical reality. Its applicability scope must be tested and verified by the experiments or exact computations. The DSMC method has stood the strict test of the experiment (see next Chapter) and has the merit of verified solution to test various approximation models. We simulated the problem of the gas flow caused by the sudden wall temperature change and the Rayleigh problem using the DSMC method with the employment of Maxwellian molecular model [26, 27] and compared the results with the exact numerical results of solution of the BGK equation [28]. Fig.6.1 shows this comparison. From this figure it is seen that near equilibrium (in 1~2 collision times) the BGK equation yields correct results (in agreement with the DSMC result), but when far from equilibrium (after 5 or more collision times from the starting of the flow) the BGK model equation is inexact (deviated from the correct DSMC simulation result).

The idea of the BGK equation considering the transition of a gas from its present state to equilibrium as a simple relaxation process is applied to the solution of the Euler equation and the Navier-Stokes equations. In section 5.2 we have seen that the zero order approximation of the Chapman-Enskog expansion in the solution of the Boltzmann equation gives the Euler equation, the first order approximation gives the Navier-Stokes equations. Employment of the Chapman-Enskog

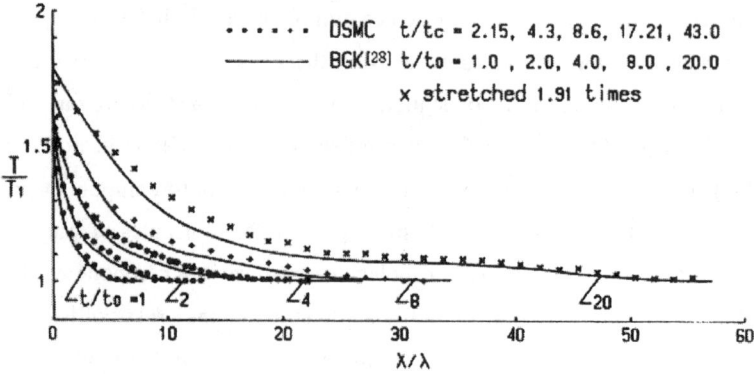

Fig. 6.1 Transients of temperature in the sudden temperature change problem. Comparison of DSMC simulation [26] and BGK calculation [28]. In BGK the expedience of modifying collision frequency is made

method to the BGK equation yields the same result. This leads Xu et al. to put forward the *gas kinetic scheme* for solving the fluid mechanics equations [29].

For the two-dimensional flows the BGK equation in the x direction can be written as

$$\frac{\partial f}{\partial t} + u\frac{\partial f}{\partial x} = \frac{g-f}{\tau}, \qquad (6.56)$$

where for convenience the symbol is changed into g to denote the equilibrium state the distribution function f approaches, τ is the reciprocal of the collision frequency v or the collision time of the particles. The equilibrium Maxwell distribution, when the thermal motion in the third dimension is considered as the internal energy, can be written as

$$g = n\left(\frac{\beta^2}{\pi}\right)^{\frac{N+2}{2}} \exp\left[-\beta^2\left((u-u_0)^2 + (v-v_0)^2 + \xi^2\right)\right],$$

where ξ is the variable of internal degree of freedom and includes the internal energy of the gas and the thermal motion in the z direction. The total number of freedom of ξ is $N = (5-3\gamma)/(\gamma-1)+1$, $(5-3\gamma)/(\gamma-1)$ is the number of freedom of the internal energy (see Eq. (4.40)).

6.4 MODEL EQUATIOINS

The BGK kinetic scheme starts from the solution of the BGK equation (6.56) at $x_{j+1/2}$ of the grid boundary surface

$$f\left(x_{j+\frac{1}{2}},t,u,v,\xi\right)=\frac{1}{\tau}\int_0^t g(x',t',u,v,\xi)e^{-(t-t')/\tau}dt' + e^{-t/\tau}f_0\left(x_{j+\frac{1}{2}}-ut\right), \quad (6.57)$$

where $x' = x_{j=1/2} - u(t-t')$ is the trajectory of the motion of the particle, and f_0 is the initial distribution function at the star of every time step ($t=0$). For obtaining f the unknown functions g and f_0 in Eq. (6.57) should be determined. For the sake of simplifying the notation in the following $x_{j+1/2}=0$ is used to represent the position of $x_{j+1/2}$.

In the early BGK scheme f_0 is supposed to be

$$f_0 = \begin{cases} g^l(1+a^l x), x \le 0 \\ g^r(1+a^r x), x \ge 0 \end{cases}. \quad (6.58)$$

g^l and g^r are the Maxwell distributions on the left hand side and the right hand side of the cell boundary. The slopes a^l and a^r are obtained from the space derivatives of the Maxwell distribution, the latter has the unique relation of correspondence with the slopes of the conservation variables. The basic assumption here is, even when there is discontinuity on the cell boundary, the gas on both sides of the discontinuity is assumed to be in equilibrium states. Such an assumption is suitable for the case when the cell size Δx is large in comparison with the thickness of the shock or the case of Euler equation. For the case when Δx is small and the shock structure can be distinguished, i.e., when deal with the Navier-Stokes equations, the initial distribution f_0 should present the deviation from the equilibrium Maxwell distribution and describe the interior structure of the shock transition zone. So in the recent gas kinetic BGK scheme method aiming at the application on equation solution the initial distribution function f_0 is assumed to be

$$f_0 = \begin{cases} g^l\left[1+a^l x - \tau(a^l u + A^l)\right], x \le 0 \\ g^r\left[1+a^r x - \tau(a^r u + A^r)\right], x \ge 0 \end{cases}. \quad (6.59)$$

The additional terms $-\tau(a^l u + A^l)g^l$, $-\tau(a^r u + A^r)g^r$ are the deviations of the non-equilibrium state obtained from the Chapman-Enskog expansion of the BGK equation from the Maxwell distribution.

The equilibrium state g to which the initial distribution f_0 approaches in the recent version [30] is supposed to have two slopes at the two sides of the cell boundary and have time derivative

$$g = g_0\left[1+(1-H(x))\overline{a^l}x + H(x)\overline{a^r}x + \overline{A}t\right], \qquad (6.60)$$

where $H(x)$ is the Heaviside function defined as

$$H(x) = \begin{cases} 0, x < 0 \\ 1, x \geq 0 \end{cases},$$

and g_0 is the local Maxwell distribution at $x = 0$.

It is noted, f_0 is discontinuous at the boundary surface $x = 0$, on each side of the boundary it is the non-equilibrium distribution function in local cell, g is continuous at $x = 0$, but has different slopes at $x < 0$ and at $x > 0$. $a^l, a^r, A^l, A^r, \overline{a^l}, \overline{a^r}$ and $\overline{A}$ characterize the deviations from the equilibrium Maxwell state in space and time. Their determination see reference [30].

Substituting thus obtained f_0 and g into Eq. (6.57) yields the distribution function $f(x_{j+1/2}, t, u, v, \xi)$ of the gas at the cell boundary. The fluxes of the mass, momentum and energy across the cell boundary can be obtained by searching the moments relative to f. The technical details in the application of the gas kinetic BGK scheme can not be explained here thoroughly (see [29, 30]). Here only confine to noting that as the distribution function is introduced in the method, at the wall the reflection model of the molecule surface interaction can be applied to obtain the changes of the mass, momentum and energy of the molecules on the boundary (suitable for the entire transitional regime). This under small Kn number corresponds to the introduction of the slip boundary condition. Besides, after ensuring the correct viscosity coefficient μ through the modification of the computation result of the heat flux the method can guarantee also the correct conductivity coefficient K, so the correct Pr number is assured.

The method has been applied to the Couette flow with temperature gradient, the shock wave structure, the two-dimensional Mach=3 step flow, the interaction of the laminar boundary layer with shock wave etc. (see reference [30]. As the efficiency of the gas kinetic scheme is higher than that of the DSMC method, validation of the accuracy of the scheme by the DSMC or experimental results in its applications to the near continuum and transitional regime is an interesting issue having practical meaning.

Li and Zhang attempted to find a *gas kinetic unified algorithm* to solve the transitional flow bridging the free molecular flow to the continuum flow [31]. The starting point is that the basic equation of the kinetic theory is replaced by the model equation, the latter adopted the so called S-model equation [32] developed by Shakhov from the BGK model. The authors introduced the discrete velocity method, adapted the Gause-Hermite method of quadrature, applied and developed the method of quadrature of Hua and Wang approximating multiple integral by single sums, and employed the equally spaced three-point composite Newton-Cotes quadrature formula, and implemented the solution of the model equation successfully. The method was applied to one-dimensional shock-tube problems, the flows past two-dimensional circular cylinder, and the three-dimensional flows around sphere and spacecraft with various Knudsen numbers. In the case where the experimental data are available (the drags of the cylinder and sphere) the agreement with experiment is fairly good. It is desirable to have more exact quantitative comparison. The transplantation of the method to the solution of the Boltzmann equation is of value.

6.5 THE FINITE DIFFERENCE METHOD

The most natural numerical method in solving the Boltzmann equation is analogous to the finite difference method in the computational fluid dynamics(CFD). The velocity distribution function is the unique dependent variable, but in general (three-dimensional and unsteady flow) it is a function of seven independent arguments, and it is a difficult task to solve it by the finite difference method. If allocate in each dimension 100 cells, then for the 3-D unsteady flow 10^{14} cells is required, and in addition the velocity space is infinite. The upper bound of the finite velocity should be chosen reasonably to make the problem manageable. This leads to even larger difficulty to the hypersonic flow, for in this case there are some molecules having velocity much larger than that of most molecules, although they constitute only a small portion of all the molecules, but have strong influence on the whole macroscopic motion.

Except the above difficulty the peculiar trouble in solving the Boltzmann equation by CFD method is of course in the computation of the collision term. The computation of the collision integral requires for each phase point the summation over all points of the other velocity space, requires finding the contribution of the reverse collision for each collision and each term should be summed over all impact parameters of the collision. So the methods of the solution of the Boltzmann equation are mainly embodied in the treatment of the computation of the collision integral.

Nordsieck and Hicks [33] and Yen et al. [34] suggested a method of Monte-Carlo quadrature of the collision term of the Boltzmann equation. Replacement of the direct numerical quadrature by this Monte-Carlo quadrature reduced the computation time by a factor of 10^{-4}. The infinite velocity space was firstly replaced by the finite velocity domain R, which was taken large enough to involve all except 0.1% molecules. The Monte-Carlo quadrature is to replace the integration over the domain R by the product of the mean of the values of the under-integral expression at N randomly chosen points and the volume of R. The error of this replacement is proportional to $N^{-1/2}$. The under-integral expression is the function of 8 arguments, i.e., c^*, c and two impact parameters. For the fluid dynamics terms on the left hand side of the equation the ordinary finite difference method was employed. Here the usual problems of dependence on the grid and the computational stability would be encountered. Nordsieck et al. by using this method solved some one-dimensional problems including the structure of shock wave problem etc., satisfactory results were obtained.

Tcheremissine, Aristov and others developed the algorithm to use the finite difference method for solving the Boltzmann equation (for relatively recent references see [36] and works cited there). In the computation of the collision integral the Monte-Carlo quadrature method is still used, but some improvements are introduced. The symmetry of the binary collision is utilized to remarkably reduce the arithmetic amount and a special projection technique is introduced for the calculation of the collision integral to ensure the conservation of mass, momentum and energy. These accelerated the solution process and enable to use the method to solve some unsteady, 2-D and 3-D problems.

6.6 DISCRETE ORDINATE METHOD

In section 2.11 of Chapter 2 on the example of Couette flow we studied in detail the 8 velocity gas model introduced by Broadwell [37], in which the velocities of molecules can assume only 8 magnitudes $c_1, c_2, \cdots, c_8$ (see Fig.2.9). Such an approach of using finite number of discrete velocities to replace the entire velocity space is called the *discrete ordinate or discrete velocity method*. In an 8 velocity gas the number density of the molecules having a certain velocity can be counted, and corresponding to 8 velocities there are 8 number densities $n_1, n_2, \cdots, n_8$. The velocity distribution function of the gas now is degenerated into the set of 8 number densities. For the molecules of class 1 Boltzmann equation is degenerated into (see Eq. (2.141))

$$\frac{\partial n_1}{\partial t} - q\frac{\partial n_1}{\partial x} + q\frac{\partial n_1}{\partial y} = 2\theta\left(n_2 n_3 - n_1 n_4\right)/n \ . \tag{6.61}$$

Analogously, the equations for molecules of classes 2, 3, and 4 can be written

$$\frac{\partial n_2}{\partial t} + q\frac{\partial n_2}{\partial x} + q\frac{\partial n_2}{\partial y} = 2\theta\left(n_1 n_4 - n_2 n_3\right)/n \ , \tag{6.62}$$

$$\frac{\partial n_3}{\partial t} - q\frac{\partial n_3}{\partial x} - q\frac{\partial n_3}{\partial y} = 2\theta\left(n_1 n_4 - n_2 n_3\right)/n \ , \tag{6.63}$$

$$\frac{\partial n_4}{\partial t} + q\frac{\partial n_4}{\partial x} - q\frac{\partial n_4}{\partial y} = 2\theta\left(n_2 n_3 - n_1 n_4\right)/n \ , \tag{6.64}$$

where q is the magnitude of the projection of velocity on the rectangular coordinate system, θ is a magnitude proportional to the collision frequency and equals to $0.7v$ (see Eq. (2.142)), n is the number density of the gas, for the case of low speed to be discussed here $n = const$. In deriving the above equations two-dimensional flow is assumed: the flow characteristics are not dependent on z, hence the symmetry condition holds

$$n_1 = n_5, \quad n_2 = n_6, \quad n_3 = n_7, \quad n_4 = n_8. \tag{6.65}$$

Now we continue to examine with the help of the 8 velocity gas the problem of *shear flow between two plates* or *the Couette flow*, i.e., two plates separated by a distance d moving in the x, z plane in opposite directions, the velocity in the y direction is 0:

$$V = \sum_{i=1}^{8} v_i n_i / n = 0.$$

Thus one has

$$n_1/n + n_5/n + n_2/n + n_6/n = n_3/n + n_7/n + n_4/n + n_8/n = \frac{1}{2}.$$

From the symmetry condition Eq. (6.65) we have

$$n_1/n + n_2/n = n_3/n + n_4/n = \frac{1}{4}. \tag{6.66}$$

This condition makes the equations (6.62) and (6.64) out of the system (6.61)~(6.64) superfluous, and Eq. (6.61) and Eq. (6.63) can be written as

$$\frac{\partial n_1}{\partial t} + q \frac{\partial n_1}{\partial y} = \frac{\theta}{2}(-n_1 + n_3), \tag{6.67}$$

$$\frac{\partial n_3}{\partial t} - q \frac{\partial n_3}{\partial y} = -\frac{\theta}{2}(-n_1 + n_3). \tag{6.68}$$

For the steady flow $\partial/\partial t = 0$, by introducing the dimensionless quantities

$$\eta = y/d, \quad \alpha = \theta d/2q,$$

the equations (6.67) and (6.68) can be written as

$$\frac{dn_1}{d\eta} = \alpha(-n_1 + n_3), \tag{6.69}$$

$$\frac{dn_3}{d\eta} = \alpha(-n_1 + n_3). \tag{6.70}$$

The boundary conditions for the Couette flow problem are

$$U\bigg|_{\eta=\frac{1}{2}} = -\frac{1}{2}U_w, \quad U\bigg|_{\eta=-\frac{1}{2}} = \frac{1}{2}U_w.$$

The mean velocity in the x direction leaving the upper plate at $\eta = 1/2$ and the mean velocity in the x direction leaving the lower plate at $\eta = -1/2$ are

$$q^2\left[-n_3\left(\frac{1}{2}\right) + n_4\left(\frac{1}{2}\right)\right] \bigg/ q\left[n_3\left(\frac{1}{2}\right) + n_4\left(\frac{1}{2}\right)\right] = -\frac{1}{2}U_w,$$

$$q^2\left[-n_1\left(-\frac{1}{2}\right) + n_2\left(-\frac{1}{2}\right)\right] \bigg/ q\left[n_1\left(-\frac{1}{2}\right) + n_2\left(-\frac{1}{2}\right)\right] = \frac{1}{2}U_w, \tag{6.71}$$

$$n_1\left(-\frac{1}{2}\right)\bigg/n = \frac{1}{8}(1 - U_w/2q),$$

$$n_3\left(-\frac{1}{2}\right)\bigg/n = \frac{1}{8}(1 - U_w/2q).$$

The solution of equations (6.69) and (6.70) satisfying Eq. (6.71) is

$$n_1/n = \frac{1}{8}\left[\frac{\alpha}{(\alpha+1)}\frac{U_w}{q}\eta + 1 - \frac{1}{2(\alpha+1)}\frac{U_w}{q}\right], \tag{6.72}$$

$$n_3/n = \frac{1}{8}\left[\frac{\alpha}{(\alpha+1)}\frac{U_w}{q}\eta + 1 + \frac{1}{2(\alpha+1)}\frac{U_w}{q}\right]. \tag{6.73}$$

The velocity of the fluid in the direction of x is obtained from the above density solution

$$U = \sum_i u_i(n_i/n) = 2q[-n_1/n + n_2/n + n_3/n - n_4/n] = -\frac{\alpha}{(\alpha+1)}U_{W\eta}. \tag{6.74}$$

The shear stress is

$$\tau_{yx} = \rho \sum_i u_i v_i (n_i/n) = 2\rho q^2 [-n_1 + n_2 + n_3 - n_4]/n =$$

$$4\rho q^2 (-n_1 + n_3)/n = \frac{1}{2(\alpha+1)} \rho q U_W. \tag{6.75}$$

Expressing τ_{ij} as $\mu(\partial U/\partial y)$, from Eq. (6.74) the expression of μ can be obtained

$$\mu = \rho q d / 2\alpha = \rho q^2 \theta.$$

Substituting $\theta = 0.70\bar{c}/\lambda$ and $q = \bar{c}/\sqrt{3}$ yields

$$\mu = 0.48 \rho \bar{c} \lambda. \tag{6.76}$$

Although this value is quite close to the viscosity expression Eq. (2.222) of the Chapman-Enskog approximation for the hard sphere model, but if examine the value of τ_{ij} in the entire transitional regime (when $Kn = \lambda/d$ is between 0.01 and 100) then the result of the discrete ordinate method of the 8 velocity approximation is not so good. This is undoubtedly the result of adopting a too simple discrete velocity model to replace the continuous velocity space.

Broadwell in [37] investigated also the Rayleigh problem using the 8 velocity model. He also tested the 6 velocity model (6 velocities with equal speed value and pointed into the positive and negative directions of the x,y,z axes) applying it to the study of shock wave structure [38]. Gatignol utilized the coplanar 6 velocity model [39] (pointed into the apexes of a right hexagon) to investigate the shock wave structure and the Couette flow, and with help of the coplanar 4 velocity model [40] discussed the boundary conditions and the H theorem in the discrete velocity model. Cabannes [41] introduced the 14 velocity model (combined from the Broadwell 8 velocity and 6 velocity models) and investigated the Couette flow problem. A result of universal meaning is that when 4 velocities in the flow plane is eliminated from the 14 velocities the result on the contrary is improved. The reason for this is that the molecules moving along the direction of the surface do not collide with the surface, are isolated from the surface, the existence of such parasitic molecules causes the errors at the wall. Gatignol [42] and Cabannes [43] introduced in general form the discrete velocity model, utilizing the set of $c_1, c_2, \cdots, c_p$ to replace the entire velocity space and n_i to represent the number density in time t at position r having velocity c_i, and write the discrete Boltzmann equation (compare Eqs. (6.61)~(6.64))

$$\frac{\partial n_i}{\partial t} + c_i \cdot \frac{\partial}{\partial r} n_i = \sum_{j=1}^{p} \sum_{(k,l)} A_{ij}^{kl} \left(n_k n_l - n_i n_j \right), \quad i = 1, 2, \cdots, p, \tag{6.77}$$

where A_{ij}^{kl} is the transition probability

$$A_{ij}^{kl} = \sigma_T c_{rij} p_{ij}^{kl}, \tag{6.78}$$

no summation convention is implied in Eq. (6.78)

$$\sum_{k,l} p_{ij}^{kl} = 1, \text{ valid for any } i, \ j. \tag{6.79}$$

p_{ij}^{kl} is the probability the pair of molecules with velocities c_i, c_j before collision changes into pair of molecules with velocities c_k, c_l after collision. An often used model is the one in which the ends of the velocity vectors are the uniformly distributed grid points in the phase space [44, 45]. Fig 6.2 shows a peculiar example of such model in planar case with the values of velocity components equal to semi-integers

$$\begin{cases} u = \left(n + \frac{1}{2} \right) \Delta v, -5 \leq n \leq 4, \\ v = \left(n + \frac{1}{2} \right) \Delta v, -5 \leq n \leq 4, \\ \Delta v / \sqrt{2kT/m} = 0.667. \end{cases} \tag{6.80}$$

The semi-integer points are chosen because the molecules with velocities parallel with the body surface would cause errors near the wall as indicated in reference

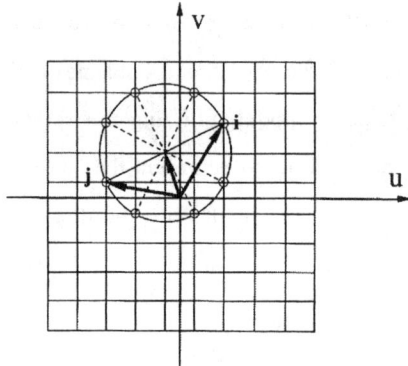

Fig. 6.2 The model with the ends of the velocity vectors as uniformly distributed grid points is shown in planar case. In the figure are shown $c_i(1.5, 2.5)$ and its certain collision partner $c_j(-2.5, 0.5)$ and the possible velocity pairs c_k, c_l after collision (shown by small circles, altogether 4 of them)

[41]. For such a model Eq. (6.77) is written for each $c_i(u,v)$ (in Fig 6.2 the case $c_i = (1.5\,\Delta x, 2.5\Delta x)$ is shown). The partner c_j of molecule c_i goes through the whole p velocity set (in the present example $p = 10 \times 10$, the figure shows the case of $c_j = (-2.5\Delta x, 0.5\Delta x)$). The conservation of momentum and energy yields the invariance of the value of the relative velocity ($|c_i - c_j|$, in the present example $= \sqrt{20}\Delta v$) and the invariance of the velocity of the mass center. So all the after collision velocity pairs lay on the circle with center at the end of the velocity vector of the mass center and diameter of the value of the relative velocity. In the figure the velocity pairs from set $c_1, c_2, \cdots, c_p$ satisfying such condition are shown by the small circles connected by dashed lines (there are 4 pairs in the present example). For hard sphere model the appearance of all after collision pairs are of equal probability. The probability p_{ij}^{kl} in Eq. (6.77) is easily calculated

$$p_{ij}^{kl} = 1/N_{ij}, \text{ valid for any } k, l, \qquad (6.81)$$

where $N_{i,j}$ is the number of all possible c_k, c_l pairs after collision resulted from the pre-collision c_i, c_j pairs. According to the discussion about the reverse collision in Chapter 2, the collision $(c_k, c_l \to c_i, c_j)$ is the reverse collision of $(c_i, c_j \to c_k, c_l)$. So we have $p_{ij}^{kl} = p_{kl}^{ij} = p_{lk}^{ji}$, this result has been employed in writing Eq. (6.77).

The above expounding of the discrete velocity model and the discrete Boltzamnn equation is only a kind of physical heuristic elucidation, the relationship of them with the Boltzmann equation still requires strict mathematical proof. In 1994 Bobylev et al. [46] proved that the uniform grid discrete velocity model [44, 45, 47] converges to the Botzmann equation and gave an estimate of the error of the quadrature by the discrete velocity method. This is an important development of the discrete velocity method. From the heuristic explanation of the method for the planar case where the ends of the velocity-vector pairs are situated on a circle and at the same time are on the nodal points of a integer grid, it is understandable why this mathematical proof is closely relevant to a congruence problem in number theory related with the presentation of a integer number by the sum of three squires. An assertion in number theory states: any number t which is not congruent to 7 relative to 8 ($t \neq 7(\mod 8)$) can be represented as the sum of three squires $t = p^2 + q^2 + r^2$, and the number of ways of presentation grows sufficiently fast with t.

Gatignol [48] discussed the boundary condition problem of the discrete velocity model in detail. Despite the above progresses on the whole the problems solved by the discrete velocity method are mainly of simple geometry and molecular models.

6.7 INTEGRAL METHODS

The Boltzmann equation (2.152) can be written in the following form

$$\frac{df}{dt} = J_1(t, r, c) - fJ_2(t, r, c), \qquad (6.82)$$

where

$$J_1 = \int_{-\infty}^{\infty} \int_0^{4\pi} f^* f_1^* \, c\sigma \, d\Omega dc_1, \qquad (6.83)$$

$$J_2 = \int_{-\infty}^{\infty} \int_0^{4\pi} f_1 c_r \sigma d\Omega dc_1 . \qquad (6.84)$$

The differentiation is carried along a certain molecular trajectory. Equation (6.82) can be considered as a ordinary differential equation relative to f

$$\frac{df}{dt} + fJ_2(t,f) = J_1(t,f) . \qquad (6.85)$$

Suppose the f in J_1 and J_2 of Eq. (6.85) is known, the formal solution of the first order linear ordinary differential equation can be written

$$f(t,r,c) = f(t_0, r - c(t-t_0), c) \times$$

$$\exp\left\{-\int_{t_0}^{t} J_2(s, r - c(t-s), c) ds\right\} + \int_{t_0}^{t} J_1(\tau, r - c(t-\tau), c) \times$$

$$\exp\left\{-\int_{\tau}^{t} J_2(s, r - c(t-s), c) ds\right\} d\tau . \qquad (6.86)$$

The right hand side of Eq. (6.86) includes unknown function f and is a integral equation relative to f. This is the Boltzmann equation in the integral form.

Vallander [49] presented the direct physical derivation of the integral equation (6.86) without citing on the Boltzmann equation, he initiated the method of solving the rarefied gas dynamics problems basing on the integral form of the basic equation, i.e., the integral method [50]. The most convenient approach of searching solution of the integral equation is of course the method of iteration. The iteration method of the integral equation is utilized to prove the existence theorem of the solution of the Boltzmann equation.

6.8 DIRECT SIMULATION METHODS

These are the methods different from those of numerical solution of the Boltzmann equation but based directly on the simulation of the physics of the flow.

They were born with the appearance of the electronic computers and are fully developed with the enhancement of the speed and the capacity of the computers. Direct simulation methods distinguish *deterministic simulation methods* and *probabilistic simulation methods*. Both categories of methods trace the motion of enormous number of molecules, their encounter with the boundaries, the collisions between themselves, the exchange of internal energies during the collisions and chemical reactions, etc. The simulation should ensure that the processes traced in the computer are able to reproduce the processes in the real flows. In the computer a time (clock) is introduced synchronously with the physical time in the flows. The positions, velocities and the internal energies of the molecules are recorded, they change due to the motion of the molecules, their interaction with the wall surfaces and the collisions between themselves. Obviously such simulation calculation is an unsteady process. The steady process is obtained as the stabilized state of the unsteady process.

The deterministic direct simulation method is the most early suggested physical simulation method put forward by Alder and Wainwright [51] in 1957, and is called the *molecular dynamics (MD) method*. In this method the probabilistic method is employed only when distribute the initial position and velocities, and when compute the molecular motion, the interaction of molecules with the boundaries and their collisions between themselves, the deterministic method is used without exception. For example, when judge the occurrence of a certain collision, the overlap of the collision cross sections of two mo lecules at the same time is considered and the impact parameters of the collision are provided by the mutual configuration of the molecules, and the velocities after the collision are determined. Such a simulation method aims at the full reproduction of the physical processes, so the size of the molecules, the number density of the molecules and the flow geometry are to be simulated. Thus the number of simulated molecules in a certain simulation region should keep entirely identical with actual situation of the physical flow.

Such a requirement is more apt to be implemented for dense gas than for dilute gas. This point can be explained through the consideration of the number of molecules in a cube with side of one mean free path

$$N_\lambda = n\lambda^3. \tag{6.87}$$

As λ is inversely proportional to the number density n of the molecules, so N_λ is inversely proportional to n^2, with the help of Eq. (2.220) N_λ can be written

$$N_\lambda \equiv n\lambda^3 = \frac{1}{2\sqrt{2}\left(\pi d^2\right)^3 n^2}.$$

If introduce the number density $n_0 = 2.687\times 10^{25} m^{-3}$ under the standard state ($p_0 = 0.101325 MPa$, $T_0 = 273K$), the above equation can be written

$$N_\lambda = \frac{1}{2\sqrt{2}\left(\pi d^2\right)^3 n_0^2}\left(\frac{n_0}{n}\right)^2.$$

For certain definite molecule d is fixed, the expression before $(n_0/n)^2$ is a constant, if take a typical value $d = 4\times 10^{-10} m$ (the VHS model gives $d = 4.17\times 10^{-10} m$ for nitrogen and $d = 4.07\times 10^{-10} m$ for oxygen at $273K$), the required number of molecules is written

$$N_\lambda = 3856\left(\frac{n_0}{n}\right)^2. \tag{6.88}$$

Thus, if study a certain problem involving N_2, O_2 or gas of alike size under standard density using the MD method, around 3856 molecules should be allocated in a cube of mean free path, which will be difficult for a three-dimensional problem with a space scope of many mean free paths. But if study the same problem under a density 100 times higher than the standard state, only less than half (0.39) a molecule is needed to be allocated in a volume of λ^3, or in a cube of side of 10λ allocate around 386 molecules. This is already a relatively easy task. Thus, the molecular dynamics method is particularly suited for the simulation of dense gas or liquid but not for simulating the dilute gas. At the same time the Boltzmann equation is appropriate for the dilute gas where binary collision prevails but not for the dense gas where some modifications are needed, e.g., the Enskog put forward the Enskog equation to take into account the influence of the dense gas. The MD method supplement an additional computational means at the particle level in the domain of dense gas and at the same time is a deterministic method, so it is readily accepted by people working in this field.

The *direct simulation Monte-Carlo (DSMC)* method was put forward by G. A. Bird and was first applied to the simulation of the relaxation in the uniform gas [52] and the shock wave structure [53] problems, and then was developed to the application in simulating two-dimensional, three-dimensional and problems with complex geometry, including cases of flows with complex physic-chemical processes (see the monograph of Bird published in 1976 [54] and the new extended edition 1994 [55]). The DSMC method as the MD method traces in the computer the motion, the collisions and the change of internal energies etc. of enormous number molecules, but its specific feature is the employment of the probability processes. Not only the allocation of the initial positions and states of the molecules, but also the judgment of the occurrence and the outcomes of the collisions (including the change of the velocities and the internal energies of molecules), are determined by the test (generation and transformation) of random functions. This is just the origination of the name of Monte-Carlo method[2]. The DSMC method traces the binary collisions in gas, and we see in the discussion at beginning of section 2.3 that the binary collisions are prevailing over the three-body collisions only for dilute gas. So the prerequisite of the probabilistic simulation of collisions is the assumption of the dilute gas. This is entirely different from the MD method suitable for simulation of dense gas. And it is not like the deterministic MD method in which the equality of the number of simulated molecules in the simulation region and the number of molecules in the actual flow is required, but relatively few simulation molecules are used to represent the enormous number of real molecules. This is another difference between the DSMC method and the MD

---

[2] Monte-Carlo – a city in Monaco dukedom, the world famous gambling city. The Monte-Carlo method is so called because the implementation of its process depends on the generation and transformation of the uniform random number as the outcome in the turntable gambling depends on where the arrow stops. Von Neumann and Ulam put forward this method when studying the reactor in 1949. The method is also called method of statistical test.

method. In the early stage of the method the product of the number density and the collision cross section is kept identical for the simulated molecules and in the real flow (so the mean free path similarity is gained) to ensure the similarity law of Kn number. This led to very huge molecular size but did not have any essential influence on the simulation. The present approach is to fix the number of actual molecules a simulated molecule represents and take this into account when calculate the collision number and the macroscopic quantities. The key point of the DSMC method is also in the decoupling of the molecular motion and collision in a time step Δt. Each molecule moves a distance according to its velocity and Δt (taking into account the interaction with the boundary and the motion after reflection) and then calculate the representative collisions corresponding to this time interval Δt. The foundation of algorithm of choosing the representative collisions to ensure the identity of the motion and collisions in the simulation and in the real flow is the kinetic theory of gases. In fact the assumption of molecular chaos and dilute gas is the prerequisite for both the Boltzmann equation and the DSMC method. And Bird proved that following the procedures of the DSMC method the Boltzmann equation can be derived [56]. Wagner et al. proved that the DSMC method converges to the Boltzmann equation [57, 58].

The *test molecule Monte-Carlo method* put forward by Haviland and Levin [59] was one of the probabilistic simulation methods developed the earliest. The Monte-Carlo method developed by Kogan in the book [60] is also a test molecule method. This kind of method requires the initial estimate of the distribution function over the grid of the flow field and the allocation of the so called target molecules. Then the trajectories of numerous so called test molecules are calculated, their collisions with the target molecules are considered and the new distribution of the target molecules is established basing on the trajectories of the test molecules. Iterate such processes until convergence is reached. This method was limited to the one dimensional steady flow owing to the necessity of starting the iteration from the supposed known initial distribution and the proportionality of the computation time to the number of trajectories of the test molecules.

Nanbu developed a version of DSMC method directly derived from the Boltzmann equation [61], the main feature of it is that only one of the two molecules involved in each collision changes its velocity, this is consistent with the fact that the dependent variable of the Boltzmann equation is the single molecule distribution function. Thus, the momentum and energy conservation in each collision is not satisfied. However, the Boltzmann equation does not require such conservation but requires the overall conservation of the physical space quantities, the momentum and the energy, and this is satisfied in the simulation. The problem is: when the molecular momentum and energy is not exactly conserved but is only conserved in the average sense, the random walk is brought about, that is, the deviation from the mean value grows with the squire root of time. This must be taken into account when dealing with small disturbance or low speed problems. Babovsky gave the strict proof of the convergence of the method of Nanbu to the Boltzmann equation [62].

REFERENCES

1. Cercignani C (1969) Mathematical Methods in Kinetic Theory. Plenum Press
2. Cercignani C (1975, 1987) Theory and Application of the Boltzmann Equation, Scottish Academic; The Boltzmann equation and Its Applications, Springer
3. Gross EP and Ziering S (1958) Kinetic theory of linear shear flow. Phys. of Fluids, 1: 213-224
4. Gross EP and Jackson EA. (1958) kinetic theory of the impulsive motion of an infinite plate. Phys of Fluids, 1: 318
5. Grad H (1963) Asymptotic theory of the Boltzmann equation. In: JA Laurmann edited Rarefied Gas Dynamics, Academic, 1:26-59
6. Cercignani C (1975, 1987) See [2] Chapter 4, sec 5
7. Sone Y, Ohwada T and Aoki K (1989) Temperature jump and Knudsen layer in a rarefied gas over a plane wall. Phys. of Fluids, A1: 363-370
8. Sone Y, Ohwada T and Aoki K (1989) Evaporation and condensations of plane condensed phase. Phys. of Fluids, A1: 1398-1405
9. Ohwada T, Sone Y and Aoki K (1989) Numerical analysis of the shear and thermal creep flows of a rarefied gas over a plane wall. Phys. of Fluids A1: 1588-1599
10. Ohwada T, Sone Y and Aoki K (1989) Numerical analysis of the Poiseuille and thermal transpiration flows of a rarefied gas between two parallel plates. Phys. of Fluids A1: 2042-2049

11. Sone Y, Takata S and Ohwada T (1990) Numerical analysis of the plane Couette flow of a rarefied gas. Eur J Mech., B/Fluids, 9: 273-288
12. Mott-Smith HM (1951) The solution of the Boltzmann equation for a shock wave. Phys. Re.v, 82:855-892
13. Shapiro AH. (1953) Compressible Fluid Flow. Ronald Press, 1: 118
14. Muckenfuss C (1962) Some aspects of shock structure according to the bimodal model. Phys. of Fluids, 5: 1325
15. Rode DL, Tanenbaum BS (1962) Mott-Smith shock thickness using the v_x^p method, Phys. of Fluids, 10: 1352
16. Deshpande SM and Narasimha R (1969) The Boltzmann collision integrals for a combination of Maxwellians. J Fluid Mech., 36: 545
17. Liu CY and Lees L (1961) Kinetic theory description of plane compressible Couette flow. In: L Talbot edited Rarefied Gas Dynamics, Academic Press, NY, 391
18. Ostmo S, Ytrehus T (1995) Strong evaporation in a dusty polyatomic gas. In: J Harvey and G Lord edited Rarefied Gas Dynamics 19, 1: 291
19. Bhatnagar P L, Gross E P and Krook M. (1954) A model for collision processes in gases. I, Phys. Rev., 94: 511-525
20. Welander P (1954) On the temperature jump in a rarefied gas. Ark Fys, 7: 507-553
21. Vincenti WG and Kruger CH Jr (1965) Introduction to Physical Gas Dynamics. John Wiley & Sons
22. Chapman S, Cowling TG (1970) The Mathematical Theory of Non-uniform Gases, 3rd ed., Cambridge Univ. Press
23. Gross EP and Jackson EA (1959) Kinetic models and the linearized Boltzmann equation. Phys. of Fluids, 2: 432-441
24. Holway LH Jr (1966) New statistical models for kinetic theory: methods of construction. Phys. of Fluids, 9: 1658
25. Boley CD and Yip S (1972) Modeling theory of the linearized collision operator for a gas mixture. Phys. of Fluids, 18: 1424-1433
26. Shen C, Xu XY, Hu ZH and Wu WQ (1994) Transient motion of rarefied gas caused by heat addition simulated by DSMC method. In: BD Shizgal and DP Weaver edited Rarefied Gas Dynamics, Progress in Astronautics and Aeronautics, 159: 234-242
27. Shen C, Yi Z (2000) Direct numerical test of the BGK model equation by the DSMC method. Acta Mechanica Sinica, 16: 133

28. Aoki K, Sone Y, Nishino K, Sugimoto H (1991) Numerical analysis of motion of rarefied gas caused by sudden change of wall temperature. In: AE Beylich edited Rarefied Gas Dynamics, VCH, 222-231
29. Xu K (1993) Numerical Hydrodynamics from Gas Kinetic Theory. Ph D thesis, Columbia University
30. Xu K (2001) A gas-kinetic BGK scheme for the Navier-Stokes equations and its connection with artificial dissipation and Godunov method. J Comput. Phys
31. Li ZH and Zhang HX, (2003) Study on gas kinetic unified algorithm for flows from rarefied transition to continuum. Acta Aerodynamica Sinica, 21: 255-266 (in Chinese)
32. Shakhov EM (1985) Kinetic model equations and numerical results. In: H Oguch edited Rarefied Gas Dynamics, Univ. of Tokyo Press, 1:137-148
33. Nordsieck A, Hicks BL (1967) Monte Carlo evaluation of the Boltzmann collision integral. In: C L Brundin edited Rarefied Gas Dynamics, Academic Press, 675-710
34. Yen SM. (1970) Monte Carlo solutions of nonlinear Boltzmann equations for problems of heat transfer in rarefied gases. International J of Heat Mass Transfer, 14: 1865-69
35. Hicks BL, Yen SM and Reiley BJ (1972) The internal structure of shock waves. J Fluid Mech., 53:85-112
36. Tcheremissine FG (1999) Solution of the Boltzmann equation for arbitrary molecular potentials. In: R Brun et al. edited Rarefied Gas Dynamics, 2: 165-172
37. Broadwell JE (1964) Study of rarefied flow by the discrete velocity method. J Fluid Mech., 19: 404-414
38. Broadwell JE (1964) Shock structure in a simple discrete velocity gas. Phys. of Fluids, 7: 1243
39. Gatignol R (1975) Kinetic theory for a discrete velocity gas and application to the shock structure. Phys. of Fluids 18: 153
40. Gatignol R (1977) Kinetic theory boundary conditions for discrete velocity gases. Phys. of Fluids, 20: 2022
41. Cabannes H (1976) Couette flow for a gas with a discrete velocity distribution. J Fluid Mech., 76: 273
42. Gatignol R (1975) Theorie Cinetique des Gas a Repartition Discrete de Vitesses. Lecture Notes in Physics, Springer, Berlin, 36

43. Cabannes H (1980) The Discrete Boltzmann Equation, Lecture Notes, Univ. of California, Berkeley
44. Goldstein D, Sturtevent B and Broadwell JE (1989) Investigation of the motion of discrete velocity gases. In Progress in Astro-Aeronautics, 118: 100
45. Inamuro T, Sturtevent B (1990) Numerical study of discrete velocity gases. Phys. of Fluids, A2: 2196
46. Bobylev AV, Palczewski A and Schneider J (1995) Discretization of the Boltzmann equation and discrete velocity models. In: J Harvey and G Lord edited Rarefied Gas Dynamics 19, Oxford Univ. Press, Oxford, 2: 857
47. Rogier F and Schneider J (1994) A direct method for solving the Boltzmann equation. Transport Theory and Statist. Physics, 23: 313
48. Gatignol R. (1997) Boundary conditions in discrete kinetic theory-application to the problem of evaporation and condensation, In: C Shen edited Rarefied Gas Dynamics 20, Peking Univ. Press, 253-262
49. Vallander SV (1960) Doklady Akad Nauk SSSR, 131 (1) (in Russian)
50. Aerodynamics of Rarefied Gases, collected works, (1963-) edited by SV Vallander, Leningrad Univ. Press, vol.1-vol.10 (in Russian)
51. Alder BJ and Wainwright TE (1957) Studies in molecular dynamics. Journal of Chem. Physics, 27: 1208-1209
52. Bird GA (1963) Approach to translational equilibrium in a rigid sphere gas. Phys. of Fluids, 6: 1518-1519
53. Bird GA (1965) Shock wave structure in a rigid sphere gas. In: J H de Lesuw edited Rarefied Gas Dynamics, 1: 216-222, Ac. Press
54. Bird GA (1976) Molecular Gas Dynamics, Clarendon Press
55. Bird GA (1994) Molecular Gas Dynamics and the Direct Simulation of Gas Flows. Clarendon Press, Oxford
56. Bird GA (1970) Direct simulation of the Boltzmann equation. Phys. of Fluids, 13: 2676-2681
57. Wagner W (1992) A convergence proof for Bird's direct simulation Monte Carlo method for the Boltzmann equation. J Stat Phys, 66: 1011
58. Pulvirenti M, Wagner W and Zavelani MB (1994) Convergence of particle schemes for the Boltzmann equation. Euro J Mech., B7: 339
59. Haviland JK and Lavin ML (1962) Applications of the Monte Carlo method to heat transfer in a rarefied gas. Phys. of Fluids, 5: 1399-1405

60. Kogan MN (1969) Rarefied Gas Dynamics. Plenum Press
61. Nanbu K (1980) Direct simulation scheme derived from the Boltzmann equation. I Multi-component gases, J Phys Soc Japan, 45: 2042-2049
62. Babovsky H (1989) A convergence proof for Nanbu's Boltzmann simulation scheme. Euro J Mech. B: Fluids, 8 (1): 41-45

7 DIRECT SIMULATION MONTE-CARLO (DSMC) METHOD

7.1 INTRODUCTION

The *direct simulation Monte-Carlo* (*DSMC*) method [1, 2] developed by G. A. Bird is a method basing directly on the physical simulation of the gas flows. Concerning its relation with the Boltzmann equation there are references [3, 4, 5] and others proving its consistence with the Boltzmann equation. In fact, both DSMC method and the Boltzmann equation are based on the same physical reasoning. In handling the molecular collisions and the molecule surface interactions both methods need to introduce physical models. But as DSMC method deals with the actions of individual molecules, it is easier for it to introduce models in agreement with the physical realities, but it is relative difficult to involve the realistic models in the mathematical solution of the Boltzmann equation. This is all the more so in treating problems of gas flows accompanied by chemical reactions and radiation. And as the DSMC method does not depend on the assumption of inverse collisions, it can be applied to such complex phenomena as the recombination reactions involving three-body collisions, which is beyond the capability of the Boltzmann equation. Thus, it is unnecessary and impossible to verify each simulation procedure involving complicated physic-chemical processes by strict proof that it is derived from the Boltzman equation. But still the DSMC method is not a method independent of the Boltzmann equation or a method which stands side by side with the Boltzmann equation. The Boltzmann equation plays fundamental roles in the formulation of the DSMC method. For example, the various molecular models are based on the relation of the collision cross section with the gas viscosity given by the Chapman-Enskog approximation in the solution of the Boltzmann equation (see section 2.4). The sampling of collision pairs and the implementation of collisions in DSMC is based on ensuring the matching of motion and collisions in simulation, keeping it consistent with the actual flow. This is achieved by the satisfaction of conditions expressed through collision cross sections depending in turn on the molecular models (see the explanation in section 7.2). On the other hand, the experimental verification played an important role for the general acceptance of the DSMC method. Both in the aspect of the global characteristics of the flow field and in the aspect of the micro-structure of the flow the DSMC method gave results in good agreement with the experimental data. The result of simulation by the DSMC obtained for the space shuttle in the transitional regime (including the ratio of drag to lift) [6] has excellent agreement with the flight experimental data. The experimental measurement of the distribution functions for the molecular velocity in the flow direction and for the velocity in the transverse direction in the strong shock wave structure was carried out as early as in 1966, but as the result obtained had poor agreement with the available at that time theoretical result (the Mott-Smith solution [7], see also section 6.3), so it had not been published until 1989, and was published in Science [8] only when the DSMC calculations were carried out and excellent agreement with the experimental result was obtained (see Fig. 7.1) and received general recognition of the scientific community. The verification of the DSMC method by the experiments enhanced its status and significance.

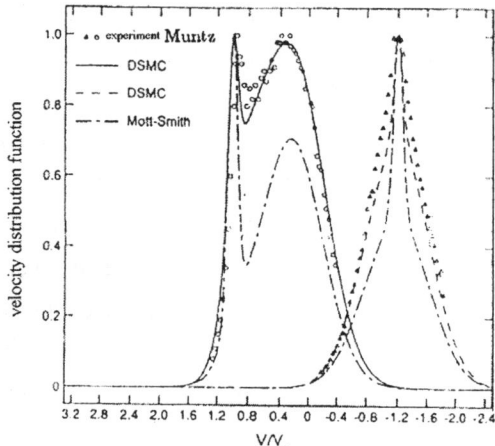

Fig. 7.1 The distribution functions of the parallel and normal velocities in the structure of normal shock wave of helium ($Ma = 25$ at $\bar{x} = 0.565$), comparison of the DSMC calculation and the experimental measurements [8]

The DSMC method employs a large number of simulated molecules to simulate the real gas. The number of simulated molecules must be large enough so that in the cells of the flow field they can fully represent the distribution of the real gas molecules. At the same time this number is much smaller than the number of real molecules, i.e., each simulated molecule represents an enormous number of real molecules. In the computer the position coordinates, the velocity components and the internal energies of each simulated molecules are stored, they are changing with the motion of molecules, their encounters with the boundaries and the collisions between themselves unceasingly. The time parameter in the simulation is identical with the physical time in the real flow. All computations are unsteady, the steady flow is obtained as the long time asymptotic state of the unsteady flow. The essential approximation of the DSMC method is: in a small time step Δt the molecular motion is decoupled with the molecular collisions. In time Δt each simulated molecule moves a certain distance according to its velocity, then the representative collisions between the molecules in time Δt are calculated. The time step Δt must be small in comparison with the local mean collision time. The flow filed is divided into cells of linear size Δr used to choose collision pairs among molecules in them, and also to sum molecular quantities to obtain the macroscopic characteristics of the flow field. The size of Δr should be small in comparison with the scale length of the gradient of the macroscopic quantities of the flow field, in general, letting Δr to be 1/3 or so of the mean free path can satisfy this condition. Corresponding to the motion in time Δt, how to choose the collision pairs and how many representative collisions are to be implemented, is a key issue in ensuring the consistency of the simulation with the processes occurred in real gas. This issue will be discussed specially in the next section.

The calculation of the motion in time Δt is a simple and direct deterministic step. Whenever a molecule encounters with the boundary the interaction of it with the boundary is taken into account, the DSMC method allows the introduction of various reflection models, about them detailed discussion has been given in Chapter 3. Except the specular reflection, which is a simple deterministic action, the implementation of the diffuse reflection, the Maxwellian reflection and the CLL model reflection requires the employment of the random fraction in obtaining the sampling of the velocity after reflection (see sections 3.2 and 3.4). Then the molecules move according to the after-collision velocities, and the new positions the molecules move to in the remaining time are obtained.

What are the velocities of the molecules after collisions depends on what kind of molecular models are being employed, this plays a determinant role in the exact modeling of the real gas flows. The DSMC method allows the introduction of the phenomenological models capable of reflecting the essential features of the flow field, in particular, the VHS model, the VSS model, the GHS model and the GSS model and their implementation in the DSMC simulations have been discussed in sections 2.4.4~2.4.7.

In section 7.2 the sampling and the counting of collisions will be discussed, in section 7.3 the implementation of the DSMC method will be explained through a simple example of Couette flow. In sections 7.4 and 7.5 the methods of handling the excitation and relaxation of the internal energies and the chemical reactions are introduced. In section 7.6 some development of the general codes in solving the complicated flow field and the position element method are introduced.

7.2 SAMPLING OF COLLISIONS

The essential idea of the DSMC method is the decoupling of the molecular motion and collisions. The correct implementation of the collision sampling, i.e., the appropriate choice of the collision pairs and the implementation of a certain number of collisions is the key issue in ensuring the matching of the collisions with motion in time Δt and the consistency of the simulation with the actual flow processes.

The way of implementation by the *time counter* (*TC*) method introduced by Bird [1] is: (1) Choose one pair randomly from all available molecules out of a cell; (2) Take

$$S(c_r)/S_{max}, \qquad (7.1)$$

as criterion, where $S(c_r) = c_r \sigma_T$, S_{max} is the maximum of $S(c_r)$ in the cell. To this apply the acceptance-rejection method to determine if this pair is selected, if not, return to (1); (3) If the pair is selected the collision is calculated according to the molecular model, the cell time is advanced a value

$$\Delta t_c = \left[\frac{nN}{2}S(c_r)\right]^{-1}, \qquad (7.2)$$

where n and N are the number density and the number of molecules in the cell; (4) the processes (1), (2), (3) are repeated until the cell time exceeds Δt.

The mean collision frequency of the molecules is (see Eq. (2.207))

$$v = n\overline{S(c_r)}. \qquad (7.3)$$

The number N_t of collisions occurred in time Δt within a cell is

$$N_t = \frac{N}{2}v\Delta t = \frac{nN}{2}\overline{S(c_r)}\Delta t. \qquad (7.4)$$

From Eq. (7.4) it is seen intuitively, that whenever a collision occurs, letting the cell time to advance a value Δt_c (see Eq. (7.2)) will ensure the correct collision frequency in the cell. There is a strict analytic proof in reference [1].

The TC method had been widely used for its high efficiency and for it can ensure the correct collision frequency when the number of molecules in a cell is large enough. But when the number of molecules in a cell is not large enough, the occasionally selected collision with very small probability ($S(c_r)/S_{max}$ very small) would make the cell time advance too long a time (surpassing several Δt) and distort the collision frequency, leading to errors. Koura [9] suggested the *nil collision* (*NL*) method, Ivanov et al. [10] suggested the *major frequency* (*MF*) method, which can overcome this defect of the TC method and does not lead to much increase of the computation time.

Bird [11] suggested the *no time counter* (*NTC*) method, which is widely used in many simulations. The NTC method is a modification of the *direct* or *Kac method*. The direct method considers all possible collision pairs in a cell, the number of them is

$$N_D = \frac{N(N-1)}{2}. \qquad (7.5)$$

The probability P_D of the occurrence of collision of the two molecules of one collision pair in time Δt equals to the ratio of the volume swept by the collision section with the relative velocity c_r to the cell volume

$$P_D = \frac{F_N S \Delta t}{V_c}, \tag{7.6}$$

where F_N is the number of the actual molecules one simulated molecule represents. The efficiency of the direct method is low, for P_D is very small, and the computation time needed for all collision pairs is proportional to the squire of the number N of molecules in the cell. The NTC method only considers a very small portion of the N_D collision pairs (N_D is multiplied by a small factor), but at the same time P_D is amplified accordingly by the reciprocal of the same proportionality. It is readily seen that when this small factor is taken as

$$F_N S_{max} \Delta t / V_c,$$

the probability of occurrence of the collision is changed into

$$P_{NTC} = S/S_{max},$$

that is, is equal to the criterion Eq. (7.1) of TC method. And, the number of the collision pairs need to be considered is changed accordingly

$$N_{NTC} = \frac{N\overline{N}}{2} F_N S_{max} \Delta t / V_c, \tag{7.7}$$

where as F_N is large, $N(N-1)/2$ was replaced by $N^2/2$, and to keep the linear relation of N_{NTC} with the random number N, N^2 was replaced by $N\overline{N}$ ($\overline{N}$ is the time or ensemble average of N). Noting that $\overline{N}F_N/V_c$ in fact is the number density n in the flow field, then according to Eq. (7.7) the number of collision pairs to be considered is actually

$$N_{NTC} = \frac{nN}{2} S_{max} \Delta t. \tag{7.8}$$

The probability that the collision pair is selected is given by Eq. (7.1), it is seen that the number of collisions that actually happened agrees with the theoretical value (Eq. (7.4)).

We suggested the *randomly sampled frequency* (*RSF*) method [12]. The way of implementation of it is: M pairs of molecules are randomly selected from all possible pairs with returning back when not selected, take

$$\overline{S}_{RSF} = \frac{1}{M}\sum_{j=1}^{M} S(c_{ri}) \tag{7.9}$$

as the estimate of $\overline{S(c_r)}$, thus the number of collisions occurred in each cell is (see Eq. (7.4))

$$N_{RSF} = \frac{nN}{2}\overline{S}_{RSF}\Delta t . \tag{7.10}$$

The pairs of molecules are chosen following the same steps as in the TC method and are accepted or rejected until the number of collisions reaches the number of Eq. (7.10). The analysis and test calculation in [12] show that, when M is chosen as 1, sufficient accuracy can be reached.

Note, the NTC method calculates the number of collisions pairs that need to be considered according to Eq. (7.8), while the RSF method calculates the number of collisions that should happen in each cell according to Eq. (7.10), all these numbers thus calculated have to be rounded-off to integers and the remainders of the number of collisions have to be stored. The concrete way of doing will be elucidated in the practical example.

7.3 EXAMPLE OF SOLUTION OF PROBLEM BY THE DSMC METHOD

In this section the details of concrete implementation of the DSMC method is presented through the example of a simple, one-dimensional problem of *Couette flow*. The space between two parallel plane plates separated by a distance $Y-LENGTH = \lambda/Kn$ is filled with gas of temperature $273K$ and pressure $0.01 atm$. The lower plate is located at $y=0$ and moves with a velocity $U\_WALL/2$ in the direction of positive x, the upper plate is located at $y = Y-LENGTH$ and moves with a velocity $-U\_WALL/2$ in the direction of negative x. The wall temperatures of the two plates are the same as the gas temperature, the gas molecules are reflecting diffusely at the wall. The FORTRAN program of the solution of the Couette flow problem is listed in APPENDIX IV.

The flow chart of the program of problem by the DSMC method is shown in Fig.7.2. This flow chart is suitable for the solution of any flow problem, and includes the cases of steady and unsteady flows. The program of the present example also follows this chart. The FORTRAN code of the Couette flow problem consists of the main program, the FUNCTION RF(IDUM) generating the random fraction and 7 subroutines. The 7 subroutines correspond to the 7 steps in the flow chart (see Fig 7.2):

subc1   set constants and initial values
subc2   set the initial velocities and positions of molecules
subc3   calculation of the motion and the reflection of the simulated molecules at the surfaces
subc4   the reordering and the indexing of the molecules
subc5   calculation of the collisions
subc6   sampling of the flow properties
subc7   the summation of the characteristics of the flow field and on the surfaces

In the main program first the adjustable number of cells (no-cell), the number of molecules in each cell (no-molecule-each-cell) and the total number of molecules (no-molecule) are specified by the PARAMETER statement. As the IP (information preservation) method introduced in the next Chapter uses the same program for illustration, the statements needed for the IP method are marked by the symbols * and **. The former marks statement that is used to replace the one before it, the latter marks the statement that needed to be added anew. For example, the IP method needs to use different number of cells (no-cell, 300 is used to replace 50) and different total number of molecules (no-molecule, 9000 to replace 1500), so the PARAMETER statements marked by * are used to specify the values required by the IP method. Then the COMMON and DIMENSION statements introduce the variables which are commented in detail in the program and are not explained here. After the assignment of the reverse rkn of the Kn number, i.e., the specification of the Kn number, the subroutines of subc1,... are called successively to implement the various steps of the DSMC simulation. Note

7.3 EXAMPLE OF SOLUTION OF PROBLEM BY THE DSMC METHOD

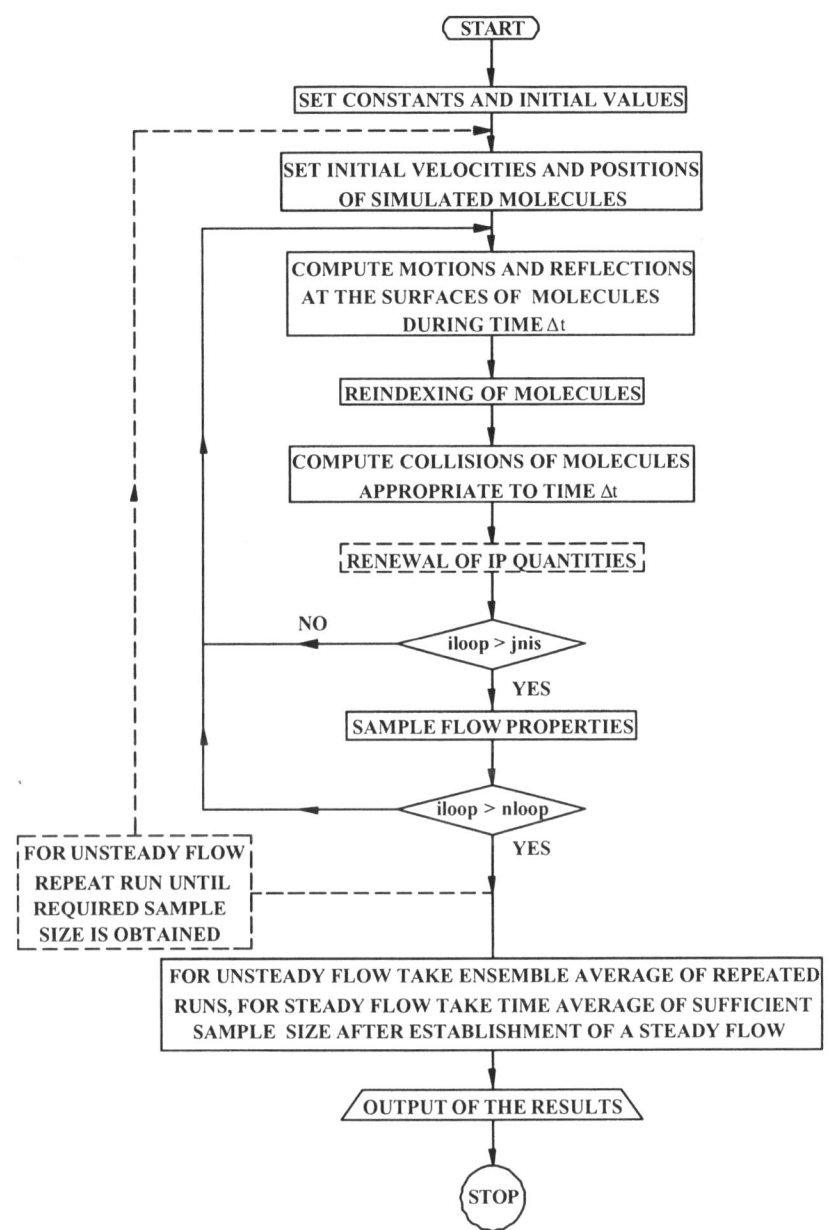

Fig. 7.2 Flow chart of program of the DSMC-IP method

in the program the following parameters are used to judge whether the code continues to proceed or goes back to the beginning of a loop.

iloop: the number of current loops; *jnis*: the number of loops at which the sampling starts; *nloop*: the total number of loops.

The statement CONTINUE labeled 100 is the position where the loop starts, after it the loop counter *iloop* adds 1 to the record. After the calculation of collisions by calling subc5 judge if the number (*iloop*) of current loops is larger than the number (*jnis*) of loops at which the sampling starts. If yes, the quantities on the wall and in the flow field are sampled, if no, provided that *iloop* is less than *nloop* go back to the beginning of the loop labeled 100. Proceed the cycle until the total number (*nloop*) of loops is reached, and then accomplish the sampling and output. The surface parameters are stored in surf-cout.dat and the flow field parameters are stored in u-cou.dat.

In the subroutine **subc1** SET CONSTANTS AND INITIAL VALUES, the ordinary constants are explained by the comment lines following them. As helium is chosen as the media and HS model is adopted DIAREF is taken as $3.659 \times 10^{-10} m$ (Table 2 of Appendix I). In the IP method DIAREF is taken as $3.963 \times 10^{-10} m$, this statement is marked with $*$. Although in the present program the HS model is adopted, but in preparing the expression for TKOM to be used in the calculation of the collision frequency the notation of VHS-coe is retained, all one has to do is let $VHS - coe = (\Gamma(5/2 - \omega))^2$, the VHS model is obtained (see the comment line before the assignment statement of TKOM). Then the initial temperature, pressure and the number density FND are set. Note, that in the DSMC method the wall velocity U-WALL is taken as $100 m/s$, but in the IP method the wall velocity U-WALL is taken as $1.0 m/s$. The molecular mean free path λ (AMDA) is calculated according to the hard sphere model (see Eq. (2.221)). The scope of the flow field Y-LENGTH is taken as λ/Kn, and the time step DTM is taken as $0.23\lambda/c_m$, where c_m is the initial most probable speed denoted by VM_INI in the program. AREA is a representative cross section area through which we observe the flow field corresponding to the number of molecules and normal to the direction (y direction) of variation of the flow field. In the whole flow field only NO-MOLECULE molecules are introduced, and the scope of the flow field in the y direction is Y-LENGTH. If we observe the NO-MOLECULE molecules through

AREA obtained in the program, then the number density in the simulation is just the initial number density FND (=NO_MOLECULE/Y_LENGTH * AREA). This AREA will be needed when calculating the molecular flux towards the surface.

In the subroutine **subc2** the initial velocities and positions of all simulated molecules are given. The initial velocity is the initial macroscopic velocity plus the thermal velocity in the equilibrium gas, the generation of the components of the latter see Eq. (3.18) and Eq. (3.19) and the discussion followed in section 3.2. In the allocation of the positions of the simulated molecules in the cells first the molecules are distributed according to the number NO_MOLECULE _EACH_CELL of molecules in each cell to various cells, and then assign randomly the locations of the NO_MOLECULE molecules in various cells. In comparison with the direct allocation of all the simulated molecules between 0 and Y_LENGTH this method has taken into account the variance reduction principle. In fact, according to Eq. (III.5) of the Appendix, in which if take $b = Y\_LENGTH, a = 0$, x would be the random position of the molecule, but the scatter-deviation would be very large. Following the variance reduction principle the sampling should have been taken according to (III.5)', and should take $m =$ NO_MOLECULE, the positions of the molecules of the initial distribution possesses the reduced variance. It is a general issue to take into account the variance reduction principle, for example, when allocate the initial positions of the oncoming molecules in three-dimensional flows, the equation (III.5)' should be repeatedly used in two directions normal to the oncoming flow to fix the initial coordinates of the molecules.

The subroutine **subc3** is used to calculate the motion of molecules and the reflection and sampling at the upper and lower plates. Detailed explanation is given in the comment lines of the program for the upper plate. Note, before the accumulative counting of the contribution of the incident molecules, the reference frame is transferred to be connected with the upper plate (moving with velocity –U_WALL/2). In the DSMC method it is required that $p(1,m) = p(1,m)$ +U_WALL/2.d0, for the IP method it is required that $vmean(1,m)$ $= vmean(1,m) + U\_WALL/2.d0$. In UP_WALL(I), I=1, 2, 3, 4, the following quantities of the incident molecules are accumulatively counted: (1) number of molecules; (2) tangential momentum; (3) normal momentum; (4) kinetic energy. For the sampling of the molecular velocities after the diffuse reflection from the plate see Eqs. (3.18), (3.19), (3.15) and (3.20) of section 3.2. In UP_WALL(I), I=5, 6, 7 the following quantities of the reflected molecules are accumulatively counted: (5) tangential momentum; (6) normal momentum; (7) kinetic energy. Then the reference frame is transferred back to be connected with the stationary system and the new positions the molecules move to are calculated. The reflection and the sampling on the lower plate are similar with those on the upper plate.

The subroutine **subc4** is used for the reordering and indexing of the molecules. The molecule with the original identifying number M moves into the cell with order number NCELL. IC(1,I) stores the number of molecules in the Ith cell, IC(2,I) stores the order number of the first molecule in the Ith cell minus 1. When recounting IC(1,NCELL), by letting $K = IC(2, NCELL) + IC(1, NCELL)$ the new order number K of the molecule with the original identifying number M is obtained. The original number M is stored in LCR(K).

The subroutine **subc5** is used to calculate the collisions. Subroutine consists of a loop which successively calculates collisions in each cell. Before the label 300 the average (TKOM * VAVER) of $S(c_r)$ (see Eq. (7.1) and below) in the cell is calculated, and the number of collisions N_{RSF} (CNOIC, see Eq. (7.10)) that should occur in the cell according to the randomly sampled frequency method is obtained. Take into account the stored remainder of the number of collisions and after rounding-off anew the number of collisions NCOLL that should actually happen is calculated. After the label 300 two molecules are chosen randomly from the cell and are been examined whether they are selected or not according to the step involving Eq. (7.1), if yes, the velocities after collision are calculated, this process continues until the number of collisions in the cell reaches NCOLL, and the collisions in the next cell are calculated. In the above the number of collisions in a cell is controlled by the randomly sampled frequency method. The NTC method is the most used method for controlling the number of collisions. It is readily to modify the present program to shift to the NTC method. For this one needs to calculate first, before the label 300, the number N_{NTC} of collision pairs to be considered in the cell (see Eq. (7.8)), where S_{max} is obtained by multiplying TKOM by YRMAN_ETA51, and then change the program segment between the statement labeled 300 and the statement labeled 140 into a program segment cy-

7.3 EXAMPLE OF SOLUTION OF PROBLEM BY THE DSMC METHOD

cling N_{NTC} times (the number N_{NTC} obtained from Eq. (7.8) should be rounded-off and the remainder of the number of collisions should be stored, etc.).

The subroutine **subc6** runs the summation of the parameters of the flow field and stores the results in FIELD(N, NO_CELL), the subroutine **subc7** outputs these parameters and stores the parameters on the surface in surf-coue.dat, the parameters of the flow field in u_coue.dat.

The program of the present section has been aimed at the problem of Couette flow, but is readily modified to be applied to the solution of the problem of the Poiseuille flow and the Rayleigh probem. *Poiseuille flow* as the Couette flow is a steady flow problem, so the structure of the main program is not changed, the changes are limited to the changes in subroutine subc 3. At the plate surfaces the condition of stationary wall is put forward. At the same time, due to the influence of the pressure gradient the velocities of the molecules during every time DTM gain increments, which can be found from the momentum conservation equation. When the change of the pressure is given by the following equation

$$p = p_0(1 + \alpha x / h),$$

where $h = Y\_LENGTH$ is the distance between the two plates, the increment of the velocity in Δt can be found to be (cf. the beginning part of section 8.5)

$$\Delta u = -(\alpha p_0 / \rho h) \Delta t . \tag{7.11}$$

This increment is to be added to the thermal velocities and the IP velocities of the simulated molecules. The molecules are retarded near the plate surfaces. The equilibrium between acceleration and the retardation enable the flow to reach the steady state.

For the *Rayleigh problem* according to its geometric condition the boundary conditions in the subroutine subc3 is certainly to be modified. The lower plate moves with velocity U_WALL in the positive direction of x. The scope of the flow field Y_LENGTH should be taken large enough, so that the disturbance originated from the lower plate could not reach the upper boundary of the flow field during the time interval we are interested in. At the same time the specular reflection boundary condition is set at the upper plane of the boundary. As the Rayeigh problem is an unsteady flow problem, so the structure of the main program must be changed. The location of the beginning of the loop, i.e., the location of the CONTINUE statement labeled 100 and the statement $illop = illop + 1$ following it, should be shifted from the location before CALL SUBC3 to that before CALL SUBC 2 (set the initial velocities and positions of the molecules). The conditional statement after CALL SUBC5 is to be cancelled, the program segment between CALL SUBC 3 and CALL SUBC 6 is to be made a loop by a DO statement to accomplish repeatedly the procedures "calculation of the molecular movement", "the indexing of the molecules", "the calculation of collisions" and "the sampling of flow properties", thus obtain the parameters of the flow field and on the surface at various time moments. The conditional statement after the loop is to allow return to the statement with label 100, so that to start the loop beginning from the initial state. This is to ensure the obtaining of large enough sampling size at various time moments utilizing the ensemble averaging, this is necessary for small storage of the micro computers. If the workstation of large enough storage is available, it is not necessary to restart the whole process from the statement labeled 100, obtaining large enough sampling size with one run is possible. This is the right means to demonstrate an unsteady process in real-time and to study the problem of instability.

7.4 THE EXCITATION AND RELAXATION OF THE INTERNAL ENERGIES

7.4.1 INTRODUCTION OF PHENOMENALOGICAL MODELS

For diatomic and polyatomic molecules, as we have seen in Chapter 1, the molecules possess as well internal energies, i.e., the rotational and vibrational energies. How to simulate the behavior of the internal energies in the direct simulation is the issue to be resolved in this section. In the equilibrium state various molecules can be assigned according to Eq. (1.94) certain rotational energy ($\zeta = 2$) and vibrational energy (ζ_v given by Eq. (1.86)). The problem of how to simulate the energy exchange between molecules in the collision is to be solved. Introduction of the traditional models that reflect the diatomic characteristics of the molecules, such as the model of dual repulsive centers, the sphero-cylindrical model, the rough-sphere model and the harmonic oscillator model, into the DSMC simulation is unpractical, for in the simulation an enormous number of collisions must be computed. And for the purpose of simulation it is essential to know, how the energies are distributed after the collisions, and to ensure, that the rate of excitation and relaxation of the internal energies be in agreement with that given by the experiments. Larsen and Borgnakke [13, 14] introduced a phenomenological model and resolved this problem. The central idea is to assume that the kinetic energy (expressed by the relative velocities) and the internal energy follow the conservation of energies, the internal energy after collision is sampled according to the equilibrium distribution of the combination of the kinetic energy and the internal energy, and the rate of relaxation process of the energy is determined by adjusting the proportion of the elastic collisions and the inelastic collisions to make it satisfy the result obtained from the experiment. Such an approach is very simple in the conceptual and implemental aspects, and needs not much computation time. Such phenomenological philosophy deserves attention in approaching other problems. In the following the Larsen-Borgnnake model will be described, firstly in the assumption of the continuum distribution of the internal energies, the case of collision between molecules of different components in a gas mixture will be discussed. When discussing the continuum distribution of the vibrational energy, the combined cumulative distribution acceptance-rejection method in handling the distributions with singularities own to $\zeta_v < 2$ is introduced. Then the implementation of the Larsen-Borgnakke model in the case of discrete vibrational energy levels is discussed. Finally, the adjustment of the relaxation time by introducing the exchange probability of the vibrational energy is discussed.

7.4.2 IMPLEMENTATION OF LARSEN-BORGNAKKE MODEL

In this section the Larsen-Borgnakke model is explained on the example of collision between molecules of different components in the gas mixture. According to Eq. (1.94) the distribution functions of internal energies of component 1 and component 2 are proportional to

$$f(\varepsilon_{i1}) \propto \varepsilon_{i1}^{\frac{\zeta_1}{2}-1} e^{-\frac{\varepsilon_{i1}}{kT}}, \tag{7.12}$$

$$f(\varepsilon_{i2}) \propto \varepsilon_{i2}^{\frac{\zeta_2}{2}-1} e^{-\frac{\varepsilon_{i2}}{kT}}, \tag{7.13}$$

respectively. Now the distribution function of the total energy of the collision pair

$$E_i = \varepsilon_{i,1} + \varepsilon_{i,2} \tag{7.14}$$

is to be found according to Eq. (7.12) and (7.13).

The probability that the internal energy of component 1 is in the interval from $\varepsilon_{i,1}$ to $\varepsilon_{i,1} + d\varepsilon_{i,1}$ and the internal energy of component 2 is in the interval from $\varepsilon_{i,2}$ to $\varepsilon_{i,2} + d\varepsilon_{i,2}$ is proportional to

$$\varepsilon_{i,1}^{\frac{\zeta_1}{2}-1} (E_i - \varepsilon_{i,1})^{\frac{\zeta_2}{2}-1} e^{-\frac{E}{kT}} d\varepsilon_{i,1} d\varepsilon_{i,2}. \tag{7.15}$$

Fix first the value of $\varepsilon_{i,1}$, then $dE_i = d\varepsilon_{i,2}$, the integration over $\varepsilon_{i,1}$ from 0 to E_i results in that the probability that the total internal energy is in the interval from E_i to $E_i + dE_i$ is proportional to

$$f(E_i)dE_i \propto \left(\int_0^{E_i} \varepsilon_{i,1}^{\frac{\zeta_1}{2}-1} (E_i - \varepsilon_{i,1})^{\frac{\zeta_2}{2}-1} d\varepsilon_{i,1} \right) \cdot e^{-\frac{E_i}{kT}} dE_i =$$

$$E_i^{\frac{\zeta_1+\zeta_2}{2}-1} \frac{\Gamma\left(\frac{\zeta_1}{2}\right)\Gamma\left(\frac{\zeta_2}{2}\right)}{\Gamma\left(\frac{\zeta_1+\zeta_2}{2}\right)} e^{-\frac{E_i}{kT}} dE_i,$$

that is

$$f(E_i) \propto E_i^{\bar{\zeta}-1} \exp\left[-\frac{E_i}{kT}\right], \tag{7.16}$$

where

7.4 THE EXCITATION AND RELAXATION OF THE INTERNAL ENERGIES

$$\bar{\zeta} = \frac{\zeta_1 + \zeta_2}{2} \tag{7.17}$$

is the mean value of the internal degrees of freedom of the two molecules.

In section 2.11 we discussed the mean value of the quantity Q in collision and obtained the expression Eq. (2.224). Expressing the relative velocity c_r by the relative translational energy $\varepsilon_t = (1/2)m_r c_r^2$, one obtains

$$\bar{Q} \propto \int_0^\infty Q \varepsilon_t^{3/2-\omega} \exp\left(-\frac{\varepsilon_t}{kT}\right) d\varepsilon. \tag{7.18}$$

That is, the *distribution function* $f(\varepsilon_t)$ of the *relative translational energy* is proportional to

$$f(\varepsilon_t) \propto \varepsilon_t^{3/2-\omega} \exp\left(-\frac{\varepsilon_t}{kT}\right),$$

where ω is the temperature power exponent of the viscosity of the (mono-component) gas.

Suppose for different components 1, 2 a temperature power exponent $\omega_{1,2}$ of viscosity can be determined, then the distribution function of the translational energy in the collision between molecules of components 1 and 2 of the gas mixture can be written as

$$f(\varepsilon_t) \propto \varepsilon_t^{3/2-\omega_{12}} \exp\left(-\frac{\varepsilon_t}{kT}\right). \tag{7.19}$$

The total energy E_c in the collision is the sum of the relative translational energy ε_t and the total internal energy E_i

$$E_c = \varepsilon_t + E_i. \tag{7.20}$$

According to Eqs. (7.16) and (7.19) the distribution function of the translational energy ε_t and the total internal energy E_i can be written as

$$\varepsilon_t^{3/2-\omega_{12}} E_i^{\bar{\zeta}-1} \exp\left[-\frac{(\varepsilon_t + E_i)}{kT}\right],$$

or

$$\varepsilon_t^{3/2-\omega_{12}}\left(E_c-\varepsilon_t\right)^{\bar{\zeta}-1}\exp\left[-\frac{E_c}{kT}\right].$$

As the effective temperature T in the collision is determined by the total energy E_c in the collision (see Eq. (7.26) to appear in the following), so the exponential term is a constant, and the translational energy in the collision is proportional to

$$\varepsilon_t^{3/2-\omega_{12}}\left(E_c-\varepsilon_t\right)^{\bar{\zeta}-1}. \tag{7.21}$$

In the collision the total energy $E_c=\varepsilon_t+E_i=\varepsilon_t+\varepsilon_{i,1}+\varepsilon_{i,2}$ remains unchanged. In the inelastic collision the translational energy ε_t^* and the internal energy after collision are sampled according to the equilibrium distribution Eq. (7.21), i.e., ε_t^*/E_c is sampled by the acceptance-rejection method from the following expression

$$f\left(\frac{\varepsilon_t^*}{E_c}\right)\propto\left(\frac{\varepsilon_t^*}{E_c}\right)^{\frac{3}{2}-\omega_{12}}\left(1-\frac{\varepsilon_t^*}{E_c}\right)^{\bar{\zeta}-1}, \tag{7.22}$$

the *relative velocity after collision* is obtained thereby equal to $\sqrt{2\varepsilon_t^*/m_r}$.

From Eq. (7.15) it is seen that the distribution of the post-collision internal energy $E_i^*=E_c-\varepsilon_t^*$ between the two molecules is sampled from the following formula by using the acceptance-rejection method

$$f\left(\frac{\varepsilon_{i,1}^*}{E_i^*}\right)\propto\left(\frac{\varepsilon_{i,1}^*}{E_i^*}\right)^{\frac{\zeta_1}{2}-1}\left(1-\frac{\varepsilon_{i,1}^*}{E_i^*}\right)^{\frac{\zeta_2}{2}-1}. \tag{7.23}$$

The description of the Larsen-Borgnakke model given here is aimed at the collisions between molecules of different components in the gas mixture. There is not any difficulty to apply the above method in the description of collisions between mono-component molecules. For the mono-component gas one has in Eq. (7.22) $\omega_{1,2}=\omega$, $\bar{\zeta}=\zeta$, and in Eq. (7.23) $\zeta_1=\zeta_2=\zeta$.

In the analysis leading to Eqs. (7.22) and (7.23) it was implicitly assumed that the internal energy includes all modes of the internal energies. In fact, the internal

7.4 THE EXCITATION AND RELAXATION OF THE INTERNAL ENERGIES

energies of various modes after collision can be distributed one by one successively with the translational energy utilizing the Larsen-Borgnakke method, the result will be the same as if to distribute all the internal energy modes together with the translational energy (and then by using Eq. (7.23) to distribute the internal energy among various modes). This can be verified by computation test: for example, the result of distribution of the total internal energy and the translational energy according to Eq. (7.22) and the later distribution of the rotational energy ($\zeta_r = 2$) and the vibrational energy (ζ_v according to Eq. (1.86)) according to Eq. (7.23) is the same as the result of distribution of the rotational energy and the vibrational energy separately with the translational energy according to Eq. (7.22). In Bird's book there is an analytical proof of this inference (see section 5.5 of [2]).

7.4.3 CASES OF DISTRIBUTIONS WITH SINGULARITIES, GENERAIZED ACCEPTANCE-REJECTION METHOD

When distribute various energies in the collisions according to the equilibrium distribution in the Larsen-Borgnakke model, there are the following cases when in the distributions appear singularities.

1. When distribute the initial vibrational energy according to the temperature of the gas, the equilibrium distribution Eq. (1.94) is used

$$f(\varepsilon_v) = \frac{1}{\Gamma(\zeta_v/2)} \frac{\varepsilon_v^{\frac{\zeta_v}{2}-1}}{(kT)^{\zeta_v/2}} e^{-\varepsilon_v/kT}, \qquad (7.24)$$

where the number ζ_v of the vibrational degree of freedom (see Eq. (1.86)) is less than 2

$$\zeta_v = \frac{2\Theta_v/T}{\exp(\Theta_v/T)-1} < 2. \qquad (7.25)$$

It is seen that, when $\varepsilon_v \to 0$, singularity appears, as the power exponent of ε_v is less than 0. Then it is impossible to use directly the acceptance-rejection method to sample energy according to Eq. (7.24). Near the singularity by using the method of taking the truncated value can not gain the exact result as well. For example by using the truncated value acceptance-rejection method according to Eq. (7.24) to distribute the vibrational energy, the average of the sampled values could not reach the accurate correct average value ($\zeta_v/2)kT$ (see Eq. (1.95)).

2. The sampling of the translational energy after the collision is done according to Eq. (7.22), when in the collision pair one molecule is monatomic molecule and only the vibrational mode of the other molecule is considered, one has $\zeta_1 = 0$, . Then $\bar{\zeta} < 1, \zeta_2 < 2$ in the equation (7.22) near $\varepsilon_t = E_c$ appears singularity.

3. When the internal energy E_i^* after collision is distributed between the two molecules (according to Eq. (7.23)), and when one or both of ζ_1 and ζ_2 is less than 2, the singularity appears at $\varepsilon_{i,1} = 0$ and $\varepsilon_{i,1} = E_i$. After the assignment of the internal energy to a molecule, the distribution between different modes is also assigned according to Eq. (7.23), then also appears the case of $\zeta_1 < 2$ or $\zeta_2 < 2$ leading to singularity. The combined consideration of vibrational energy and the rotational energy can make number of the degrees of freedom of the internal energy to be $\zeta_i = \zeta_r + \zeta_v$ to avoid the sampling from the singular distribution. But in the initial distribution of the vibrational energy and the distribution of the internal energy after collision between the vibrational and the rotational modes, due to the fact of $\zeta_v < 2$, the singularity is inevitable. In the equation (7.22) and equation (7.24) the possible singularity is the single point singularity, but in equation (7.23) it is possible to have singularities both at $\varepsilon_{i,1} = 0$ and $\varepsilon_{i,1} = E_i$, i.e., it is a double singular distribution.

We developed *the generalized acceptance-rejection method*, or *the combined cumulative distribution acceptance-rejection method*, and resolved the problem of random sampling from the single or double singular distribution [15, 16]. The explanation and the example of application of the method is given in the Appendix III.3.

In handling the vibrational mode of energy besides the problem of appearance of singularities in the distribution another problem raised is how to determine the *effective temperature in the collision* to be used for giving the vibrational degrees of freedom through Eq. (7.25). The total energy in the collision of component 1 is

$$\varepsilon_t + \varepsilon_{i,1}.$$

7.4 THE EXCITATION AND RELAXATION OF THE INTERNAL ENERGIES

The number of degrees of freedom of the relative translational energy in the collision with component 2 is (from comparison of Eq. (7.19) and Eq. (1.94))

$$\zeta_t = 5 - 2\omega_{12}.$$

The number of rotational degrees of freedom is $\zeta_{r,1}$, the number $\zeta_{v,1}$ of vibrational degrees of freedom is given by Eq. (7.25). Thus (compare with Eq. (1.95))

$$\varepsilon_t + \varepsilon_{i,1} = \frac{1}{2}\left(5 - 2\omega_{12} + \zeta_{r,1} + \frac{2\Theta_{v,1}/T_1}{\exp(\Theta_{v,1}/T_1)}\right) kT_1,$$

from where the effective temperature in a collision of component 1 is obtained

$$T_1 = \frac{(\varepsilon_t + \varepsilon_{i,1})/k}{\frac{5}{2} - \omega_{12} + \zeta_{r,1}/2 + \frac{\Theta_{v,1}/T_1}{\exp(\Theta_{v,1}/T_1) - 1}}. \tag{7.26}$$

This temperature is determined from the method of iteration, after this the number of degrees of freedom of component 1 is determined from Eq. (7.25).

7.4.4 LARSEN-BORGNAKKE METHOD FOR DISCRETE ENERGY LEVELS

In section 1.3 of Chapter 1 we have seen that the continuum distribution Eq. (1.94) (i.e., Eq. (7.24), where ζ_v is given by Eq. (1.86), i.e., Eq. (7.25)) of the vibrational energy is actually derived from the discrete distribution. And as the spacing between the vibrational energy levels is quite large (see section 1.1), it is appropriate to characterize the vibrational energy of the molecules directly by the discrete energy levels. Haas et al. [17] and Bergemann and Boyd [18] expanded the Larsen-Borgnakke method to the discrete energy levels. The implementation of it in the DSMC simulation is direct and simple, and is introduced here.

The harmonic oscillator model of the quantum mechanics gives the vibrational energy of level n (see Eq. (1.16), Eq. (1.77))

$$\varepsilon_{v,n} = nh\nu = nk\Theta_v, \quad n = 0,1,2,\cdots, \tag{7.27}$$

where

$$\Theta_v = h\nu/k.$$

The equilibrium distribution function of the vibrational energy is Eq. (1.68)', where Q_v is given by Eq. (1.79). Substituting the values of Eq. (7.27) and introducing the Dirac δ function, one can write this distribution in the following form

$$f(\varepsilon_{v,n}) = \frac{1}{kT}\left[1 - \exp\left(-\frac{\Theta_v}{T}\right)\right] \times \exp\left(-\frac{\varepsilon_v}{kT}\right)\delta(\varepsilon_v - nk\Theta_v). \qquad (7.28)$$

Now consider the redistribution of the energy levels after collision of a molecule of vibrational energy level n. The total energy in the collision in such case is $E = \varepsilon_t + \varepsilon_v = \varepsilon_t + nk\Theta_v$. The distribution of ε_t is given by Eq. (7.19), the distribution function of the combination of ε_t and ε_v is proportional to

$$(E_c - \varepsilon_v)^{3/2-\omega_{12}} \exp(-E_c/kT)\delta(\varepsilon_v - nk\Theta_v). \qquad (7.29)$$

The assignment of the energies after collision follows the distribution

$$(E_c - \varepsilon_v^*)^{3/2-\omega_{12}} \exp(-E_c/kT)\delta(\varepsilon_v^* - n^*k\Theta_v). \qquad (7.30)$$

Because E_c is a constant, so the exponential term is constant. Normalizing the above formula by its maximum value (at $\varepsilon_v = 0$), one obtains the distribution after the collision

$$\left(1 - \frac{\varepsilon_v^*}{E_c}\right)^{3/2-\omega_{12}} \delta(\varepsilon_v^* - n^*k\Theta_v). \qquad (7.31)$$

Or write it in a form of discrete value

$$\left(1 - \frac{n^*k\Theta_v}{E_c}\right)^{3/2-\omega_{12}}. \qquad (7.31)'$$

To this distribution apply the acceptance-rejection method to determine, whether the energy level n^* after collision is selected or not. n^* is an integer selected from the interval 0 to n^*_{max} with uniform probability, where

7.4 THE EXCITATION AND RELAXATION OF THE INTERNAL ENERGIES

$$n^*_{max} = \lfloor E_c/(k\Theta_v) \rfloor. \tag{7.32}$$

The symbol $\lfloor \ \rfloor$ represents rounding-off.

7.4.5 RELAXATION COLLISION NUMBER AND VIBRATIONAL EXCHANGE PROBABILITY

The above discussion concerns inelastic collisions. If in the DSMC simulation all collisions without limitation were implemented as inelastic collisions, the process of change from one state to anther of the internal (vibrational ad rotational) energies would be too fast, i.e., the rate of change would be different from the rate of the actual physical relaxation process. In general, the relaxation time τ_i is introduced to characterize the rate of change of the state, this is the time needed for the deviation of the state function (temperature) to decay to $1/e$ of the initial deviation when tends to equilibrium. This time is usually several times large than the collision time

$$\tau_i = Z_i/\nu . \tag{7.33}$$

Z_i is called the *relaxation collision number*, e.g., Z_v is the vibrational relaxation collision number, Z_r is the rotational relaxation collision number. There are a number of works investigating the vibrational and rotational relaxation times and relaxation collision numbers [19, 20]. For the general purpose of computation (basing on the comparison with the experimental data) usually adopt $Z_r = 5$ and $Z_v = 50$, and in the DSMC calculation the proportion of the elastic collisions to the inelastic collisions controlled as $(1-1/Z_r -1/Z_v):(1/Z_r):(1/Z_v)$ can maintain roughly the relaxation rates of rotation and vibration.

One can introduce the exchange probability ϕ_v of the vibrational energy in the collision to characterize the vibrational relaxation process. The average collision probability P_v in each collision can be expressed through the *exchange probability* ϕ_v of the vibrational energy

$$P_v = \frac{1}{\tau_v \nu} = \int_0^\infty \phi_v(E_c) f\left(\frac{E_c}{kT}\right) d\left(\frac{E_c}{kT}\right), \tag{7.34}$$

where ϕ_v is supposed to be dependent on the total energy E_c in the collision, where ν is the collision frequency, f is the equilibrium distribution of the energy in the collision (see Eq. (1.94), Eq. (7.26))

$$f(E_c/kT) = \frac{1}{\Gamma\left(\frac{5}{2} - \omega + \zeta\right)} \left(\frac{E_c}{kT}\right)^{\frac{3}{2} - \omega + \zeta} \exp\left(-\frac{E_c}{kT}\right) \tag{7.35}$$

The vibrational energy exchange probability originally was assumed to be the function of the relative velocity c_r and proportional to $\exp(-c^*/c_r)$ (see references [21, 22]). Boyd [23, 24] followed this thread and developed the energy exchange model between translational and vibrational modes, treating the vibrational energy either as continuum distributed or as a set of discrete vibrational energy levels, and letting τ_v follow the experimental data of Millikan and White [19]. As ϕ_v is assumed to be the function of c_r, then the instantaneous ϕ_v manifests the tendency of preferential transfer from vibrational to translational modes for high relative velocity, leading to the violation of the energy equipartition. Boyd used the method of averaging the instantaneous probabilities for all collisions during one time step in one cell and achieved energy equipartition (see [21, 22]). In reference [16] on citing a series of references we showed that the vibrational translational exchange probability not only depends on c_r, but also on the rotational and vibrational energies in the collision, and on this basis ϕ_v is supposed to be dependent on E_c (see Eq. (7.34)), and have the following form

$$\phi_v(E_c) = \frac{1}{Z_0} E_c^\beta \exp\left(-\frac{S^*}{\sqrt{E_c}}\right), \tag{7.36}$$

where Z_0, β, S^* are determined by substituting into Eq. (7.34) and comparison with the experimental data, and τ_v is given by the experimental correlation of the relaxation time given by Millikan and White [19]

$$p\tau = nkT\tau = \exp\left(A/T^{1/3} + B\right). \tag{7.37}$$

For nitrogen and $p = 1 atm$, $A = 220, B = -24.8$. Substituting Eq. (7.35), Eq. (7.36) into Eq. (7.34) after some mathematical manipulations (see [16]), the expressions of S_*, β, Z_0 are obtained

$$S^* = 2\sqrt{(A/3)^3 k},$$

$$\beta = 2 + \omega + 0.5\zeta,$$

$$Z_0 = \left[2\sqrt{\frac{2}{3}} \frac{\Gamma\left(\frac{5}{2} - \omega\right)}{\Gamma\left(\frac{5}{2} \cdot \omega + \zeta\right)} \exp B\left(\frac{S^*}{2}\right)^{2+\zeta} / \sqrt{\mu} \left| \sigma_{ref} \times \left[\left(\frac{5}{2} - \omega\right) kT_{ref}\right]^{\omega - 1/2} \right. \quad (7.38)$$

The vibrational exchange probability in the form of Eq. (7.36) is applied to DSMC simulation of the relaxation process from the initial zero vibrational energy to higher equilibrium translational energy, the relaxation process of the vibrational temperature is the same as the result of the relaxation process given by the τ_v of the Millikan-White correlation. Introduction of Eq. (7.36) in the simulation of the exchange of vibrational and translational energies in the quiescent gas reveals the satisfaction of the energy equipartition principle and that the vibrational energy follows the equilibrium Boltzmann distribution.

7.5 SIMULATION OF CHEMICAL REACTIONS

7.5.1 CHEMICAL REACTION RATE COEFFICIENT

In general bimolecular chemical reaction is expressed by the chemical formula

$$A + B \rightleftharpoons C + D. \quad (7.39)$$

A, B, C, D are different molecular components. In the molecular gas dynamics and in DSMC simulation the chemical reaction rate is most conveniently expressed through the change of number densities of molecules, then the rate equation of Eq. (7.39) can be written

$$-\frac{dn_A}{dt} = k_f(T)n_A n_B - k_r(T)n_C n_D, \tag{7.40}$$

$k_f(T)$ and $k_r(T)$ denote the *direst* and *reverse reaction rate constants*, they are only functions of the temperature. Usually the reaction rate constants can be expressed in the following form

$$k(T) = aT^b \exp(-E_a/kT). \tag{7.41}$$

This form of dependence on the temperature T in conformity with the multitudinous experimental data was put forward by Kooij [25] in 1892, a,b are constants, E_a is called the *activity energy*: when $b = 0$

$$k(T) = a\exp(-E_a/kT), \tag{7.42}$$

is called the *Arrhenius formula*. For different chemical reactions a,b,E_a in the reaction rate constant expressed in the form Eq. (7.41) are determined by the experimental data. Eq. (7.41) is called *Kooij* or *Arrhenius-Kooij formula*.

7.5.2 PHENOMENOLOGICAL CHEMICAL REACTION MODEL OF BIRD

When a molecule of component A collides with a molecule of component B, chemical reaction happens with a certain probability. Usually the reaction cross section σ_R is introduced, the ratio σ_R/σ_T of it to the total collision cross section σ_T represents the probability with which the elastic collision leads to chemical reaction and is called *sterical factor*. More over, for the occurrence of the reaction, the total energy E_c in the collision, i.e., the sum of the kinetic energy ε_t in the mass center coordinate system and the internal energy must exceed the activation energy E_a. In the previous section we have obtained the equilibrium distribution $f(E_c/kT)$ of the energy in collision (see Eq. (7.35)). Thus, the probability of occurrence of reaction from the collision between A,B is

$$\int_{E_a/kT}^{\infty} \frac{\sigma_R}{\sigma_T} f(E_c/kT) d(E_c/kT). \tag{7.43}$$

7.5 SIMULATION OF CHEMICAL REACTIONS

Bird [26] introduced a phenomenological model of chemical reaction, the idea is to ensure the reaction rate coefficient realized in the simulation by introducing the appropriate σ_R to be in conformity with that given by the experimental data in the form of Eq. (7.41). Bird supposes that the reaction cross section σ_R depends on E_c and E_a and has the form

$$\begin{cases} \sigma_R = 0, & \text{when } E_c < E_a, \\ \sigma_R = \sigma_T C_1 (E_c - E_a)^{C_2} (1 - E_a/E_c)^{\bar{\zeta}+3/2-\omega_{AB}}, & \text{when } E_c > E_a. \end{cases} \quad (7.44)$$

The total number of collisions between molecules A and B in unit time and unit volume is according to Eq. (2.254)

$$N_{cAB} = n_A \nu_{AB}, \quad (7.45)$$

where ν_{AB} is the collision frequency of an A molecule with the B molecules and is given by Eq. (2.265). The probability of occurrence of reaction in these collisions is Eq. (7.43), in which $f(E_c/kT)$ is given by Eq. (7.35). Thereby the rate of the forward reaction is obtained

$$-\frac{dn_A}{dt} = 2\pi^{1/2} (d_{ref})^2_{AB} n_A n_B \left(\frac{2kT_{ref}}{m_r}\right)^{1/2} \times \left(\frac{T}{T_{ref}}\right)^{1-\omega_{AB}} \frac{C_1}{\Gamma\left(\frac{5}{2} - \omega_{AB} + \zeta\right)} \times$$

$$\int_{E_a/kT}^{\infty} (E_c - E_a)^{C_2} (1 - E_a/E_c)^{\bar{\zeta}+\frac{3}{2}-\omega_{AB}} \left(\frac{E_c}{kT}\right)^{\bar{\zeta}+\frac{3}{2}-\omega_{AB}} \times \exp\left(-\frac{E_c}{kT}\right) d\left(\frac{E_c}{kT}\right).$$

The integral can be expressed through the gamma function. By comparison with Eq. (7.40) the forward reaction rate constant can be written as

$$k_f(T) = \frac{2C_1 \sigma_{ref}}{\varepsilon \pi^{1/2}} \left(\frac{2kT_{ref}}{m_r}\right)^{1/2} \frac{\Gamma\left(\bar{\zeta}+\frac{5}{2}-\omega_{AB}+C_2\right)}{\Gamma\left(\bar{\zeta}+\frac{5}{2}-\omega_{AB}\right)} \times \frac{k^{C_2} T^{C_2+1-\omega_{AB}}}{T_{ref}^{1-\omega_{AB}}} \exp\left(-\frac{E_a}{kT}\right), \quad (7.46)$$

where ε is the symmetry factor, $\varepsilon = 1$ for different components, and $\varepsilon = 2$ for $A = B$. Comparison of Eq. (7.46) with Eq. (7.41) determines C_1 and C_2 in the definition of σ_R

$$C_1 = \frac{\varepsilon \pi^{1/2} a}{2\sigma_{ref}} \frac{\Gamma\left(\overline{\zeta} + \frac{5}{2} - \omega_{AB}\right)}{\Gamma\left(\overline{\zeta} + b + \frac{3}{2}\right)} \left(\frac{m_r}{2kT_{ref}}\right)^{1/2} \frac{T_{ref}^{1-\omega_{AB}}}{k^{b-1+\omega_{AB}}},$$ (7.47)

$$C_2 = b - 1 + \omega_{AB}.$$

Substitution of these constants into Eq. (7.44) yields the expression of the reaction probability

$$\frac{\sigma_R}{\sigma_T} = \frac{\varepsilon a \pi^{1/2} T_{ref}^b}{2\sigma_{ref}\left(kT_{ref}\right)^{b-1+\omega_{AB}}} \frac{\Gamma\left(\overline{\zeta} + \frac{5}{2} - \omega_{AB}\right)}{\Gamma\left(\overline{\zeta} + b + \frac{3}{2}\right)} \left(\frac{m_r}{2kT_{ref}}\right)^{1/2} \frac{(E_c - E_a)^{b+\overline{\zeta}+\frac{1}{2}}}{E_c^{\overline{\zeta}+\frac{3}{2}-\omega_{AB}}}.$$ (7.48)

When E_c approaches E_a, this probability should be finite, this requires that $b > -1/2 - \overline{\zeta}$. When $b < -1 + \omega_{AB}$, the reaction probability tends to 0 together with $E_c \to \infty$.

This reaction model of Bird is a phenomenological model. Assigning σ_R the form of Eq. (7.44) is mainly for integrating the expression, but is not based on physical considerations. From the above derivation it is seen, that the gas should be in the equilibrium state (the distribution $f(E_c/kT)$ of the form Eq. (7.35) is used), and many times of collisions is needed. However, the practice has shown, for gas in highly nonequilibrium state and with the consideration of only few collisions, this model can provide the correct order of the value of reaction rate. This model received quite wide application in practice.

7.5.3 A STERICALLY DEPENDENT CHEMICAL REACTION MODEL

We put forward a sterically dependent chemical reaction model. It started from a microscopic criterion of occurrence of the dissociation or exchange reaction as the result of the break down of the chemical bond of a diatomic molecule colliding with another particle, and derived the chemical reaction rate constant in the Arrhenius-Kooij form [27]. Fig. 7.3 shows the case of collision of a diatomic molecule CD with another particle A. v_r is the velocity of A relative to CD, θ_1

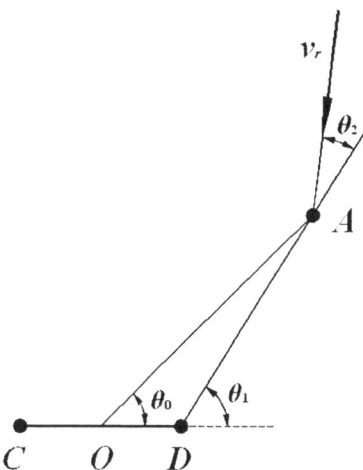

Fig. 7.3 Schematic of collision of a diatomic molecule CD with another particle A

is the angle between CD and DA, θ_2 is the angle between DA and v_r. The microscopic criterion of occurrence of the dissociation or exchange reaction as the result of collision between CD and A is

$$F_w + F_v \geq F_{max}, \qquad (7.49)$$

where F_w is the impact stress occurred in the sphero-cylinder model molecule CD due to collision [28]

$$F_w = (k_e m_{CD})^{1/2} v_r \cos(\theta_1 + \theta_2). \qquad (7.50)$$

k_e is the force constant, m_{CD} is the mass of CD. F_v is the stress acting on CD originated from the vibration

$$F_v = (2k_e \varepsilon_v)^{1/2} \cos\phi, \qquad (7.51)$$

ε_v is the vibrational energy, ϕ is the phase angle of CD. F_{max} is the dynamic breaking factor of the chemical bond CD. Suppose F_{max} is different from the static breaking factor

$$F^s_{max} = (2k_e \varepsilon_D)^{1/2}, \tag{7.52}$$

where ε_D is the dissociation energy, but

$$F_{max} = a_d F^s_{max}, \tag{7.53}$$

where a_d is called the dynamic factor. Substituting Eq. (7.50), Eq. (7.51) and Eq. (7.53) into Eq. (7.49) yields

$$\sqrt{\frac{1}{2} m_{CD}} v_r \cos(\theta_1 + \theta_2) + \cos\phi \sqrt{\varepsilon_v} \geq a_d \sqrt{\varepsilon_D}. \tag{7.54}$$

The factor a_d is introduced because the configuration of collision may influence the energy structure of the system and change the breaking stress of the chemical bond. This is not contradictory with the fact that the chemical bond energy is a constant. After collision the energy of A and the separated atoms C, D is equal to the total energy of CD and A before collision minus the dissociation energy ε_D. This assumption is in agreement with the theory that in the process of the chemical reaction of two atoms with one collision partner there exists a transitional state [29].

One can suppose that θ_1 and θ_2 are small, for only then the collisions lead to the occurrence of the chemical reaction, so one has

$$\cos(\theta_1 + \theta_2) \approx \cos\theta_1 \cos\theta_2 \approx \cos\theta_0 \cos\theta_2,$$

θ_0 is the angle between OA and CD (O is the middle point of CD), then the equation (7.54) can be written as

$$v_r \cos\theta_0 \geq \frac{a_d \sqrt{\varepsilon_d} - \cos\phi \sqrt{\varepsilon_v}}{\sqrt{\frac{1}{2} m_{CD} \cos\theta_2}}. \tag{7.55}$$

This is the microscopic criterion of occurrence of the chemical reaction. If the factor a_d can be fixed and assume the uniform distribution of ϕ, it is possible to implement the simulation of chemical reaction in the DSMC calculation according to this criterion. But here starting from this criterion we derive the expression of the chemical reaction rate constant

In section 2.11 it is obtained, that the fraction of collision pairs with the kinetic energy, corresponding to the component of the relative velocity along the apse line of the centers of the colliding molecules, exceeding a certain amount of energy ε_0, is $\exp(-\varepsilon_0/kT_t)$ (see Eq. (2.233), for HS model). This is to say, the proportion of collisions in which $v_r \cos\theta_0$ exceeds the right hand side (denoted by v_0) of Eq. (7.55) is $\exp(-m_r v_0^2/2kT_t)$. Basing on this result, integrating over all possible ϕ, θ_2 and ε_v, one obtains the probability of occurrence of the chemical reaction

$$P_r = \frac{2}{\pi^2} \int_0^{\pi/2} \int_0^{\pi} \int_0^{\varepsilon_D} \exp\left[-\frac{m_r}{m_{CD}} \frac{\left(a_d\sqrt{\varepsilon_D} - \cos\phi\sqrt{\varepsilon_v}\right)^2}{\cos^2\theta_2 kT_t}\right] f_{\varepsilon_v} d\varepsilon_v d\phi d\theta_2, \qquad (7.56)$$

where f_{ε_v} is the distribution function of the vibrational energy ε_v, which we obtained in Chapter 1 as Eq. (1.94). Thus, we obtained the probability of the chemical reaction

$$P_r = \frac{2}{\pi^2 \Gamma(\zeta/2)} \int_0^{\pi/2} \int_0^{\pi/2} \int_0^{t_*} t^{\frac{\zeta}{2}-1} \times \exp\left[-t - \gamma \frac{\left(a_d t_*^{\frac{1}{2}} - \cos\phi t^{\frac{1}{2}}\right)^2}{\cos^2\theta_2}\right] dt d\phi d\theta_2, \qquad (7.57)$$

where $t = \varepsilon_v/kT_v$, $t_* = \varepsilon_D/kT_v$, $\gamma = m_A T_v/(m_A + m_{CD})T_t$, ζ is the number of vibrational degrees of freedom, T_v is the vibrational temperature, T_t is the translational temperature. In general case when the temperature is not too high (for N_2 when $T < 25000K$), one has $t_* \gg 1$. Then the integral Eq. (7.57) can be calculated approximately. The integral relative to t is (for simplicity, the lower index of θ_2 is eliminated)

$$I = \int_0^{t_*} t^{\frac{\zeta}{2}-1} \exp\left[-t - \gamma \frac{\left(a_d t_*^{\frac{1}{2}} - \cos\phi t^{\frac{1}{2}}\right)^2}{\cos^2\theta}\right] dt. \qquad (7.58)$$

This integral can be found by the *method of the steepest descend*. Assume

7 DIRECT SIMULATION MONTE-CARLO (DSMC) METHOD

$$Z = -t - \gamma \frac{\left(a_d t^{\frac{1}{2}} - \cos\phi t^{\frac{1}{2}}\right)^2}{\cos^2\theta}. \tag{7.59}$$

It is readily seen that when

$$t = t_m \equiv \frac{a_d^2 t_*}{\left(\frac{1}{\gamma}\frac{\cos^2\theta}{\cos\phi} + \cos\phi\right)^2}, \tag{7.60}$$

one has

$$\frac{dZ}{dt} = 0.$$

Besides, it is easy to obtain the values of Z_0 and d^2Z/dt^2 at $t = t_m$

$$Z_0 = Z\big|_{t=t_m} = -\frac{a_d^2 t_*}{\left(\frac{1}{\gamma}\cos^2\theta + \cos^2\phi\right)},$$

$$Z_0 = Z\big|_{t=t_m} = -\frac{a_d^2 t_*}{\left(\frac{1}{\gamma}\cos^2\theta + \cos^2\phi\right)}.$$

According to the method of the steepest descend one has

$$I = t_m^{\frac{\zeta}{2}-1} \int_0^{t_*} \exp\left[Z_0 + \frac{1}{2}\frac{d^2Z}{dt^2}\bigg|_{t=t_m}(Z-Z_0)^2\right] dt =$$

$$t_m^{\frac{\zeta}{2}-1} \exp\left(-\frac{a_d^2 t_*}{\cos^2\phi + \frac{1}{\gamma}\cos^2\theta}\right) J, \tag{7.61}$$

where

$$J = \int_0^{t_*} e^{-\beta^2(t-t_m)^2} dt, \tag{7.62}$$

$$\beta = \frac{1}{2}\left[\frac{\cos^2\theta + \gamma\cos^2\phi}{a_d^2 t_* \cos^2\theta}\right]^{1/2}\left(\frac{1}{\gamma}\frac{\cos^2\theta}{\cos\phi} + \cos\phi\right).$$

From the definition of γ and Eq. (7.60) it is seen

$$t_m \ll t_*,$$

from where one has

$$J \approx \pi^{1/2}/\beta. \tag{7.63}$$

Then the chemical reaction probability Eq. (7.57) can be written

$$P_r = \frac{4}{\pi^{1/2}\Gamma(\zeta/2)}\left(a_d^2\frac{\varepsilon_D}{kT_v}\right)^{\frac{\zeta-1}{2}}\int_0^{\pi/2}\int_0^{\pi/2}$$

$$\exp\left[-\frac{a_d^2\dfrac{\varepsilon_D}{kT_v}}{\cos^2\phi + \dfrac{1}{\gamma}\cos^2\theta}\right]\frac{\cos\theta(\gamma\cos\phi)^{\zeta-1}}{(\gamma\cos^2\phi + \cos^2\theta)^{\zeta-\frac{1}{2}}}d\phi d\theta. \tag{7.64}$$

Denote the right hand side integral by M, it can be written by applying the *generalized theorem of mean value* as

$$M = \exp\left[\frac{a_d^2\dfrac{\varepsilon_D}{kT_v}}{\cos^2\overline{\phi} + \dfrac{1}{\gamma}\cos^2\overline{\theta}}\right]K(\gamma,\zeta), \tag{7.65}$$

where

$$K(\gamma,\zeta) = \int_0^{\pi/2}\int_0^{\pi/2}\frac{\cos\theta(\gamma\cos\phi)^{\zeta-1}}{(\gamma\cos^2\phi + \cos^2\theta)^{\zeta-\frac{1}{2}}}d\phi d\theta. \tag{7.66}$$

The integrals M and K can be evaluated by the numerical quadrature. Define the value of $(\cos^2\phi + \frac{1}{\gamma}\cos^2\theta)$ at some mean value point $(\overline{\phi},\overline{\theta})$ that satisfies Eq. (7.65) as

$$L(\gamma,\zeta,a_d^2 t_*) \equiv \cos^2\overline{\phi} + \frac{1}{\gamma}\cos^2\overline{\theta}, \tag{7.67}$$

and call it the *exponential factor of the reaction rate constant*, then one can write P_r as

$$P_r = \frac{4\left(a_d^2 \dfrac{\varepsilon_D}{kT_v}\right)^{\frac{\zeta-1}{2}}}{\pi^{1/2}\Gamma(\zeta/2)} \exp\left[-\frac{a_d^2}{L(\gamma,\zeta,a_d^2 t_*)}\frac{\varepsilon_D}{kT_v}\right] K(\zeta,v). \tag{7.68}$$

Note, that when γ,ζ,t_* and a_d are given, K and L are quantities that can be evaluated. Then the reaction rate constant $k_f = \overline{\sigma_T c_r} P_r$ can be written

$$k_f = aT_v^b \exp\left[-\frac{a_d^2}{L(\gamma,\zeta,a_d^2 t_*)}\frac{\varepsilon_D}{kT_v}\right], \tag{7.69}$$

in which

$$a = \overline{\sigma_T c_r} \frac{4}{\pi^{1/2}\Gamma(\zeta/2)} K(\gamma,\zeta)\left(a_d^2 \frac{\varepsilon_d}{k}\right)^{\frac{\zeta-1}{2}},$$

$$b = \frac{1-\zeta}{2}.$$

$K(\gamma,\zeta)$, $L(\gamma,\zeta,a_d^2 t_*)$ are defined by the equation (7.66) and equation (7.67).

The chemical reaction rate constant Eq. (7.69) has the Arrhenius-Kooij form of dependence on the temperature given by the experiments (see Eq. (7.41)), all the parameters appeared in it are determined by the physical properties of the components (*CD* and *A*) taking part in the reaction, with the only exception that a_d is determined by comparison of Eq. (7.69) with Eq. (7.41) given by the experimental data. The reaction of the dissociation of nitrogen occurred from its collision with argon

$$N_2 + Ar \rightarrow 2N + Ar, \tag{7.70}$$

was provided with the measured data [30, 31] of the reaction rate constant. When in Eq. (7.69) take $a_d = 1.45$ and 1.49, respectively, the rate constants have excellent agreement with the experimental data of [30] and [31].

From both the microscopic criterion Eq. (7.54) and the form Eq. (7.69) of the reaction rate constant k_f derived from it, it can be clearly seen that the dissociation is dependent on the vibration relaxation. The ratio $k_f(T_v, T)/k_f(T, T)$ obtained from the calculation of the present model is in good agreement with the experimental results [32, 33]. For the detailed comparison concerning the above two issues one can refer to [27].

The success of the chemical reaction model described in this subsection is also in the introduction of the exponential factor of the rate constant. This enables the explanation of the fact that the value of T_s in the *exchange reaction rate constant* $k_f = aT^b \exp(-T_s/T)$ obtained experimentally is usually much less than ε_d/k (see [34, 33]). And this fact seems haven't received reasonable explanation before. Consider the following exchange reactions

$$N_2 + O \rightarrow NO + N, \tag{7.71}$$

$$NO + O \rightarrow O_2 + N. \tag{7.72}$$

Under not too high temperature ($T = 1000K \sim 6000 K$) the values of $a_d^2 \varepsilon_d/(L \cdot k)$ (see Eq. (7.69)) at $a_d = 1$ are really much less than ε_D/k and are very close to the values of T_s provided by the experiments (see Table 7.1).

Table 7.1 The T_s values in the exchange reaction rate constant provided by the experiments compared with the T_s values calculated from the present model

Reaction	$\frac{\varepsilon_D}{k}$ [34]	$\frac{\varepsilon_D}{k}$ [33]	T_s [34]	T_s [33]	$\frac{\varepsilon_D}{Lk}$ (when $T = 3000K$)
(7.71)	113000K	113200K	37500K	38370K	37800K
(7.72)	76500K	75500K	19700K	19450K	26000K

7.6 COMPUTATION OF COMPLICATED FLOW FIELDS

The space vehicles already developed by mankind (the space ships and the space shuttles) and the space vehicles that will be developed – the space shuttles of various countries and various space earth transportation systems – will more and more fly in the transitional regime and utilize the aerodynamic force gained in it to do the maneuver flight. The understanding of the aerodynamic force and heating on the vehicle in the transitional regime becomes more and more important. The experimental means in this regime is not complete, there is not such a high enthalpy facility that could simulate completely the non-equilibrium effects (satisfying the binary scaling low, i.e., the equality of total enthalpy and $\rho L = cons$, see Eq. (0.3)). The numerical simulation of the aerodynamic force and heating becomes very important. Since 90ties of the 20th century the DSMC method has been applied to solve the flow fields of AFE (Aero-assisted Flight Experiment) vehicle, intersected blunt wedges, the plate with incidence, delta wing, sphere, full scale space shuttle and planet vehicles (Viking and Magellan), the satellites, SSTO (single stage to orbit) vehicle and other configurations.

For the demands in exploring the space it is necessary to continuously develop space vehicles of various configurations satisfying various purposes of the investigation, from the point of view of the users it is desirable to have a high efficiency general program that is readily applied to various complicated configurations and capable of providing exact reliable results. Corresponding to such requirements two types of DSMC simulation programs has been developed, one is the program with unstructured body-fitted grid, the other is the program with the Cartesian coordinate grid.

The body-fitted grid program is developed in the direct simulation of the flow fields of AEF vehicle [35] and delta wing [36,37] by Celenligil and Moss et al., the hexahedron cells are used, each cell is further divided into 6 (or 5) tetrahedron subcells. Such cells are readily combined with the triangle cells characterizing the body surface and can ensure the exact representation of the configuration and the implementation of the correct boundary conditions, and the alignment of the cells are in conformity with gradients of the flow field characteristics. The problem is that the algorithm of tracing the molecules is rather complicated and time consuming.

The performance of the body fitted grid program was improved by Wilmoth et al. [38]. Preprocessing software was directly used to the unstructured, surface and volume grid generation files, the cell face geometrical quantities were pre-computed, and the accuracy in tracking the molecules through the grid was improved. The improved code was used to the computation of the SSTO vehicle.

The work of systematic use of the Cartesian grid code to solve the problem of flow around complicated configurations was started by the position element method of Bird [39]. The flow field was divided into multilevel Cartesian cells, the most fine level was called the position elements the size of which represents the accuracy of the locations of molecules and boundaries, the surface elements allotted from them determined the surface of the body. Although the surface was of stepped form, but as the direction cosines of the surface were stored, the surface appeared with smooth characteristics. This general code of the position element concept had efficiency of the level of the specific code. The agreement of the lift to drag ratio of the space shuttle thus obtained with flight data [6] demonstrated the bright prospective of the DSMC method in prediction of the flow characteristics in the transitional regime. Rault [40] developed the Cartesian grid code with flexible cell self adaptation, which was readily applied to various configurations, such as the triangle wing, space shuttle, wave rider, AEF vehicle, planet vehicle and the high altitude satellite etc. For the simulation of the high density region of the windward side of the hypersonic vehicle, the local body fitted grid was embedded into the Cartesian grid [41], to describe properly the normal gradients near the wall. In reference [38] a Cartesian grid algorithm was developed called DAC (DSMC Analysis Code), in which the local refinement of the grid was allowed to satisfy the requirement of improving the space resolution. Meanwhile, measures had been taken to ensure the approximate equality of the number of molecules in each cell to enhance the efficiency of computation. Two codes had been utilized to solve the hypersonic rarefied gas flow field of the SSTO vehicle, results in excellent agreement had been obtained for surface quantities, flow field quantities and global aerodynamic characteristics (lift drag ratio, the center of pressure, the pitching moment etc.). The Cartesian grid required less preparation work and was more efficient, in general judging by the computation time per molecule or per time step, the Cartesian grid software was found 2~10 times faster than the body fitted code.

We suggested a new version of the position element algorithm of using DSMC method to calculate the three dimensional flow in transitional flow regime [42, 43]. This is a general code capable to simulate rarefied gas flow around multiple complicated configurations. The configuration of the vehicle was marked not only by the most frontier and the most rear position element cubes, but all the position element cubes that intersect with the body surface were marked as surface elements. Whether the body was presented analytically or by digital data, the points of intersection of the arrises of the surface element cubes with the body surface were exactly determined, and the area ΔS of the body surface (consisted of several triangles) stretched on each surface element was calculated and the directional cosines of its normal were determined and stored. This was a kind of extension of the 2-D scheme 'rectangular subcells with adaptive body fitted cells' [2] to the three dimensions. ΔS was used to calculate the flux characteristics such as heat transfer, pressure and shear, and also to judge on which surface element a molecule is reflected, i.e., the so called probable criterion of collision of molecule with the surface element was used (see [42, 43], the probability of glancing of molecules was taken into account).

In the code the real numbers were used to record the position of molecule, the surface position element was used only to present the form of the body, record ΔS of the surface element and the direction cosines of the normal to the surface, and thereby determine the reflection of the molecules at the surface. Thus, the collision of the moving molecule with the body surface is a deterministic event, the utilization of the above probability criterion of collision could lead to errors. So in [44, 45] a deterministic criterion of collision of molecule with surface element was developed, the reflection of molecule at the surface element was accurately determined.

For either analytical or digital means of presentation of the body configuration, the compiling of the surface element marking program in the position element algorithm of the DSMC is very time consuming. The accomplishment of it by individual user for specific configuration is a very complicated and arduous task.

Reference [45] has made a successful attempt of marking the surface elements by using a general code. This code met some difficulties in treating the nose of the cone, the wingtip of thin aerofoil and other configurations, the apex might be truncated, the tip of the aerofoil might be combined into a single surface. Aiming at such circumstances, in [46] a new method of marking the surface elements is suggested: The body surface is presented by small triangles with apexes located on the plane templates normal to the body axis and equally spaced along the axis of the body; the relations of the surface element-cubes with the body surface triangles are determined; the reflections of the molecules on the surface are accurately tracked by the deterministic criterion of molecular reflection. The position element program embedded with this new surface marking code has been applied to the computation of flows around sphere, the impingement of the 3-D plume onto plate and the force action and the pollution problem of the discharge of residual liquid of a space vehicle of complicated shape [46, 47], showing that the software is convenient and prompt in treating the flow problems of different complicated configurations.

REFERENCES

1. Bird GA (1976) Molecular Gas Dynamics. Clarendon Press, Oxford
2. Bird GA (1994) Molecular Gas Dynamics and Direct Simulation of Gas Flows, Clarendon Press, Oxford
3. Bird GA (1970) Direct Simulation of the Boltzmann equation. Phys. of Fluids, 13: 2676
4. Wagner W (1992) A convergence proof for Bird's direct simulation Monte Carlo method for the Boltzmann equation. J Stat Phys, 66: 1011
5. Pulvirenti M, Wagner W and Zavelani MB (1994) Convergence of particle schemes for the Boltzmann equation. Euro J Mech., B7: 339
6. Bird GA (1990) Application of the DSMC method to the full shuttle geometry. AIAA paper, 90-1692
7. Mott-Smith HM. (1951) The solution of the Boltzmann equation for a shock wave. Phys. Rev., 82:885
8. Pham-Van-Diep G, Erwin D and Muntz EP (1989) Nonequilibrium molecular motion in a hypersonic shock wave. Science, 245: 624
9. Koura K (1986) Null-collision technique in the DSMS method. Phys. of Fluids, 29: 3529
10. Ivanov MS, Rogazinskii SV (1988) Comparative analysis of algorithms of DSMC in rarefied gas dynamics. Comput. Math and Math Phys, 23 (7): 1058

11. Bird GA (1989) Perception of numerical methods in rarefied gas dynamics. Progress in Astro & Aeronautics, 118: 211
12. Fan J, Shen C (1992) A new algorithm in DSMC method –the randomly sampled frequency method. In: The theory, methods and applications of CFD, Science Press, Beijing, p.127 (in Chinese)
13. Larsen PS and Borgnakke C. (1974) Statistical collision model for simulating polyatomic gas with restricted energy exchange. In: M Becker and M. Fiebig edited Rarefied Gas Dynamics, A. 7-1, DFVLR-Press
14. Borgnakke C, Larsen PS (1975) Statistical collision model for Monte Carlo simulation of polyatomic gas mixture. J Comput. Phys, 18: 405
15. Shen C, Wu WQ, Hu ZH, Xu XY (1991) The direct statistical simulation of excitation, relaxation of internal energy and chemical reaction. Acta Aerodynamica Sinica 9: 1-7 (in Chinese)
16. Shen C, Hu ZH, Xu XY and Fan J (1995) Monte Carlo simulation of vibrational energy relaxation in rarefied gas flows. In: J Harvey and G Lord edited Rarefied Gas Dynamics 19, Oxford Univ. Press, 1: 469-475
17. Haas BL, McDonald JD and Dagum L (1993) Models of thermal relaxation mechanics for particle simulation methods. J Comput. Phys, 107: 348
18. Bergemann F and Boyd ID (1995) New-discrete vibrational energy model for the direct simulation Monte Carlo method. Progress in Astro and Aeronautics, 158: 174
19. Millikan RC, White DR (1963) Systematics of vibrational relaxation. J Chem. Phys, 39: 3209
20. Belikov AE, (1989) et al. Preprint 183-88, Institute of Thermophysics, Siber. Branch, Ac Science USSR; Belikov AE, Sukkhinin GI, Sharafutdinov RG (1988) Nitrogen rotational relaxation time measured in free jets. Rarefied Gas Dynamics, Progr. in Astro. And Aeronautics, 117: 40
21. Zener C (1931) Low velocity inelastic collisions. Phys Rev, 38: 277
22. Landau L and Teller E (1936) Theory of sound dispersion. Phys Z Sowjet, 10: 34
23. Boyd D (1990) Rotational and vibrational nonequilibrium effects in rarefied hypersonic flow. J Thermophysics 4: 478
24. Boyd ID (1991) Analysis of vibrational-translational energy transfer using the DSMC method. Phys. of Fluids, A3: 1785
25. Kooij DM, Z (1893) Phys Chem, 12:155
26. Bird GA (1981) Simulation of multi-dimensional and chemically reacting flows. In: R Campargue edited Rarefied Gas Dynamics, 1: 365, Bird GA (1979) Monte Carlo simulation in an engineering context. Progress in Astro and Aero, 74: Part 1, 239
27. Fan J and Shen C (1995) A sterically dependent chemical reaction model, In: J Harvey and G Lord edited Rarefied Gas Dynamics 19, Oxford Univ. Press, 1: 448
28. Kelsey H (1978) Stress Waves in Solid, Oxford Univ. Press; Miklowitz J (1953) The Theory of Elastic Waves and Wave Guides, North Hollard
29. Laider KJ (1987) Chemical Kinetics (3rd ed), Harper and Row
30. Roth P Thielen K (1986) Measurements of N atom concentrations in dissociation of N_2 by shock waves, Proc. on 15th Shock Waves and Shock Tubes, 245
31. Appleton JP (1968) Shock-tube study of nitrogen dissociation using vacuum ultraviolet light absorption. J Chem. Phys, 48: 599
32. Hansen CF (1991) Dissociation of diatomic gases. J Chem. Phys, 95: 7226
33. Park C (1989) A review of reaction rates in high temperature air. AIAA paper 89-1740
34. Schexnayder CJ Jr and Evans JS (1974) AIAA Jornal,12:805-811

35. Celenligil MC, Moss JN and Blanchard RC (1989) 3-D flow simulation about the AFE vehicle in the transition regime. AIAA paper 89-0245
36. Celenligil MC and Moss JN (1990) Direct Simulation of hypersonic rarefied flow about a delta wing. AIAA paper, 90-0143
37. Celenligil MC and Moss JN (1992) Hypersonic rarefied flow about a delta wing-direct simulation and comparison with experiment. AIAA Journal, 30: 2017
38. Wilmoth RC, LeBeau GJ and Carlson AB (1996) DSMC grid methodologies for computing low-density, hypersonic flow about reusable launch vehicles. AIAA paper 96-1812
39. Bird GA (1990) Application of the DSMC method to the full shuttle geometry. AIAA paper 90-1692
40. Rault DFG (1994) Efficient 3-D DSMC for complex geometry problems. In: BD Shizgal and DP Weaver edited Rarefied Gas dynamics, Progress in Astro & Aeronautics, 159: 137
41. Rault DFG. (1994) Aerodynamics of the shuttle orbiter at high Altitudes. J. of Spacecraft & Rockets, 31: 944
42. Shen C, Fan J, Hu ZH, Wu XY (1996) A new version of position element space discretization in direct statistical simulation of three-dimensional flow in transitional regime. Acta Aerodynamica Siniea 14: 295-303 (in Chinese)
43. Shen C, Fan J, Hu ZH and Xu XY (1997) A new version of position element algorithm of DSMC in calculation of 3-D transitional flows. In: C Shen edited Rarefied Gas Dynamics 20, Peking Univ. Press, 162
44. Fan J, Liu HL, Shen C, Chen LM (2000) A molecular reflection deterministic criterion used in the position element algorithm of direct statistical simulation. Acta Mechanica Sinica 18: 180-187 (in Chinese)
45. Fan J, Peng SL, Liu HL, Shen C, Chen LM (1999) Program marking of surface element and the determination of molecular surface reflection in position element algorithm of DSMC method. Acta Mechanica Sinica 31: 671-676 (in Chinese)
46. Liu HL, Li Z W, Fan J and Shen C Computation of supersonic rarefied gas flows past a sphere .Submitted to Acta Mechanica Sinica (English Series)
47. Fan J, Liu HL, Jiang JZ, Peng SL, Shen C (2004) Analysis and simulation of discharging residual rocket propellants in orbit, Acta Mechanica Sinica 36:129-139 (in Chinese)

8 MICROSCALE SLOW GAS FLOWS, IP METHOD

8.1 INTRODUCTION

The research on the rarefied gas flows carried out by mankind commenced at the beginning of the 20th century from the study of the low speed flows of micro scale. In the middle of the 20th century owing to the demands in the aerospace exploration the interest of the rarefied gas dynamics was concentrated mainly on the flows around bodies flying with hypersonic speed. At the end of the 20th century microscale low speed gas flows rekindled the interest of the rarefied gas dynamics community. But after near a hundred years the motivation of the study has transferred from the research on the basic problems of science into the application study related to the manufacture and the prediction of the performances of the micro-machines. There are tremendous changes in the complexity of the flow patterns and the tools of analysis of the problems.

At the beginning of the 20th century there were the experimental research of Knudsen on the mass flow rate of gas flowing through tiny tubes [1] and the experimental study by Millikan, Knudsen and Weber on the drag of small sphere in the air [2, 3, 4]. All these were important basic research topics. Knudsen obtained the result that the normalized mass flow rate through the tube had a minimum in the transitional regime (the *Knudsen paradox* or *Knudsen minimum*), Millikan measured the velocity of the charged oil drop and with the help of the formula of the low speed drag of the small sphere in the air determined exactly the electric charge of the electron (for this he won the Nobel prize in 1923).

Richard Feynman in his lecture 'There's plenty of room at the bottom' at the 1959 annual meeting of the American Institute of Physics envisaged the possibility of manufacture of micromachines by the chip processing technology, analyzed the difficulties that might encountered with the manipulation and control of micromachines, and even offered a reward of 1000 U.S. dollars for the manufacture of a micro motor with diameter less that 1/64 inch ($400\mu m$). The reward was won by McLellan in November 1960 for making a small motor which was rather an art work than a machine. In the late 1980s commenced the fabrication and the research of the flows in the *micro-electro-mechanical systems* (*MEMS*). By the 1990s the micromachinary fabrication techniques became mature, including the combined surface-bulk silicon micro machining, EDM (electro discharge machining) and LIGA (abbreviation of German words Lithographie Galvanoformung Abformung, i. e., lithographic electroforming) etc. The size of most tiny micro motors at present day is $1 \sim 10 nm$. The *Journal of Microelectromechanical Systems* for paying deserved honor to R.P. Feynman for his farsightedness and insight republished this 1959 lecture in the Journal's initial issue [5]. (R. P. Feynman's main contribution, of course, is his fundamental research in quantum electrodynamics for which he won the physics Nobel prize in 1965)

The micro electro mechanical systems (MEMS) fabricated by these techniques are complicated systems in which simultaneously occur the motion of the working media, the perception of the sensors and the retroaction controlled by the electronic components. The full system simulation of MEMS is beyond the scope of this book: out of the three functions of MEMS only the motion of the working media is concerned. And in the following sections only the case of gas media will be addressed in detail. For the case of liquid media, as it was explained in section 6.8, the MD method is the appropriate means of simulation. The MD simulations revealed strong density fluctuation of the liquid molecules along the normal direction near the wall [6] which is the result of layered structure of the liquid molecules that have a tendency to arrange in rows parallel to the wall. The layering phenomenon of the liquid molecules near the wall is the basis of the origination of the slip boundary condition in liquid (see Thompson and Troian [7]) and the anomalous diffusion (the diffusion coefficient in the vicinity of the wall decreases or increases by a large portion in comparison with that in the bulk of the liquid). Also, in the liquid such phenomena as the wetting, adsorption and electro-kinetics (the accumulation of ions near the dielectric surfaces that can be driven by the voltage difference) closely related to the surface effects become prevailing.

To have an idea of the typical spatial and temporal sizes and the flow parameter (such as the Knudsen number etc.) ranges of the micro devices let's have a close look at the modern Winchester hard disc drive [8, 9] and the micromachined channel fabricated both by the UCLA- CALTECH [10, 11, 12] and MIT [13, 14] groups.

In a Winchester-type hard disc drive the write/read head floats approximately $50 nm$ above the spinning platter surface. The head, the platter and the gas layer between them together form a thin film slider air bearing. The characteristic length (the height) is less than the mean free path ($\sim 65 nm$) of molecules in air at STP. The typical Knudsen number is about 1.3. The platter speed is typically about $25 m/s$ (at radius $5 cm$ of the magnetic disc rotating at 4800 revolutions per minutes), corresponding to a Mach number of ~0.07 and a Reynolds number of ~0.12 (see Eq. (0.27)). To enhance the recording capacity the gap is expected to further decrease resulting in a further increase of the Kn number, and with the increase of the revolution the Mach number and Reynolds number can also increase. The typical length of the slider bearings in disc drives is 1 mm, i. e. , 20,000 times the gap at the rear edge of the head, and the width is usually $1/10 \sim 1/3$ of the length.

The UCLA- CALTECH group first proposed and fabricated an integrated micro-channel/pressure sensor system using the combined surface-bulk silicon micro-machining. The microchannels are formed by silicon with a $1.2 \mu m$ layer of wet oxide on the silicon substrate then bulk-etched with HF to obtain a channel with straight vertical walls of a height of $1.2 \mu m$. Surface micromachining also enables them to make micro sized pressure sensors integrated with the flow system. The second generation micro channel is $40 \mu m$ wide and $1.2 \mu m$ high with 11 pressure sensors uniformly distributed along $4000 \mu m$ length of the channel with intervals of $400 \mu m$, the reading from the other two end sensors are omitted for the end effects. Both helium and nitrogen are used as working media. When nitrogen is used, as the mean free path of nitrogen molecules is almost the same as that of the air, the Knudsen number at the outlet of the channel under STP is ~0.055, but when helium is used, as the mean free path is inversely proportional to the squire root of the molecular mass (see Eq. (2.222)), the Knudsen number is ~0.16, the flow is surely beyond the slip flow regime. Reference [12] showed that under the conditions of the experiment the Reynolds number is less than 0.07 for nitrogen and less than 0.009 for helium, corresponding to Mach number of

~0.0026 and 0.00089 respectively. The MIT micro channel is fabricated approximately the same way with a height of 1.33, width of 52.25 and length of $7500\mu m$. To measure exactly the flow rate the modified accumulation techniques have been developed and the thermal stability requirements have been decreased by five orders of magnitude [13,14]. Nitrogen, argon, carbon dioxide and helium have been used, the flow of argon has a Knudsen number of 0.05 at the exit at an atmospheric pressure, that of helium has a $Kn \sim 2.5$ at the exit at a low pressure of $6.5 \times 10^3 Pa$.

The silicon micromachining fabrication technology has manufactured besides micro channels also micro nozzles, micro valves, micro accelerometers, micro pumps, micro motors and other micro devices. The gas flows in them owing to the micro scale of the devices usually enter into the slip flow regime, and the flows in micro channel, micro pump, micro valve, micro nozzle and the hard disc drive slider bearing enter the transitional flow regime. Thus for simulation of the gas flows in MEMS the methods of molecular gas dynamics or rarefied gas dynamics must be invoked. The objects of study in comparison with circular pipes and sphere studied in the beginning of 20th century are much more complicated. As for the tools of solution various methods developed in the transitional regime and elucidated in chapter 6 can be utilized. In MEMS the flow is usually very slow, the information to noise ratio is very small, thus leads to difficulties in statistical simulation. In the next section some methods of solution of the rarefied flow problems, such as the method of linearized Boltzmann equation, the Lattice Boltzmann method (LBM), the slip Navier-Stokes solution and the direct simulation Monte Carlo (DSMC) method, will be examined from the point of view of utilization for simulation of the flows in MEMS. In particular the unfeasibility of LBM in simulation of transitional flow is shown by comparison with the DSMC results. A method developed by Fan and Shen called the information preservation (IP) method allows the simulated molecules to carry the macroscopic information of the enormous number of molecules one simulated molecule represents, uses it to obtain the macroscopic characteristics, and in principle has found the way to overcome difficulties of large noisy to useful information ratio. The IP method will be introduced in section 8.3, with a general description and some validation of the method and a program demonstrating the method. In section 8.4 the results of

IP simulation for the unidirectional flows are described. The specific features of low flow speed and large length to height ratio of flows in MEMS pose a problem of elliptic nature with boundary conditions set far apart and requiring to be specified in the process of solution, leading to the issue of mutual influence of the inlet and outlet boundary conditions and the need to regulate them. The resolving of the boundary condition regulation problem by using the conservative scheme of continuity equation and the super relaxation method is illustrated on the example of flow in long micro channels in section 8.5. The thin film air bearing problem is solved in section 8.6. By using the same scheme and method the IP simulation of the flow of authentic length of the hard disc drive is described and compared with the result of the Reynolds equation. The use of the degenerated Reynolds equation is suggested by the author to solve the microchannel flow and to serve as a criterion with the merit of strict kinetic theory to test various methods intending to solve the transitional internal MEMS flows. The method, the comparison with the experimental data and the IP calculation and the test of the LBM by it is given in section 8.7. Finally, some review and summary are given in section 8.8.

8.2 METHODS FOR SOLVING THE RAREFIED GAS FLOWS IN MEMS

In the previous section we have seen that the gas flows in MEMS typically are in the slip and transitional flow regimes. The method of Navier-Stokes equation plus slip boundary condition, the method of linearized Boltzmann equation, the Lattice Boltzmann method and the direct simulation Monte Carlo (DSMC) method will be examined in this section from the point of view of utilization for simulation the flows in MEMS.

The solution of the rarefied gas dynamics problems by using the *Navier-Stokes equation with slip boundary conditions* (see Chapter 5) can make advantage of the mature and efficient methods of the Computational Fluid Dynamics (CFD). Karniadakis and Sherwin developed high order finite element (spectra element) method [15] to solve the compressible and incompressible Navier-Stokes equations with the first and higher order slip boundary condition, and by using the so called μ Flow code solved many interesting MEMS flow problems which were also reported in [16]. There is no doubt about the appropriateness of this method in solving problem in the slip flow regime. Kardiadakis and Beskok extended the method for use in micro flows with Kn as high as 0.5. This seems to be a kind of extrapolation beyond the reasonable application range. But it brings into full play the high efficiency of the continuum model in treating the complex geometries. Still one should be cautious relative to the results of the extrapolation. Here we cite two examples with one showing the necessity of the caution and the other showing the success of the extrapolation. The first example is the calculation of mass flow rates through short micro channels [17]. The flow rates obtained by the Navier-Stokes equation with slip boundary conditions and the DSMC methods (see Fig. 8.1) differs significantly as $Kn > 0.1$, and the slip Navier-Stokes solution can not yield the flux minimum predicted first by Knudsen [1] experimentally (for more detailed account of the Knudsen minimum see section 8.4). The second example is the flow in the air bearing between the read/write head and the hard disc drive platter. The slip corrected Reynolds equation can provide result in fair agreement with the DSMC result for Knudsen number as high as 4.2 (see [18]). But the calculation by the generalized Reynolds equation based on the solution of the linearized Boltzmann equation for the flow rate of Poiseulille flow by Fukui and Kaneko [19] is in excellent agreement with the result of DSMC. This latter success of course must attribute to the employment of the Boltzmann equation that

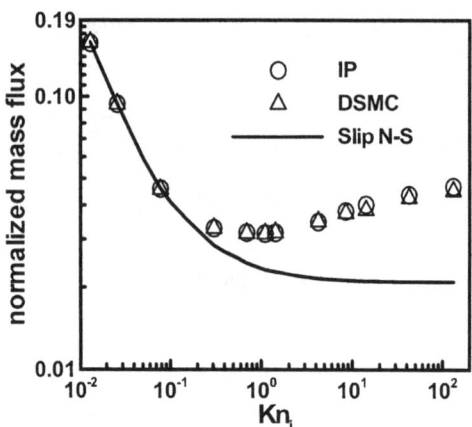

Fig. 8.1 The variation of the mass flux through short channels calculated by the IP, DSMC methods and the Navier-Stokes equation with slip boundary conditions [17]

is appropriate in the entire transitional regime. (For more detailed account of the thin film air bearing problem see section 8.6.)

The linearized Boltzmann equation (see section 6.2) is suitable for solution of low speed problems in MEMS, and can serve as the criterion for testing other methods. At the same time the linearized Boltzmann equation can be used to solve the flow field with temperature variation which is the typical case in MEMS. It is an actual task to develop the solution of the linearized Boltzmann equation to complex geometry. Some times the equation being linearized is not the Boltzmann equation but the BGK equation, in which case the solution is much simpler. But to make the solution of the BGK equation corresponding to the physical reality, some modification of the kind of parameter regulation is needed. And there are still differences between such solution and the solution based on real molecule models.

At the second half of 1980 years Frish et al. developed *the lattice gas method* [20], in which particles are allocated at lattice consisting of equilateral triangles with velocities either along the sides of the triangles or equal zero. Every time step the particles move a cell length (except the particles with zero velocity), and it is shown that Navier –Stoke equation can be obtained from such lattice gas. The shortcomings of such lattice gas are: the amount of work increases with the increase of Reynolds number, and it can only simulate incompressible fluid under small Mach number and the statistical noise is large. The first two shortcomings are tolerable for small speed micro flows. The latter shortcoming is essential and is resolved by introducing the *lattice Bolotzmann method* (LBM, see [21] and [22], and the literature cited in the latter). Lattice Boltzmann method integrates the kinetic theory equation (Boltzmann equation or its simplified version) at the location of each lattice along each discrete velocity. The arithmetic operations of this method are simple, and it is easy to treat arbitrary complex geometry and implement parallel computation. It seems especially suitable for treating micro scale flows. Recently Nie, Doolen and Chen [23] simulated the flows in microchannels under large Knudsen numbers in the transitional regime using the LBM and obtained results of the pressure distribution etc. The microchannel flows under the same parameters are simulated in [24, 25] using both Nie et al.'s LBM method and DSMC method to examine the feasibility of the LBM method in the transitional regime. The simulation results show that for small Knudsen number ($Kn = 0.0194$) the LBM and DSMC methods agree fairly well. For $Kn = 0.194$, the velocity profiles of the LBM and the DSMC (as well the IP) methods differ slightly, but the pressure distribution results have apparent difference (see Fig. 8.2). In the transitional regime, when $Kn = 0.388$, the DSMC simulation results do not verify the negative deviation of the pressure from the linear distribution predicted by the LBM method, and the results of the LBM and DSMC differ significantly in magnitudes (see Fig. 8.3). This shows clearly that this version of LBM is not able to simulate the MEMS flows in transitional regime.

The direct simulation Monte Carlo (DSMC) method (see Chapter 7) is an appropriate method to treat gas flows in MEMS and is able to simulate flow problems in regimes from free molecular to continuum. The simulation results of DSMC for bench mark problems can be used as criteria for other methods and it is able to treat problems abundant physical contents, including chemical vapor deposition, plasma processing and the flow field with temperature variations. But utilization of DSMC method in MEMS flows encounters with the problems of the excessively high demands to the storage and computation time of the computer. Take the micro channels with embedded pressure sensors fabricated by the global processing techniques [10, 11] as example, the size being $1.2 \times 40 \times 3000 \mu m^3$.

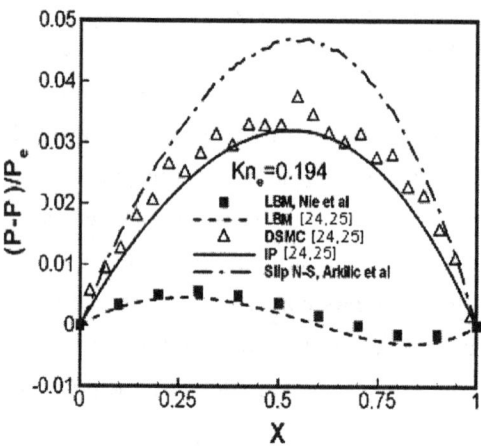

Fig. 8.2 Comparison of the deviation of the stream-wise pressure from a linear pressure distribution given by LBM, DSMC, IP, and slip Navier-Stokes equation, for the case of $Kn = 0.194$ [24,25]

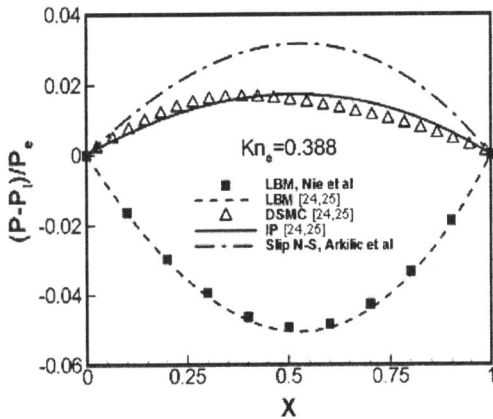

Fig. 8.3 Comparison of the deviation of the stream-wise pressure from a linear pressure distribution given by LBM, DSMC, IP, and slip Navier-Stokes equation, for the case of $Kn = 0.388$ [24,25]

When the cell dimension Δr is taken of the order of the mean free path λ, even treating the problem as two-dimensional (neglect the span wise variation), 6×10^5 cells must be allocated. If distribute 20 molecules in each cell, about 10^7 molecules must be followed in the simulation. The macroscopic velocity of the gas flow in the experiments of [10-14] is $0.2 \sim 0.5 m/s$, the time for transiting the channel is about $10^{-2} s$, or $10^8 \Delta t$ (the time step Δt is taken as the order of the collision time $10^{-10} s$). This makes un-accomplishable the task of gradual regulating the inlet and outlet boundary conditions of the channel to gain the steadiness of the flow (this requires multiple transit times). The difficulty of simulating the low speed flow in MEMS also lies in huge statistical scatter on the DSMC results. The order of the useful information is of the order $U = 0.2 m/s$, and the background noise under room temperature $c_m = \sqrt{2kT/m}$ is of the order of $10^3 m/s$. Only when the sample size N is as big as 10^8, the standard deviation $c_m/\sqrt{N}$ could be small enough, and this is an excessive requirement for the computation time. This makes some researchers think that DSMC is not suitable for simulating gas flows inside MEMS [26]. In fact there have been many experimental results of the micro-channel flows [10-14], at the same time the DSMC simulation of the micro channel flow has been limited to the high speed and even hypersonic cases [27, 28].

Recently the results of the DSMC method with fluctuations have been filtered by using the flux-corrected transport (FCT) method [29] as filter. It is shown, that when the flow velocities are much smaller than the thermal velocity and the number of the real molecules is much larger than the number of simulated molecules, FCT can extract smooth solution from the noisy solution of DSMC with the high frequency statistical fluctuations eliminated. But verification by experiment or exact solution is needed to judge whether the filtered solution is in exact agreement with physically real solution.

8.3 INFORMATION PRESERVATION (IP) METHOD

8.3.1 THE DESCRIPTION OF THE METHOD

Fan and Shen proposed a particle-based method, called the *information preservation (IP) method* [30, 31], to treat the problems encountered by the DSMC method of the huge ratio of the noise to the useful information and the demand of extremely large sample size. This is a method imbedded in the DSMC method in which each simulated molecule is assigned two velocities: thermal velocity c and information velocity u_i. The former is just the molecular velocity c in the DSMC method and is used to calculate the motion, collision and the reflection of molecules at the surfaces following the same algorithms and models as the DSMC method. Besides c we suppose that each molecule carries the so called *information velocity (IP velocity)* u_i to record the collective velocity of the enormous number of real molecules represented by each simulated molecule. The IP velocities do not produce any influence on the motion of molecules, and are used only for summation to obtain the macroscopic velocities, the primitive information is taken from the oncoming flow and the body surface. When the molecules reflect from the surface, collide with each other, experience force action and enter from boundary, the IP velocities attain new values [30-33, 17]:

1. For simulated molecules diffusely reflected from a wall, the reflected IP velocity u_i has the same velocity as the wall. If the wall has a tangential momentum accommodation coefficient of value σ the reflected molecule with a probabil-

ity of σ has an IP velocity the same as the wall, and with a probability $(1-\sigma)$ retains its tangential velocity before incidence.

2. For two simulated molecules colliding each other, the post-collision IP velocities satisfy the momentum conservation

$$u_{i,1}^* = u_{i,2}^* = \frac{m_1 u_{i,1} + m_2 u_{i,2}}{m_1 + m_2}, \tag{8.1}$$

where superscript $*$ denotes post-collision quantities.

3. If there are external forces acting on a cell, acceleration $a = F/\rho \Delta V$ will contribute an velocity increment $a\Delta V$ to each simulated molecule during a time step Δt, where F is the sum of the external forces, ρ and ΔV are the density and volume of the cell, respectively.

4. For simulated molecules entering the computational domain from boundaries, u_i is set to satisfy the boundary condition.

5. In general under the isothermal assumption (which is valid for slow subsonic micro flows without heating) the IP velocity u_i of the simulated molecule and the IP velocity U and IP density ρ (or n) of the cell are introduced which obey the mass conservation and momentum conservation equations

$$\iiint \frac{\partial \rho}{\partial t} dV = \iint \rho U \cdot l dS, \tag{8.2}$$

$$\iiint \rho \frac{dU}{dt} dV = -\iint p l dS, \tag{8.3}$$

where the integrals are taken on the volume and surfaces of a cell, l is the external normal vector of the surface. It is noted that in the right hand side of the momentum equation only a non-viscous term is retained. In fact the IP quantities are governed by a general momentum equation

$$\iiint \rho \frac{dU}{dt} dV = -\iint \sigma l dS \equiv -\iint p l dS + \iint \tau l dS, \tag{8.4}$$

where σ is the pressure stress tensor and τ is the viscous stress tensor. But as the IP quantities are carried along by the simulated molecules of the DSMC process which migrate across the cell surface in the positive and negative direction and implement the action of viscous transport, so although the IP quantities are written formally as governed by an inviscid momentum equation, but as the IP process is embedded in the DSMC process, the IP quantities are factually governed by a more complete conservation law including the viscous transport. After a time step Δt the cell IP density attains increment according to Eq. (8.2)

$$\Delta \rho = -\frac{\Delta t}{\Delta V} \iint \rho U \cdot l dS, \qquad (8.5)$$

from where the density and pressure are also renewed: $p = nkT$. The increment of the IP velocity of the cell is, according to Eq. (8.3),

$$\Delta u = -\frac{\Delta t}{\rho \Delta V} \iint p l dS, \qquad (8.6)$$

and is added to the IP velocity of the simulated molecules in the cell. The renewed quantities are used for the next step calculation. This step of the renewal of the IP quantities is conducted after 'calculation of collisions corresponding to time Δt' (see Fig.7.2 Flow Chart of program of the DSMC-IP method). The calculation of the macroscopic quantities should employ the information velocities introduced, for example, the macroscopic velocity of a cell is obtained from the averaging of the IP velocities of molecules in the cell

$$u_0 = \frac{1}{N_c} \sum_{k=1}^{N_c} u_{i,k} \qquad (8.7)$$

where N_c is the number of simulated molecules in the cell; k is the index of the molecules in a cell. The shear stress on a surface element with area ΔA_w is given by

$$\tau_w = \frac{\sum_{j=1}^{N_t} m(u_{t,j}^i - u_{t,j}^r)}{t_s \Delta A_w} \qquad (8.8)$$

where N_t is the total number of molecules incident to the element during the sampling time t_s, subscript t denotes the tangential direction of the element, and superscripts i and r denote the incident and reflecting values of the IP velocities, respectively.

8.3.2 THE VALIDATION OF THE METHOD

Now we validate the reflection rule of the IP procedure [34]. For simplicity we validate the case of diffuse reflection, namely, the statement 'for simulated molecules diffusely reflected from a wall, the reflected IP velocity u_i has the same velocity as the wall' in step 1). The extension to the case of incomplete diffuse reflection does not pose any principle difficulty. As a simulated molecule represents an enormous number of real molecules, we trace the velocities of the numerous real reflected molecules and obtain the IP value by averaging. An individual molecule after *diffuse reflection from a stationary surface* would have velocity with the components

$$u = -(\ln(ranf))^{1/2}/\beta, \tag{8.9}$$

$$v = V\cos\theta, \tag{8.10}$$

$$w = V\sin\theta, \tag{8.11}$$

where

$$\beta = (2\frac{k}{m}T_w)^{-1/2}, \tag{8.12}$$

$$V = -(\ln(ranf))^{1/2}/\beta, \tag{8.13}$$

$$\theta = 2\pi ranf, \tag{8.14}$$

and *ranf* is a random fraction uniformly distributed between 0 and 1 (see Eqs (3.20), (3.15), (3.12), (3.18) and (3.19) of section 3.2). In the DSMC procedure one records these individual components (with concrete values of *ranf*) and then uses them to obtain the macroscopic quantities only in the step 'sampling of the flow properties'. In the IP procedure we record the averaged values of u, v and w already at this stage of reflection. From the derivation of Eq. (3.20) (Eq. (8.9)) and the practice of the DSMC procedure, one sees that thus sampled velocity components u in the whole guarantees the correct value of the mass flux of diffusely reflected molecules and yields no macroscopic velocity in the normal to surface direction. So after averaging all u, the zero macroscopic velocity component is obtained:

$$\overline{u} = 0. \tag{8.15}$$

The averaging of v yields:

$$\overline{v} = \overline{V\cos\theta} = V\overline{\cos\theta} = 0, \tag{8.16}$$

as v and $\cos\theta$ are independent variates and

$$\overline{\cos\theta} = 0$$

according to Eq. (8.14). Similarly one has

$$\overline{w} = 0. \tag{8.17}$$

If the surface is not stationary but has certain velocity, the velocity components would have been added to u, v, w, and after averaging this velocity would be obtained as the IP velocity after diffuse reflection. So the statement 'for simulated molecules diffusely reflected from a wall, the reflected IP velocity u_i has the same velocity as the wall' is verified.

Next we validate the collision rule of the IP procedure [34, 35]. The components of the post-collision velocities of the two collision partners (with velocity components u_1, v_1, w_1 and u_2, v_2, w_2 before collision) have been found already in section 2.4.5 (see Eq. (2.112)):

$$u_1^* = \frac{m_1 u_1 + m_2 u_2}{m_1 + m_2} + \frac{m_2}{m_1 + m_2}\sin\theta\cos\phi c_r^*,$$

$$u_2^* = \frac{m_1 u_1 + m_2 u_2}{m_1 + m_2} - \frac{m_1}{m_1 + m_2}\sin\theta\cos\phi c_r^*,$$

$$v_1^* = \frac{m_1 v_1 + m_2 v_2}{m_1 + m_2} + \frac{m_2}{m_1 + m_2}\sin\theta\sin\phi c_r^*,$$

$$v_2^* = \frac{m_1 v_1 + m_2 v_2}{m_1 + m_2} - \frac{m_1}{m_1 + m_2}\sin\theta\sin\phi c_r^*,$$

$$w_1^* = \frac{m_1 w_1 + m_2 w_2}{m_1 + m_2} + \frac{m_2}{m_1 + m_2}\cos\theta c_r^*,$$

$$w_2^* = \frac{m_1 w_1 + m_2 w_2}{m_1 + m_2} - \frac{m_1}{m_1 + m_2}\cos\theta c_r^*. \tag{8.18}$$

8.3 INFORMATION PRESERVATION (IP) METHOD

where ϕ is a variate uniformly distributed between 0 and 2π, and $\cos\theta$ is a variate uniformly distributed between -1 and 1. Here we understand $u_1, v_1, w_1, u_2, v_2, w_2$ as one set of the velocity components of many individual real molecules the two colliding simulated molecule represent. In the IP procedure we are not interested in recording the individual $u_1^*, v_1^*, w_1^*, u_2^*, v_2^*, w_2^*$ but are intending to record (preserve) the averages of velocity components of the enormous number of molecules. For example, we have from the first equation of Eqs. (8.18)

$$\begin{aligned}\overline{u_1^*} &= \overline{\frac{m_1 u_1 + m_2 u_2}{m_1 + m_2}} + \overline{\frac{m_2}{m_1 + m_2}\sin\theta\cos\phi c_r^*} \\ &= \overline{\frac{m_1 u_1 + m_2 u_2}{m_1 + m_2}} + \frac{m_2}{m_1 + m_2}\overline{\sin\theta\cos\phi c_r^*} \\ &= \overline{\frac{m_1 u_1 + m_2 u_2}{m_1 + m_2}},\end{aligned} \qquad (8.19)$$

as the $\sin\theta$ and $\cos\phi$ are independent variates and ϕ is uniformly distributed between 0 and 2π. Analogously we have

$$\overline{u_2^*} = \overline{u_1^*} = \overline{\frac{m_1 u_1 + m_2 u_2}{m_1 + m_2}}, \qquad (8.20)$$

$$\overline{v_1^*} = \overline{v_2^*} = \overline{\frac{m_1 v_1 + m_2 v_2}{m_1 + m_2}}, \qquad (8.21)$$

$$\overline{w_2^*} = \overline{w_1^*} = \overline{\frac{m_1 w_1 + m_2 w_2}{m_1 + m_2}}. \qquad (8.22)$$

Thus, the IP collision rule, Eq. (8.1) of step 2), has been validated. As the IP procedure uses the already averaged values to obtain the macroscopic quantities (see Eq. (8.7)), it is natural that the sample size needed for convergent IP averaging is much less than that needed in the DSMC procedure.

8.3.3 PROGRAM DEMONSTRATING THE METHOD

In section 7.3 a FORTRAN program is given to demonstrate the solution of Couette problem by the DSMC method (see Appendix IV), it is also used to demonstrate the IP method. In this program VMEAN(I, NO-MOLECULE), $I = 1,2,3$, are introduced to denote IP velocities. The statements in the program used to implement the changes in the IP method are marked with * and **, the statement marked with * signify that it is used to replace the statement before it, those statements marked with ** signify that they are the statements needed to be added anew. The above described cases of changes of the IP velocities and the procedure of obtaining the macroscopic quantities from the IP velocities are shown in the program (see section 7.3 and the statements in the program in Appendix IV marked with * and **).

When employing the IP method, another change should be introduced as well. This is the change in the collision cross sections of the molecules. In section 2.4 the expression of viscosity coefficient μ for various molecular models has been given according to the Chapman-Enskog transport theory in the kinetic theory (see Eq. (2.71)), and the diameters of molecules have been determined (for HS model, see Eq. (2.77), for VHS and VSS models, see Eq. (2.234), where the reference diameter of molecules is given by Eq. (2.235)). In the IP method, when assigning the IP velocities after collision, we stipulate they follow the macroscopic momentum conservation law, Eq. (8.1), but in the DSMC method the velocities after collision are assigned according to the momentum conservation in each collision (the detailed conservation), and this is the condition implied in obtaining the expressions of d for various models. The difference in the assignment of the post-collision velocities leads to the necessity of modification of the collision cross section in the IP method to obtain the correct value of viscosity μ. The concrete method is to obtain the correct experimental value of μ by varying d, in employing the IP method to solve the Couette problem under small Kn number (see [31]). Take the HS model as example. The collision diameter (see Eq. (2.77))

$$d_{HS} = (\frac{5mc_m}{16\sqrt{2\pi}\mu})^{1/2}, \qquad (8.23)$$

is used as the initial value of the collision diameter in the IP method, the shear stress τ_{xy} of each cell can be calculated according a formula analogous to Eq. (8.8), from where the value of μ, $\mu = \tau_{xy}\Delta y / \Delta u$, of the cell is obtained, the viscosity μ is obtained as the average of the μ values in various cells (except the cells in the Knudsen layer). The diameter of HS model is modified according to the difference between this value and the experimental μ value (the increase of d_{HS} makes μ decreasing), until the correct experimental μ value is obtained. The d_{HS} thus fixed is the value to be used in the IP method. The reference diameters of the VHS model can be obtained by analogous method [31]. Some values of d_{HS} and reference diameters d_{ref} of the VHS model are listed in Table 3 of Appendix I. In the example program demonstrating the IP method, the collision diameter of the Ar molecule, when employing the IP method, is replaced by $d_{HS} = 3.963 \times 10^{-10} m$ (see the first statement with * in subroutine subc1 in the program).

8.4 UNIDIRECTIONAL FLOWS

The Couette flow is a steady flow of gas occurred between two parallel plates moving with velocity U_w in opposite directions along their own planes. The velocity profiles and the shear stress profile obtained by using the IP method in simulating the Couette flow [30, 31] are given in Fig. 8.4 and Fig. 8.5. The velocity profiles are given under three Knudsen numbers, $Kn = 0.1128$, $Kn = 1.128$ and $Kn = 11.28$, and are compared with the solutions of the linearized Boltzmann equation [36], of the moment method [37] and of the Navier-Stokes equation plus the slip boundary condition (see section 5.4.1, Eq. (5.65)). The velocity profiles of the IP method under the small, medium and large Kn numbers are all in good agreement with the solution of the linearized Boltzmann equation of Sone et al., but the agreement between the moment method of the second order approximation by Gross and Ziering and the result of the linearized Boltzmann equation (and the IP method) is not so good, especially for medium Knudsen number. The solution of the Navier-Stokes equation plus the slip boundary condition yields relatively

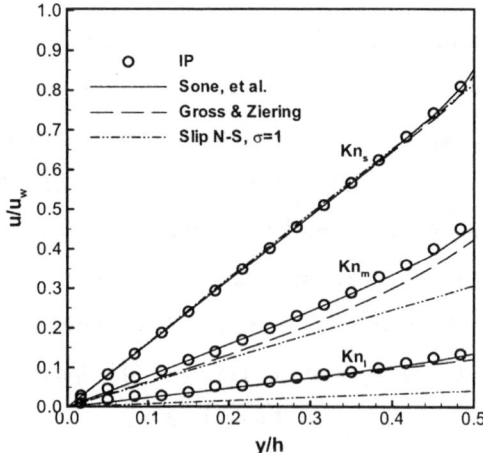

Fig. 8.4 Velocity profiles of the Couette flow for $Kn_s = 0.1128$, $Kn_m = 1.128$ and $Kn_l = 11.28$. Comparison of the IP method [30], the linearized Boltzmann equation [36], the moment method [37] and the Navier-Stokes equation plus slip boundary condition

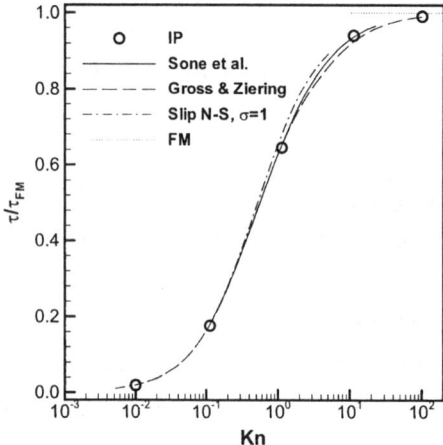

Fig. 8.5 Variation of the shear stress with Kn in the Couette flow. Comparison of the IP method [30], the linearized Boltzmann equation [36], the moment method [37], the Navier-Stokes equation plus slip boundary condition and the free molecular flow theory

good result only for small Knudsen numbers. The shear stress in the Couette problem solved by the Navier-Stokes plus slip boundary condition (see Eq. (5.64) in Chapter 5) is a case incidentally appropriate for the entire transitional regime. The solutions of the IP method, of the linearized Boltzmann equation, of the moment method and of the Navier-Stokes equation plus slip boundary condition are in good agreement in the whole transitional regime (see Fig. 8.5, in which $\tau_{FM} = \rho c_m U_w / \sqrt{\pi}$, see Eq. (4.54), note, the wall velocity U_w here constitutes only half of the velocity U in section 4. 5). It is noted that the IP method is in exact agreement with the solution of the linearized Boltzmann equation, and simultaneously agrees with the theoretical solution of the free molecular flow in the collisionless limit.

In section 7.3 it has been mentioned that the DSMC program aimed at the Couette flow problem can be amended to be used in solving the Poiseuille flow and the Rayleigh problem, the same is true for the IP method. The results of solution of the planar Poiseuille flow and the Rayleigh problem by the IP method were presented in detail in [30, 31], here only the mass flow rate of the Poiseuille flow and the velocity profile and the shear stress in the Raylegh problem will be discussed.

The mass flow rate of the Poiseuille flow has been calculated by the Navier-Stokes equation with slip condition on the boundary in section 5.4.2 (see Eq. (5.74)). This $Q_{P,SL}$ is a monotonically descending function of Kn. But at the beginning of the 20th century Knudsen [1] discovered through experiments that there appears a minimum of the mass flow rate in transitional regime, this is the so called *Knudsen minimum* or the *Knudsen paradox*. This result was confirmed in the later experiments [38] for many gases (air, helium, hydrogen, carbon dioxide and Freon-12). Fig. 8.6 shows the comparison of various methods and the experiment ($u^* = \alpha c_m$, α is a pressure gradient factor, see Eq. (7.11)). The result of the IP method agrees with Eq. (5.74) (for $\sigma = 1$) under small Kn numbers, yields the Knudsen minimum under medium Kn numbers and agrees with the numerical solution of the linearized Boltzmann equation [39] and the experimental data, demonstrating the ability of the IP method in predicting the fine flow characteristics in the transitional regime.

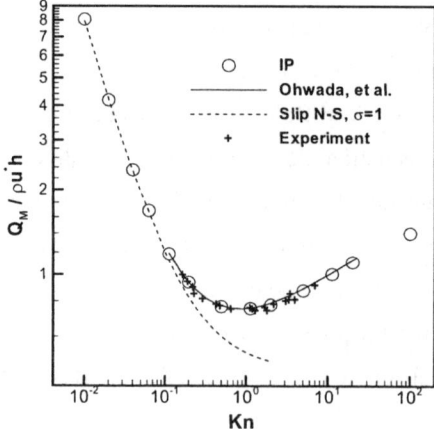

Fig. 8.6 The variation of the dimensionless mass flux in the Poiseuille flow with Kn number (comparison of IP [30] and linearized Boltzmann equation [39] with the experimental data [38].)

The velocity profiles in the Rayleigh problem obtained by the IP method in the initial stage of motion $t = 0.01\tau_c$ (τ_c collision time) agrees well with the result of free molecular flow (see Eq. (4.99) in Chapter 4), and after many collision times ($t = 100\tau_c$) agrees well with the solution of the slip Navier-Stokes equation (Eq. (5.88) in Chapter 5) (for detailed account see [30, 31]). In the transitional regime $t = 5\tau_c$, as there is no numerical solution of the Rayleigh problem by the linearized Boltzmann equation, the calculation by the DSMC method has to be employed to check the result of the IP method (see Fig. 8.7)). From the comparison it is seen that the agreement is excellent. But for the case of $U_w = 1m/s$, the DSMC method has to employ enormous sampling size 2×10^8 to reduce the statistical scatter, in the result the computational time spent is 3×10^4 times of the IP method. In Fig. 8.8 the comparison of the results of various methods of the shear stress (normalized by the value in free molecular flow $\tau_{FM} = \rho c_m U_w / 2\sqrt{\pi}$, see Eq. (4.101)) of the Rayleigh problem is given. Except the results of the IP method, the DSMC method [31], the slip Navier-Stokes equation and the FM theory, also shown is the result of the moment method [40]. The agreement of the IP method with the collisionless solution in the free molecular flow limit ($t \ll \tau_c$), with the

DSMC result in the transitional regime ($t \sim \tau_c$) and with the Navier-Stokes slip solution in the slip flow regime ($t \gg \tau_c$) is uniformly good.

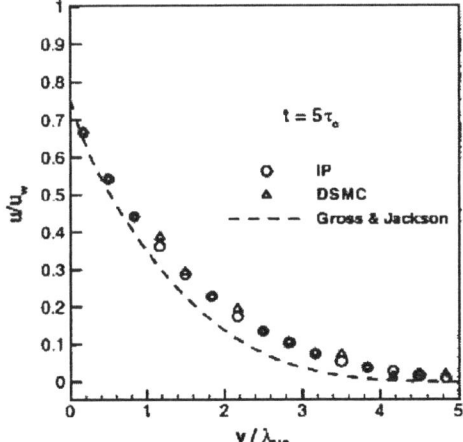

Fig. 8.7 The velocity profiles in the Rayleigh problem at $t = 5\tau_c$ obtained by the IP method, the DSMC method [30, 31] and the moment method [40]. The sample size of the IP method is 6×10^3, of the DSMC method is 2×10^8

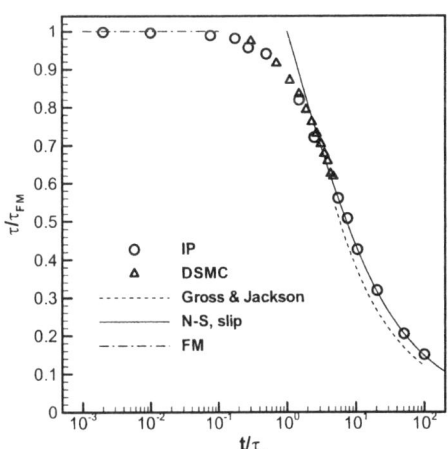

Fig. 8.8 The variation of the shear stress (normalized by $\tau_{FM} = \rho c_m U_w / 2\sqrt{\pi}$) of the Rayleigh problem with Kn number

8.5 THE MICROCHANNEL FLOW PROBLEM

In treating the unidirectional motions utilization of the steps 1), 2), 3) and 4) given in section 8.3 was sufficient for renewal of the IP velocities. Only when treat the Poiseuille flow the expression of the IP velocity increment (Eq. (8.6) of step 5) was used. When the pressure variation is expressed as

$$p = p_0(1 + \alpha x/h), \qquad (8.24)$$

the velocity increment can be found from Eq. (8.6) as

$$\Delta u = -(\alpha p_0 / \rho h)\Delta t . \qquad (8.25)$$

In fact, Eq. (7.11) in section 7.3 has been obtained in this way. In the two-dimensional and three-dimensional cases the method of renewal of the cell IP velocities U, the IP density ρ and the molecular IP velocities u described in step 5) should be used systematically.

Various two-dimensional problems have been solved by the IP method, including the microchannel flow [41, 42, 43, 17], the flow around the plane plate [44, 45, 47], the flow around the airfoil [46, 48], the cavity flow [49], non-circular Poiseuille flow [50], the flow in membrane filter [51], etc.

Microchannel is the basic constituent of the MEMS devices, the geometric form is regular and simple (see Fig. 8.9), but can reveal the specific features of the low speed micro internal flows, i. e., the issue of the mutual influence of the boundary conditions at the inlet and the outlet caused by the elliptic nature of the problem. For the DSMC-IP procedure it is necessary to prescribe the values of the pressure p and the velocity distribution U over the cross sections at the inlet and the outlet of the channel to start any simulation. But fixing all p and U at the inlet and the outlet simultaneously would over determine the boundary conditions: The arbitrarily chosen p and U would be contradictory to each other. The correct values of p and U at inlet and outlet must be obtained in the process of solution of the problem. A method of fixing p as the same of the prescribed (experimental) condition and allowing U change continuously and finally reach the steady solution is adopted here [41]. Thus the process of the DSMC-IP solution is always one of gradual adjustment towards a steady state. It is very critical that the conservative form of the mass conservation equation must be employed

8.5 THE MICROCHANNEL FLOW PROBLEM   339

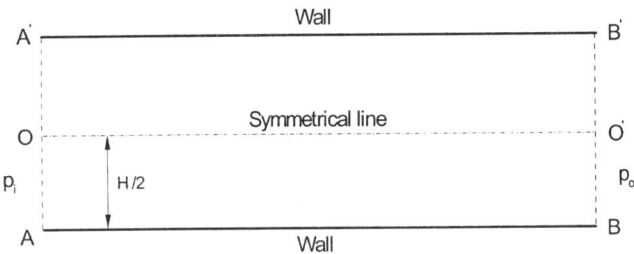

Fig. 89 Computational domain of gas flow in a microchannel

$$\frac{\partial \rho}{\partial t} + \frac{\partial \rho U}{\partial x} + \frac{\partial \rho V}{\partial y} = 0, \qquad (8.26)$$

its second order central difference scheme yields the density increment

$$\Delta \rho = \Delta t \left(\frac{\rho_{i-1,j} U_{i-1,j} - \rho_{i+1,j} U_{i+1,j}}{2\Delta x} + \frac{\rho_{i,j-1} V_{i,j-1} - \rho_{i,j+1} V_{i,j+1}}{2\Delta y} \right). \qquad (8.27)$$

This density increment expression can be obtained from the integral form mass conservation equation (8.5) directly by using an integration domain ABCD ($2\Delta x \times 2\Delta y$) with point (i, j) in the center (see Fig. 8.10). The adoption of the conservative form of the continuum equation or the integral form of conservation equation guarantees that the mass flux flown from the adjacent domain of area ABCD will flow without any numerical error into the integral area and vice versa

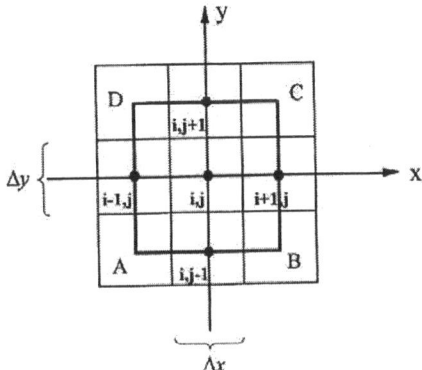

Fig. 8.10 The control surface ABCD of the conservation equation and the cell central points (i, j) in the IP method

and avoids the accumulation of numerical errors from the non-conservative scheme. This is an issue that must be taken into account whenever solving a slow rarefied channel flow or any other slow internal rarefied gas flows.

The increments $\Delta\rho$ and Δu from Eq. (8.27) and Eq. (8.6) allow one to obtain the renewed fields of ρu and ρv which are unfortunately with large fluctuations and are smoothened by a averaging technique to avoid the amplification of the errors which would influence the stability of the calculation. The increment $\Delta\rho$ obtained from Eq. (8.27) is only of the order of 10^{-9} of ρ with time step Δt being of the collision time for slow gas motion in long micro-channels [10-14]. For all simulations in such channel flows Δt has been taken as 1/2 average collision time at the inlet. Direct employment of this $\Delta\rho$ to achieve the steady (convergent) state is too time-consuming. A *super-relaxation technique* is employed to speed up the convergence process

$$\rho_{i,j}^{t+\Delta t} = \rho_{i,j}^{t} + \omega\Delta\rho_{i,j}^{t}, \qquad (8.28)$$

where ω is a *super relaxation factor*. In practical calculations ω is taken to be between 100 and 2000 and trends to 1 when convergence is achieved[1].

The necessity of using the conservative form of the mass conservation is illustrated on one of the flow cases under the experimental conditions considered [12]. Helium flows through a $1.2\times 40\times 4000\mu m^3$ micro-channel with an inlet pressure of 19.0 psig into the atmosphere (outlet pressure 0 *psig*). Fig. 8.11 and Fig. 8.12 show the evolution of mass fluxes at all cross sections along the channel by the IP calculation. The slip Navier-Stokes solution is adopted as the initial pressure distribution. It is different from the real distribution since the flow is in the transitional regime. This resulted in a non-uniform mass flux distribution along the channel length at the initial stage of simulation (at $1\sim 2\times 10^3 \Delta t$, the black triangles in Fig. 8.11 and Fig. 8.12). By using the conservative scheme, Eq. (8.27), and the super-relaxation technique, Eq. (8.28), a steady state is approached after about 2×10^5 time steps (the hollow spheres in Fig. 8.11). And the averag-

---

[1] For short channels and not slow flow speed, $\Delta\rho_{i,j}^{t}$ might be the same order of $\rho_{i,j}^{t}$, and an ω less than 1 is suggested to be used to stabilize the convergence process.

ing-smoothing process gives a relatively smooth and almost uniform mass flux distribution (solid line in Fig. 8.11). If the non-conservative scheme were used, the mass flux would remain non-uniform. Fig. 8.12 shows the mass flux distribution after 2×10^5 time steps (the hollow spheres, the solid line being the averaged smoothened data) by using the non-conservative form of the continuity equation. The mass fluxes at various cross sections have not been regulated by the simula-

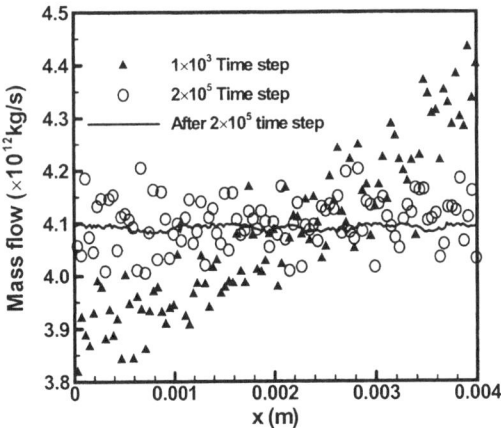

Fig. 8.11 Evolution of mass flux distribution in IP simulation of the micro-channel gas flow of [12], while the conservative form of the mass conservation equation is used

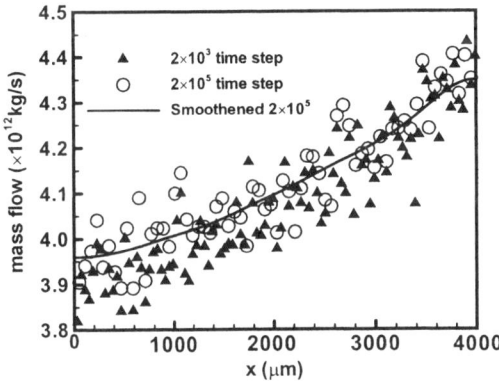

Fig. 8.12 Evolution of mass flux distribution in IP simulation of the micro-channel gas flow of [12], while a non-conservative form of the mass conservation equation is used

tion relaxation process, for the adjusting act of the mutual influence of the inlet and outlet boundaries have been damped by the numerical errors inherent in the non-conservative scheme.

The effect of the acceleration action of the super-relaxation technique is illustrated on another experimental case [11] of nitrogen flowing in a $1.2 \times 30 \times 3000 \, \mu m^3$ channel with inlet pressure of $15 \, psig$ into the atmosphere. Figure 8.13 shows the evolution of the density ρ at the center of the cross section located at $2500 \, \mu m$ from the inlet in the IP calculation by using a super-relaxation factor ω of 1, 100 and 1000, respectively. While ρ approaches the steady value of $1.39 \, kg/m^3$ in about 6×10^4 time steps with a relaxation factor ω of 1000, the evolution for $\omega = 100$ is further than halfway apart from the steady state after 6×10^4 time steps, and the value of ρ remains almost the same when no super-relaxation is employed (with $\omega = 1$). The maximum value of ω allowed in simulation is dependent on the smoothing technique of mass fluxes in the whole flow field: the smoother the mass flux, the larger value ω is allowed. But exaggerated smoothing would distort the flow field. A simple averaging from adjacent three points,

$$M(i,j,n) = \big(M(i-1,j,n-1) + M(i,j,n-1) + M(i+1,j,n-1)\big)/3,$$

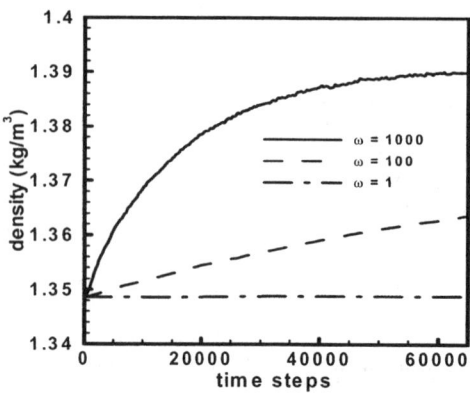

Fig. 8.13 The evolution processes of density ρ at a point located at $2500 \, \mu m$ apart from the inlet under experimental conditions of [11] with different super-relaxation factors $\omega = 1, \omega = 100, \omega = 1000$, respectively

where n is the number of iterations, is used. The iterated averaging for $n=15$ has the desirable effect of smoothing, and retains the local trend of mass flux evolution. Then the value $\omega = 2000$ can be employed and the calculation remains stable. It is noted that when the steady value of ρ is being approached the value of ω and the smoothing procedure has little consequence on the final evolution result, so after having experience one can prescribe ω a varying process from say 2000 to 1 to reach the steady state, and the final result is entirely not effected by the varying process. This is satisfactory for the purpose of the calculation, for it is the final result but not the evolution process that is concerned.

In the practice of general IP method the DSMC simulated molecules move and carry the IP quantities, the DSMC process determines the IP process and the IP process has no reverse influence on the DSMC process. In the solution of channel flow and other internal flow cases, where the macroscopic quantities on the boundaries are to be regulated during the simulation, there is another specific feature, that is, the varying IP velocities on the boundaries are used to continuously adjust the boundary conditions of the DSMC–IP procedure. This influences the DSMC simulation and enables the DSMC finally to have the correct value on the boundaries. Pure DSMC process is carried out by individual molecules and the adjustment of boundary conditions is very slow and DSMC needs sufficient sample size to allow definite boundary values of U to emerge, while the IP process is a global one: the changes of IP values happen simultaneously over the whole domain of calculation and the adjustment is quick and not limited to the boundary but spreads over all the channel length. Although the approach to the steady state requires quite a long time in the example calculation under condition [11] (120 hours CPU time on a Pentium III 450, or 98.7% of the entire computation time), but during this time the global DSMC quantities are also regulated. After arriving at the steady state the sampling time required for yielding the final IP convergent data is quite short (1.6 hour CPU time, or 1.3% of the computation time).

In micro-channel experiments [10-14] the width ($40 \sim 50 \mu m$) is much larger than the height ($1.2 \sim 1.33 \mu m$). This made the span-wise influence negligible, and the flows can be simplified as two-dimensional (the midline velocity profile and the maximum velocity remains almost the same for rectangular cross section channels with a width to height ratio larger than 5, but the flow rate is influenced in some minor degree by the slow down of the flow near the side wall even for large width to height ratio, see [50]). As we have seen in section 8.1, the experimental conditions [10-14] are in the slip and transition regimes.

An orthogonal coordinate system is employed with the origin located at point O, and x and y axes along OO' and OA, respectively (see Fig. 8.9). Since the flows are symmetric about OO', a computational domain of $OO'BA$ is considered only. Each of the uniform rectangular cells is sub-divided into a set of uniform rectangular sub-cells within which collision pairs are selected. The number of cells is around 400×15 to 700×30 and there are 5×2 sub-cells in each cell. The cell size is much smaller in the cross sectional direction than in the stream-wise direction, so is the sub-cell size. As shown by Nance et al. [28], the flow field is insensitive to the stream-wise cell size because of a relatively small velocity gradient in this direction. The test calculations observe that the smaller stream-wise cell and sub-cell sizes provide the same results as the present sizes being employed. For all cases the molecular interaction is described by the VHS model. The reference collision diameter in VHS appropriate to the IP method has been determined for common gases [31].

A specular reflection is used along the symmetrical boundary OO'. The channel surfaces are assumed to be diffusely reflecting with a tangential momentum accommodation coefficient σ (see Eq. (3.23)). Arkilic et al. [13, 14] developed a modified accumulation technique to measure the mass flux through micro-channels. Comparing the measured mass flow rate with the slip Navier-Stokes solution, they extracted σ for the micro-channel surfaces of single-crystal silicon in their system. The values appeared to be 0.80 ± 0.01 for argon and 0.88 ± 0.01 for nitrogen. The same means was also utilized by Shih et al. [11] to extract σ for their micro-channel surfaces, yielding 0.9905 for nitrogen and 1.1620 for helium. However, as we have seen in section 8.1, the microchannel helium flow has a Knudsen number of 0.16 at the outlet and is in the transitional regime. So extracting σ from the slip Navier-Stokes solution became improper. And the value 1.162 is beyond the physically realistic range of σ. In contrast, the nitrogen flow is in the slip regime and the value of $\sigma = 0.9905$ is reasonable. This shows that the micro-channel surfaces in the UCLA system is close to the full diffuse reflection. The values of σ used in simulation for nitrogen and helium flows of [11, 12] are both 1.0 and for argon flow of [13] is 0.8 (the Knudsen number at the exit of the argon channel flow is 0.05, and the σ value extracted from the slip Navier-Stokes solution is valid).

In the case of the channel flow of nitrogen [12] the density increment obtained by the conservative form of mass equation (8.27) and the super-relaxation method Eq. (8.28) make the mass fluxes at various sections tend to be the same (see the hollow spheres in Fig 8.11), at the same time the pressure distribution is adjusted to the actual configuration. For the inlet pressure of $19.0\, psig$ the mass fluxes at various cross sections are all equal to about $4.1 \times 10^{-12} kg/s$. This result is in good agreement with the experimental result in [12].

Figure 8.14 compares the stream-wise pressure distributions given by the IP method with experimental data of [11] with nitrogen as the working media for the inlet pressures of 5, 10, 15, 20 and 25 $psig$, with the error bars showing the measured confidence limits. Because of the small height of $1.2\mu m$, the velocity gradient in the normal direction is quite large that leads to a strong viscous effect which is clearly demonstrated by the non-linearity of the pressure profiles. The pressure loss is subject to the local shear stress of the micro-channel surfaces that becomes sensitive to the Knudsen number as $Kn > 0.01$. For the same outlet pressure of the atmosphere, the increase of the inlet pressure results in a more significant stream-wise variation of Kn and therefore corresponds to a more obvious non-linear pressure profile. Fig. 8.15 shows the stream-wise pressure distributions at three different inlet pressures of 8.7, 13.6 and 19.0 $psig$ given by IP and

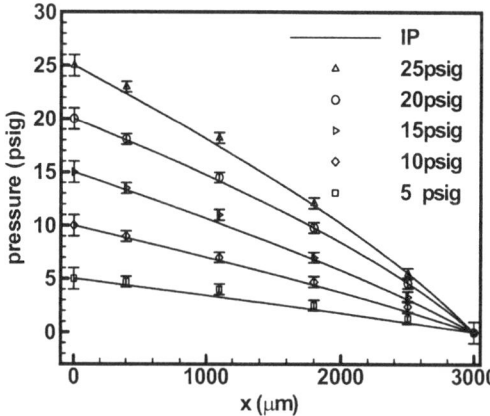

Fig. 8.14 Comparison of stream-wise pressure distributions of nitrogen flow given by IP with experimental data of [11]. $Kn_o = 0.055$. The pressure values indicated are the inlet pressures.

experiment for helium with an exit Knudsen number of 0.16, which also agree with each other.

In Fig. 8.16 the simulated mass flux by the IP method at inlet pressures of 9.5, 15, 20, 26, 30 *psig* is compared against measured data of Shih et al. for nitrogen [12]. Fig 8.17 shows the mass flux calculated by the IP method and measured by Arkilic for argon [13]. The flows are in the slip flow regime and one can see, there is a remarkable agreement between the IP and the experimental results.

Arkilic [14] has undertaken experiments under "extreme" flow conditions to investigated flows in the transition regime. The inlet pressures of helium range from 133kPa to 413kPa (with Kn_i between 0.117 and 0.04), while the helium exhausts to a low pressure of 6.5kPa that results in an outlet Knudsen number of 2.5. Therefore, a significant portion of the channel lied well beyond the slip flow regime. Arkilic defined the *flow conductance* as the ratio $C = Q/\Delta p$ of mass flow to the differential pressure across the channel length and used it to check the validity of the slip Navier-Stokes model. Arkilic obtained the slope of the measured flow conductance was approximately 11% greater than the slip Navier-Stokes prediction [14] showing an obvious breakdown of the slip flow model. The data of the flow conductance has been used in [17] to check the performance of the IP method in the transitional flow regime. The value of flow conductance versus mean pressure, $\overline{P} = (P_i + P_o)/2$, given by the IP method and the experimental data of [14], except at the largest mean pressure range, where a difference of about 5%

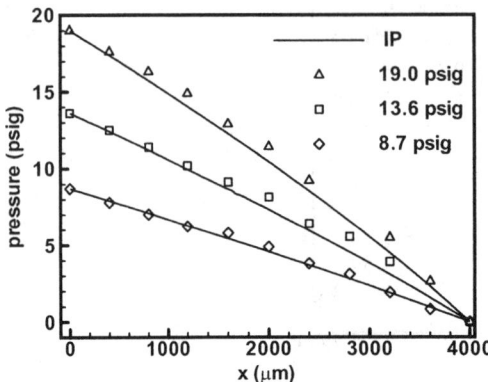

Fig. 8.15 Comparison of stream-wise pressure distributions of helium flow given by IP with experimental data of [12]. $Kn_o = 0.16$. The pressure values indicated are the inlet pressures

8.5 THE MICROCHANNEL FLOW PROBLEM   347

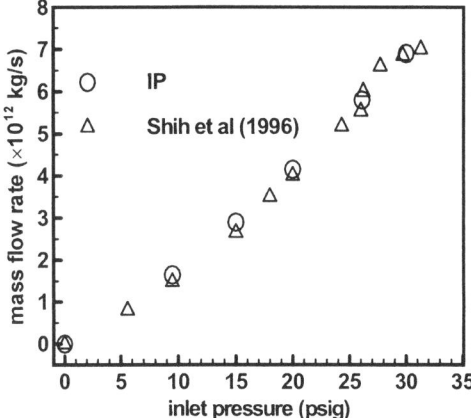

Fig. 8.16 Relation of mass flux versus the inlet pressure for the helium flow. Comparison of the IP simulation with the experimental data [12]. $Kn_o = 0.16$

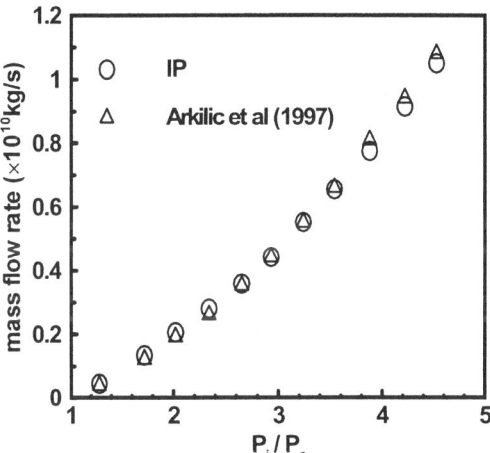

Fig. 8.17 Relation of mass flux versus the inlet pressure for the argon flow. Comparison of the IP simulation with the experimental data [13]. $Kn_o = 0.055$

appears, are generally in good agreement (see Fig.8.18). This is the first time that the result of a method appropriate for the entire transitional flow regime is compared with the experimental results of long microchannel ($1.33 \times 52.3 \times 7490 \mu m^3$) flow at rather large Knudsen numbers ($Kn_o \approx 2.5$).

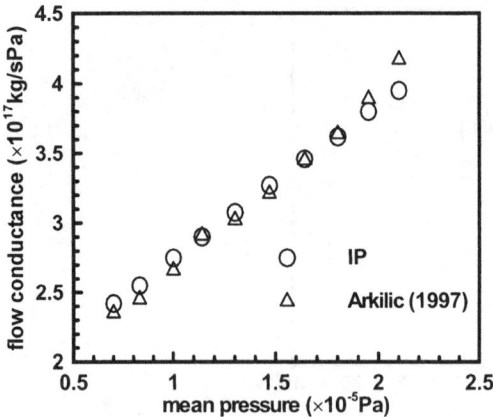

Fig. 8.18 Relation of flow conductance to mean pressure for helium flow in the transition regime. $Kn_o = 2.5$, $\sigma = 0.85$. Comparison of the IP simulation with the experimental data of [14]

8.6 THIN FILM AIR BEARING PROBLEM

The general dimensional and flow characteristics of the modern Winchester-type hard disc drive were described in section 8.1. The squeezed air bearing problem may be schematically modeled as a lower plate (the surface of the spinning platter) moving in its own plane with a velocity of U under the upper stationary tilted plate (the read/write head, see Fig. 8.19). The thin film air flow between the plates is most appropriately described by the *Reynolds equation*, which is a differential equation relating the pressure p, density ρ, platter velocity U and the height h of the gap, firstly developed by Reynolds for continuum fluid [52]. The

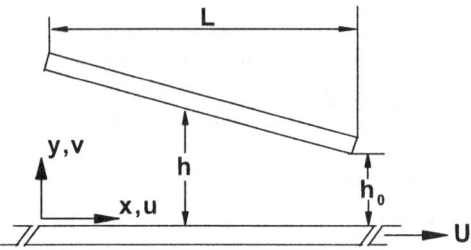

Fig. 8.19 A schematic model of the thin film air bearing flow

8.6 THIN FILM AIR BEARING PROBLEM

equation has been modified to include a number of rarefied gas dynamics effects but is still called Reynolds equation. It is essentially a mass conservation relation applied not to a fluid element but to the cross sections of the squeezed air flow and is obtained from the continuity equation by integrating it over the vertical direction with the employment of the momentum equation. Burgdorfer [53] introduced the velocity slip correction to the Reynolds equation, Fukui and Kaneko [54] developed a generalization of the equation suitable for the transitional regime.

The derivation of the Reynolds equation in the continuum regime is enlightening and can be easily extended to the slip flow and transitional flow cases so is given here. For simplicity the two-dimensional assumption is made, as the head width W is much large than the height h_o so the span wise motion can be neglected.

Writing the continuity equation

$$\frac{\partial \rho}{\partial t} + \frac{\partial \rho u}{\partial x} + \frac{\partial \rho v}{\partial y} = 0 \tag{8.29}$$

in the form

$$\frac{\partial \rho v}{\partial y} = -(\frac{\partial \rho u}{\partial x} + \frac{\partial \rho}{\partial t}), \tag{8.30}$$

and integrating it over y across the whole flow region yields

$$\int_0^h d(\rho v) = -\int_0^h (\frac{\partial \rho u}{\partial x} + \frac{\partial \rho}{\partial t}) dy. \tag{8.31}$$

The left hand side of Eq. (8.31) vanishes, as there is no fluid flown into or out of the walls. Interchanging the integration and differentiation gives

$$\frac{\partial}{\partial x} \int_0^h \rho u\, dy + \frac{\partial}{\partial t}(\rho h) = 0. \tag{8.32}$$

For thin film flow with the inertial terms neglected the steady momentum Navier-Stokes equation has the form

$$\frac{\partial p}{\partial x} = \frac{\partial}{\partial y}(\mu \frac{\partial u}{\partial y}). \tag{8.33}$$

Integration across the gap with the non-slip boundary conditions

$$u\big|_{y=0} = U, \quad u\big|_{y=h} = 0 \tag{8.34}$$

yields the solution of the stream wise velocity component u

$$u = U(1-\frac{y}{h}) - \frac{h\partial p/\partial x}{2\mu} y(1-\frac{y}{h}). \tag{8.35}$$

Substituting Eq. (8.35) into Eq. (8.32) and accomplishing integration over y, the following equation is attained

$$\frac{\partial}{\partial x}(\frac{h^3 \rho}{\mu}\frac{\partial p}{\partial x}) = 6\left[2\frac{\partial}{\partial t}(\rho h) + \frac{\partial}{\partial x}(\rho h U)\right]. \tag{8.36}$$

This is the general form of the Reynolds equation for the two-dimensional case. By introducing $X = x/L, H = h/h_o, P = p/p_o$ and the *bearing number*

$$\Lambda = 6\mu U L / p_o h_o^2, \tag{8.37}$$

Eq. (8.36) for steady and two-dimensional case can be written in the normalized form [18]

$$\frac{d}{dX}(H^3 P \frac{dP}{dX}) = \Lambda \frac{d}{dX}(PH). \tag{8.38}$$

The first term of Eq. (8.35) is the slip-less solution of the velocity in the Couette flow when the upper plate is stationary and the lower plate moves towards the right with velocity U (see section 5.4.1, compare with Eq. (5.63) with $\zeta = 0$), the second term is the slip-less solution of the velocity in the Poiseuille flow when the axis x is aligned along the lower plate (see section 5.4.2, the second term of Eq. (8.35) can be obtained from Eq. (5.69) by a simple translation of the ordinate y). The equation (8.38) shows that the flow rate across any cross section is the sum of the flow rate of the Couette flow and the Poiseuille flow and this rate does not change from one cross section to another in steady flow.

In section 5.4.2 we have seen that the flow rate of the Poiseuille flow with slip boundary condition surpasses that of the slip-less case by a factor

$$\frac{Q_{P,SL}}{Q_{P,C}} = (1 + 6\frac{2-\sigma}{\sigma}Kn) \qquad (8.39)$$

see Eq. (5.73). As for the Couette flow the flow rates have a specific feature and are identical in slip-less case and the slip case (and even in the transitional flow case) and have the following value independent of the Knudsen number owing to the symmetry of the flow (see Fig.8.20):

$$Q_C = \rho U h / 2 . \qquad (8.40)$$

From the flow rate expressions (8.39) and (8.40) for Poiseuille and Couette flows in the slip flow case one can conclude, that in the slip flow regime the following Reynolds equation is obtained in place of Eq. (8.38)

$$\frac{d}{dX}\left[(1+6\frac{2-\sigma}{\sigma}Kn)H^3 P\frac{dP}{dX}\right] = \Lambda \frac{d}{dx}(PH), \qquad (8.41)$$

where $Kn = \lambda / h$ is local Knudsen number.

When the slip boundary conditions

$$u\Big|_{y=0} = \varsigma\frac{du}{dy}, \quad u\Big|_{y=h} = -\varsigma\frac{du}{dy}, \quad \varsigma = \frac{2-\sigma}{\sigma}\lambda \qquad (8.42)$$

instead of the non-slip boundary condition (8.34) is employed in solving the momentum equation (8.33), and the resulted velocity profile is substituted into the mass conservation relation (8.32), one would arrive at the same slip corrected Reynolds equation (8.41) [53, 18].

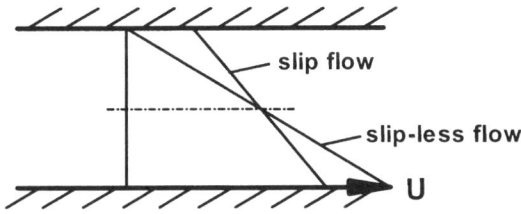

Fig. 8.20 Velocity profiles and the flow rates of the slip-less and slip Couette flow, the transitional flow is not shown but it has the same flow rate owing to the symmetry of flow

Fukui and Kaneko [19] showed that the solution of the linearized Boltzmann equation for the thin film bearing problem can be decomposed into the solutions of the plane Couette flow and the plane Poiseuille flow [55]. On this basis they derived the generalized Reynolds equation for the thin film air bearing problem by employing the flow rates of the fundamental Poiseuille and Couette flows solved by the linearized Boltzman equation. This generalized Reynolds equation in the isothermal case can be written as [19]

$$\frac{d}{dX}\left[\overline{Q}_{P,TR}(Kn)H^3 P \frac{dP}{dX}\right] = \Lambda \frac{d}{dx}(PH) \qquad (8.43)$$

where $\overline{Q}_{P,TR}(Kn)$ is the flow rate in transitional regime (normalized by the slip-less value $Q_{P,C}$) calculated from the linearized Boltzmann equation for Poiseuille flow and is shown to be the same as solved by Cernignani and Daneri [55]. A tabled database of the calculated values of $\overline{Q}_{P,TR}(Kn)$ for $\sigma=1$, $\sigma=0.9$, $\sigma=0.8$ and $\sigma=0.7$ is provided in [56], and a fitted formula approximation for diffuse reflection ($\sigma=1$) by Robert is recorded in [18] (there the second term on the right hand side is misprinted as $6A\sqrt{\pi Kn}$)

$$\overline{Q}_{P,TR}(Kn) = 1 + 6AKn + \frac{12}{\pi}Kn\log(1+BKn), \qquad (8.44)$$

where $A=1.318889$ and $B=0.387361$. Alexander, Garcia and Alder [18] used the DSMC method to simulate the short head length air bearing problem ($L=5\mu m$, $h_o=50nm=0.05\mu m$, $U=25m/s$, $\sigma=1$), and found excellent agreement of the DSMC simulation with the generalized Reynolds equation (8.43) and Eq. (8.44). Note, their description of the latter as continuum hydrodynamic Reynolds equation corrected for slip is misleading. As we have shown, the generalized Reynolds equation is a global mass conservation relation applied to the cross section of the air bearing flow with the flow rate calculated by the Boltzmann equation which is appropriate for transitional regime. The comparison made in [18] for the cases (the ratio of the inlet to outlet heights is kept as 2:1)

$L=1.5\mu m$, $h_o=15nm=0.015\mu m$, $U=153.9m/s$, $\sigma=1.0$;

$L=5\mu m$, $h_o=50nm=0.05\mu m$, $U=25m/s$, $\sigma=0.7$;

8.6 THIN FILM AIR BEARING PROBLEM

$L = 5\mu m$, $h_o = 50 nm = 0.05\mu m$, $U = 307.8 m/s$, $\sigma = 1.0$;

showed good agreement of the generalized Reynolds equation with the results of DSMC simulation, this just confirms that the generalized Reynolds can be used to solve the air bearing problem in the entire transitional flow regime and can be used to test other methods intended to solve the problem, say for longer bearing head length (the authentic length of the Winchester disc drive read/write head is $\sim 1000\mu m$, but the DSMC method was able to solve only short length ($\sim 5\mu m$) problems).

The thin film air bearing problem is solved by the IP method in [57]. The rectangular area (from $x = 0$ to $x = L$, and from $y = 0$ to $y = h$ ($x = 0$)) is divided into 200×10 uniform cells for short ($L = 5 \sim 25\mu m$) length write/read head and into 1000×10 cells for long ($L = 1mm = 1000\mu m$) head. Some of them are incised by upper surface into two parts, only the one under the upper surface is within the computational domain. The cell of this part is called incomplete cell. The smallest volume of the incomplete cells is only a very small portion of that of the standard complete cell. During the process of the IP calculation, all the incomplete cells are combined with their lower adjacent complete cells. It is found that the time step is better to be kept different for the DSMC part and the IP part of the simulation process: for DSMC the usual size of the time step of the order of collision time is sufficient, but for the IP simulation a smaller time step would ensure obtaining real macroscopic quantities of the solution without much increase of the computation time. With the employment of appropriate super-relaxation factor steady convergent results can be obtained. Fig.8.21, Fig.8.22 and Fig.8.23 show the comparison of the pressure distributions for the cases of $L = 5\mu m$, $L = 25\mu m$ and $L = 1000\mu m$ of the IP results and the results of the generalized Reynolds equation. For $L = 5\mu m$ the comparison with the DSMC simulation is given as well. One can see the excellent agreement of the IP results with the generalized Reynolds equation. This can be considered as a verification of the IP method by a criterion with the merit of the strict kinetic theory. As the generalized Reynolds equation is applicable only to a certain class of problems, where as the IP method has the flexibility and the ability to treat problems with complex geometry, this verification encourages people to use IP method to treat various complicated flow problems encountered in MEMS.

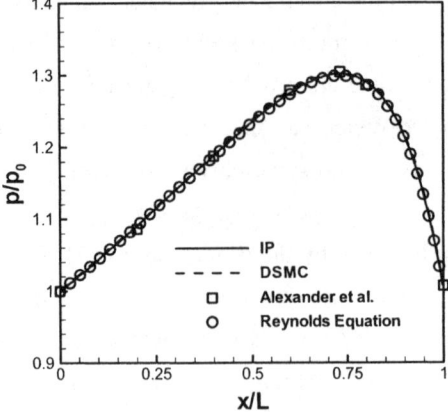

Fig. 8.21 Pressure distribution in the disc driver bearing for $Kn_o = 1.25, L = 5\mu m$, comparison of IP, DSMC and the generalized Reynolds equation results [57], also shown is the DSMC result of Alwxander et al. [18]

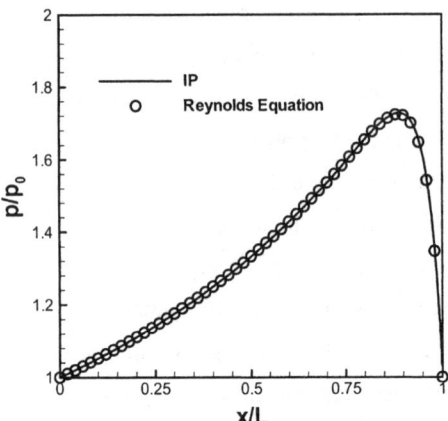

Fig. 8.22 Pressure distribution in the disc driver bearing for $Kn_o = 1.25, L = 25\mu m$, comparison of IP and the generalized Reynolds equation results [57]

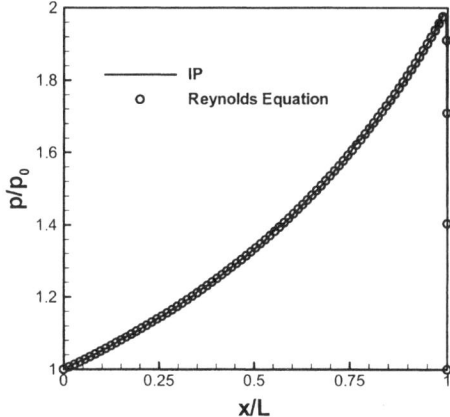

Fig. 8.23 Pressure distribution in the disc driver bearing for $Kn_o = 1.25, L = 1000 \mu m$, comparison of IP and the generalized Reynolds equation results [57]

8.7 USE OF DEGENERATED REYNOLDS EQUATION IN CHANNEL FLOW

The generalized Reynolds equation (8.43) originally is derived for application in the thin film air bearing problem with the lower plate moving with a velocity U and the upper plate tilted. Shen [58] suggests degenerate this Reynolds equation and use it to solve the microchannel flow problem. In the microchannel the lower plate is stationary and the upper plate is parallel to the lower one. Owing to the steadiness of the lower plate the right hand side term vanishes, as $U = 0$ and $\Lambda = 0$, there is not any contribution of the Couette flow. Owing to the parallelity of the two plates the value H is a constant and also can be dropped from the equation. So the generalized Reynolds equation suggested for application to the microchanel problems is degenerated to the form

$$\frac{d}{dX}\left[\overline{Q}_{P,TR}(Kn)P\frac{dP}{dX}\right] = 0. \qquad (8.45)$$

The values of P on the inlet and outlet of the channel are to be specified to make the microchannel problem solvable. This degenerated Reynolds equation is suggested be used to solve the microchannel flow in transitional flow regime provided the flow rate of the local Poiseuille flow $\overline{Q}_{P,TR}(Kn)$ in transitional regime (normalized by the slip-less value $Q_{P,C}$) is known from the strict kinetic theory. There are many works devoted to the solution of the Poiseuille flow providing the database for the flow rates at different Knudsen numbers and for different boundary conditions at the surface. With the database incorporated the degenerated Reynolds equation is valid for any surface conditions of the plates and can be integrated numerically. For example, the incomplete diffuse reflection cases with tangential accommodation coefficient $\sigma=1$, $\sigma=0.9$, $\sigma=0.8$ and $\sigma=0.7$ were calculated in [56] with tabled database of the values of $\overline{Q}_{P,TR}(Kn)$ provided under these boundary conditions. If practice has the needs, even situation with two plates having different accommodation coefficients could be considered. But for the illustrative purpose only the case of complete diffuse reflection $\sigma=1$, is expounded here. For the case of diffuse reflection, the fitted formula approximation of $\overline{Q}_{P,TR}(Kn)$, Eq. (8.44), can be used, and the degenerated Reynolds equation attains the form

$$\frac{d}{dX}\{[1+6AKn+\frac{12}{\pi}Kn\log(1+BKn)]P\frac{dP}{dX}\}=0, \qquad (8.46)$$

For the ease of integration the local Knudsen number Kn is most conveniently expressed through P, e. g., for HS model it can be written as

$$Kn=\frac{\lambda}{h}=\frac{C}{P}, \qquad (8.47)$$

where

$$C=\frac{\mu}{p_0 h}\sqrt{\frac{\pi RT_0}{2}}=\lambda_0/h=Kn_{out}, \qquad (8.48)$$

for we have for hard sphere

$$\lambda=\frac{\mu}{p}\sqrt{\frac{\pi RT}{2}}, \qquad (8.49)$$

see Eq. (2.222). p_0 is the pressure at the outlet, T_0 is the temperature of the gas, μ is the viscosity of the gas at T_0. The constant C has the physical

8.7 USE OF DEGENERATED REYNOLDS EQUATION IN CHANNEL FLOW 357

meaning of the Knudsen number at the outlet of the channel (see Eq. (8.47), at outlet $P=1$). Substituting Eq. (8.47) into Eq. (8.46), one arrives at

$$[P+6AC+\frac{12}{\pi}C\log(1+\frac{BC}{P})]\frac{dP}{dX}=D, \qquad (8.50)$$

where D is an unspecified constant to be determined from the integration and has the physical meaning of the flow rate across the channel normalized by the slip-less flow rate value.

To illustrate the use of the degenerated Reynolds equation in solving the microchanel problem we calculate the pressure distribution for nitrogen in the $1.2\times40\times3000\,\mu m^3$ channel [11] and helium in the $1.2\times40\times4000\,\mu m$ channel [12].

For $T_0=294K$ the value of C for helium is 0.15579, and for nitrogen is 0.052325. Equation (8.50) is integrated under the following boundary condition

$$P|_{X=0}=p_{in}/p_{out}, \text{ and } P|_{X=1}=p_{out}/p_{out}=1, \qquad (8.51)$$

with p_{in} provided by the experimental data in [11, 12]. The results of integration are presented in Fig. 8.24 and Fig. 8.25. It is seen from the figures that the results of the degenerated Reynolds equation agree well with the experimental data, and the IP simulation results have excellent agreement with those of the degenerated Reynolds equation, especially for the pressure distribution in the microchannel with dimension $1.2\times40\times4000\,\mu m^3$ for helium (the two curves almost coincide with each other).

In section 8.2 we have shown the unfeasibility of using LBM in simulating transitional flows in MEMS by comparison with the DSMC calculations [24, 25]. Here the LBM results [23] are compared (see [58]) with the calculations by using the degenerated Reynolds equation to attain the same conclusions as in [24, 25], but this time the conclusion is confirmed by a test stone with the merit of strict kinetic theory. Equation (8.50) is integrated under the following conditions for a short $1\times100\,\mu m^2$ microchannel that have been considered by LBM in Nie, Doolen and Chen [23]:

1. $C = 0.194, P|_{X=0} = 2, P|_{X=1} = 1$,

2. $C = 0.388, P|_{X=0} = 2, P|_{X=1} = 1$.

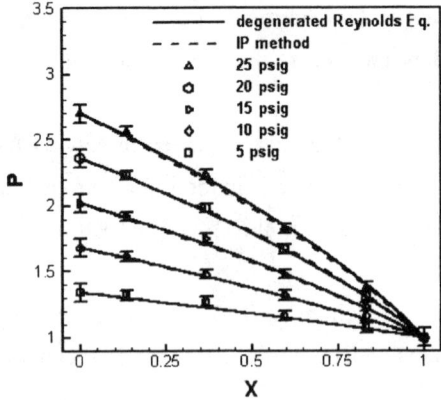

Fig. 8.24 The pressure distribution in a $1.2 \times 40 \times 3000 \mu m^3$ microchannel for nitrogen. Comparison [58] of the degenerated Reynolds equation (8.50) (solid line), the IP method (dashed line) and the experimental data [11]

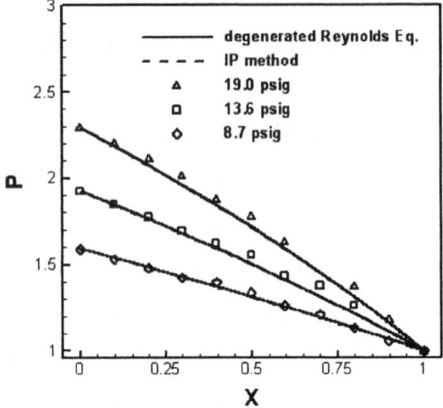

Fig. 8.25 The pressure distribution in a $1.2 \times 40 \times 4000 \mu m^3$ microchannel for helium. Comparison [58] of the degenerated Reynolds equation (8.50) (solid lines), the IP method (dashed lines, note that the solid lines and the dashed lines almost coincide) and the experimental data [12]

8.7 USE OF DEGENERATED REYNOLDS EQUATION IN CHANNEL FLOW

The results of comparison of the integration of the degenerated Reynolds equation (8.50) with the LBM, the DSMC and IP results are shown in Fig. 8.26 and Fig. 8.27. It is seen that the degenerated Reynolds equation, the DSMC method and the IP method are in excellent agreement with each other but they are in appa-

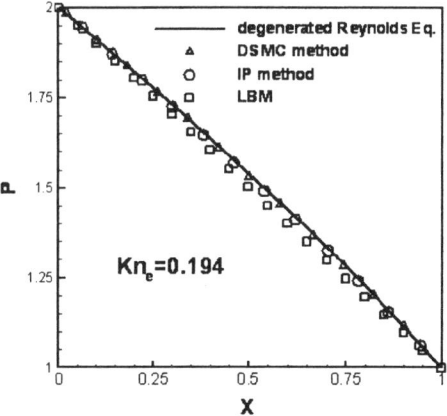

Fig. 8.26 Pressure distribution in a microchannel with $Kn = 0.194$ at outlet ($h/L = 100$). Comparison [58] of the degenerated Reynolds equation, DSMC, IP methods and the LBM method

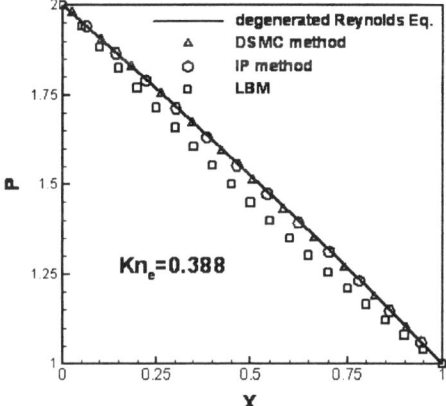

Fig. 8.27 Pressure distribution in a microchannel with $Kn = 0.388$ at outlet ($h/L = 100$). Comparison [58] of the degenerated Reynolds equation, DSMC, IP methods and the LBM method

rent disagreement with the LBM results. The LBM is shown to be unfeasible to simulate the transitional flow again, but this time by a method having the merit of kinetic theory

The generalized Reynolds equation (8.45) degenerated for application to the micro-channel problems with the flow rate $\overline{Q}_{P,TR}(Kn)$ provided by the linearized Boltzmann equation is appropriate for solving the microchannel flow problems in the entire transitional regime. It can provide the pressure distribution, the flow rate but not the detailed flow field such as the velocity profiles. But its significance lies in that it can be used as criterion of strict kinetic theoretical merit to test various methods aimed to solve the microchannel problems in transitional regime.

From the degenerated Reynolds equation (8.50) for the specific case of diffuse reflection it is seen that the microchannel rarefied gas flow is entirely specified by the inlet and outlet pressure P_{in} and P_{out}, and the Knudsen number at the outlet $C = Kn_{out}$, the length of the channel does not enter as a determining factor.

Besides the air bearing problem and the microchannel flow problem the Reynolds equation can also model the gas damping problem in micromechanical accelerometers [59]. Database for the flow rates of the Poiseuille flow with various combinations of possible surface properties calculated on the basis of linearized Boltzmann equation or other rigorous kinetic theory is desirable for the solution of microchannel flow, thin film air bearing problem and also the damping problem in the micromechanical accelerometers, especially in the form of fitting formulas.

8.8 SOME ACTUAL PROBLEMS AND CONCLUDING REMARKS

When there are temperature gradients along the MEMS or channel surfaces there occur the phenomena of thermal creep, thermal transpiration, thermal stress slip flow and the temperature stress convection etc. (see chapter 5). The Knudsen compressor in use of MEMS is worked out on the basis of thermal transpiration [60]. So it is of significance to extend the IP method to the case of temperature variation. Some useful attempt and exploration have undertaken in this aspect [32, 35, 46]. The difficulty encountered in extending the IP method to the case of temperature variation is that the average energy flux of monatomic molecules in a static gas through a surface element is $2kT\Gamma_n$, where Γ_n is the molecule number flux, but the average energy carried by a single molecule is $(3/2)kT$, the IP process can not satisfy the global energy balance across an interface. Sun [46] put forward a model of additional energy transfer and a method of assignment of the post collision IP temperature by which the IP method was able to simulate the flow between two plates with different temperatures and obtain temperature result in agreement with that of the DSMC method. But the density distribution is in some minor difference with the DSMC method. In [35] the temperature components in three directions are introduced and new method of assignment of the post collision temperature is adopted, the agreement of the temperature and density distribution with the DSMC is obtained in the flow between two plates. But these constitute only partial success for such models can not provide general method of simulation of the rarefied gas motion caused by the temperature variation. It is a challenge to modify and develop the present IP algorithm to adapt it for employment in the case of temperature variation.

The micro scale flow is usually a low speed flow as well and the flow problem is of the elliptical nature. For the external flows the size of the flow field involved in the simulation as a rule is much bigger than the body itself, thus most part of the flow field can be described by the continuum equations. For internal flows the region near the walls is described by the particle method, the region far from the walls can be described by the continuum method. The hybrid continuum/particle approach can make use of the advantages of both methods and can save enormous computation time and thus has gained extensive attention. At the interface of the continuum flow and the particle simulation boundaries (usually movable and regulated unceasingly) information must be interchanged at each time step. For ordinary particle methods owing to the huge statistical fluctuations, it is very time-consuming to pose definite boundary conditions for the continuum flow. At the same time, as the IP method preserves the macroscopic information, it is quite easy to pose boundary conditions for the continuum flow. Sun et al. [61] used hybrid IP method with the Navier-Stokes equation plus slip boundary condition to solve the flow around the plane plate and Couette flow problems and obtained smooth solutions with enhanced efficiency.

Finally, we make some concluding remarks of this chapter.

The linearized Boltzmann equation method and the DSMC method are appropriate for solving microscale rarefied gas flow problems and can be used as criteria for testing various methods intending to solve the transitional flows, the latter encounters the problem of huge statistical fluctuation for slow rarefied gas flows. The IP method preserves the averaged information of the enormous number of molecules that a simulation molecule represents, overcomes this difficulty and for low speed cases saves the computation time by a factor of $10^2 \sim 10^4$ and can treat more easily problem of complex configuration in comparison with the linearized Boltzmann equation.

The difficulty of regulating the inlet ad outlet boundary conditions of the internal flow problem is overcome by the use of conservation scheme and supper-relaxation method in the IP method. The method is validated, and on the examples of unidirectional flow, the channel flow and thin film bearing problems is checked by comparison with the experimental results, the linearized Boltzmann equation, DSMC method, the generalized Reynolds equation and the its degenerated version (with the flow rate of Poiseuille problem calculated by the Boltzmann equation). In regulating of the inlet and outlet boundary conditions the effect of super-relaxation is different from the amplification of the time interval, the requirement of sufficient small time steps remains in force to guarantee through DSMC process the true trend of the variation of the flow quantities. The super-relaxation factor amplifies the true trend and accelerates the approaching of the true solution. The regulated IP values are used as the current boundary conditions at the inlet and outlet of the DSMC process so the entire process of convergence is quickened.

The generalized Reynolds equation is appropriate to treat the thin film air bearing problem in the entire transitional regime. Example calculations and comparison with DSMC, IP and experimental results show the success of the suggestion of using the degenerated Reynolds equation to solve the transitional microchannel flow problem. Degenerated Reynolds equation with the Poiseuille flow rate calculated by the linearized Bolktzmann equation can serve as a test stone of the merit of strict kinetic theory, in particular it gives an undoubted confirmation of the unfeasibility of LBM in simulating the transitional flows. On the example of microchannel flow it provides a solid verification of the IP method.

REFERENCES

1. Knudsen M (1909) Ann Physics, 28:75
2. Millikan RA (1911) The isolation of an ion, a precision measurement of its charge, and the correction of Stokes' law. Phys Rev, 32:349
3. Millikan RA (1923) The general law of fall of a small spherical body through a gas, and its bearing upon the nature of molecular reflection from surfaces. Phys Rev, 22:1
4. Knudsen M, Weber S (1911) Ann Phys, 36:981
5. Feynman RP (1959) There's plenty of room at the bottom. Lecture on the 1959 annual meeting of AIP, in Journal of Microelectromechanical Systems, (1992) 1:60-66; also in Trimmer W (1997), Micromechanics and MEMS, Classic and Seminar Papers to 1990, Wiley, New York
6. Koplik J, Banavar J, Willemson J (1989) Molecular dynamics of fluid flow at solid surfaces. Phys. of Fluids A 1:781-794
7. Thomson P, Troian S (1997) A general boundary condition for liquid flow at solid surfaces. Nature, 389:359-362
8. Deckert KL (1990) Computer-aided design of slider bearings in magnetic disc files. IBM J Res Devel 35:660
9. Tagawa N (1993) State of the art for flying head slider mechanisms in magnetic recording disc storage. Wear 168:43
10. Liu JQ, Tai YC, Pong KC, Ho CM (1993) Micromachined channel/pressure sensor systems for micro flow studies. The 7th International Conference on Solid State Sensors and Actuators, Tranducers, 995-998
11. Pong KC, Ho CM, Liu JQ, Tai YC, (1994) Non-linear pressure distribution in uniform microchannels. ASME, FED 197, Application of Microfabrication to Fluid Mechanics, 51-56
12. Shin JC, Ho CM, Liu JQ, Tai YC, (1996) Monatomic and polyatomic gas flow through uniform microchannels. ASME-DSC 59, 197
13. Arkilic EB, Schmidt MA, Breuer KS (1997) Measurement of the TMAC in silicon microchannels, in C Shen edited Rarefied Gas Dynamics, Peking Univ. Press, 983-988
14. Arkilic EB (1997) Measurement of the mass flow and the TMAC in silicon microchannels. Ph D Thesis, MIT, DFRL TR 97-1
15. Karniadakis GE, Sherwin S (1999) Spectral/hp Element Method for CFD. Oxford Univ. Press, New York
16. Karniadakis GE, Beskok A (2001) Micro Flows, Fundamentals and Simulation. Springer-Verlag, New York, Berlin, Heidelberg
17. Shen C, Fan J, Xie C (2003) Statistical simulation of rarefied gas flows in micro-channels. J Comp Physics 189:512-526
18. Alexander FJ, Garcia AL, Alder BJ (1994) Direct simulation Monte Carlo for thin film bearings. Phys. of Fluids 6:3854-3860
19. Fukui S, Kaneko R (1988) Analysis of ultra-thin gas film lubrication based on linearized Boltzmann equation: first report-derivation of a generalized lubrication equation including thermal creep flow. J Tribology 110:253-262
20. Frish U, Hasslacher B, Pomeau Y (1986) Lattice gas automaton for the Navier-Stokes equation. *Phy. Rev Let*, 56: 1505
21. Qian Y, D'Humieres D, Lallemand P (1992) Lattice BGK models for Navier-Stokes equation. Europhysics Letters, 17:479-482

22. Chen S, Doolen GD (1998) Lattice Boltzmann method for fluid flows. Ann Rev Fluid Mech., 30:329
23. Nie X, Doolen GD, Chen S (2002) Lattice Boltzmann simulation of fluid flows in MEMS. J Stat Phys, 107:279
24. Shen C, Tian DB, Xie C, Fan J (2003) Examination of LBM in simulation of the micro-channel flow in transitional regime. Proceedings of the 1st Intern. Conference on micro and mini-channels, ASME, 405-410, see also Shen C, Tian DB, Xie C, Fan J (2004) Examination of the LBM in simulation of microchannel flow in transitional regime. Microscale Thermophysical Engineering, 8:423-432
25. Tian DB, Shen C, Xie C, Fan J (2003) The testing of the feasibility of the LBM in simulation of the transitional flow regime. Proceedings of the 2003 seminar on frontier problems of the aerodynamics (in Chinese).
26. Oran ES, Oh CK, Cybek BZ (1998) Direct simulation Monte Carlo: recent advances and applications. Ann Rev Fluid Mech 30:403
27. Oh CK, Oran ES, Sinkovits RS (1997) Computation of high-speed, high Knudsen number microchannel flows. J Thermophysics and Heat Transfer 12:447.
28. Nance RP, Hash D, Hassan HA (1997) Role of boundary conditions in Monte Carlo simulation of MEMS devices. J Thermophysics and Heat Transfer ,11:497
29. Kaplan CR, Oran ES (2002) Nonlinear filtering of low velocity gaseous micro-flows. AIAA Journal, 40:82
30. Fan J, Shen C (1999) Statistical simulation of low-speed unidirectional flows in transitional regime. Rarefied Gas Dynamics, edited by R Brun, R Campargue, R Gatignol, J C Lengrand, Cepadues Editions, 2:245-252
31. Fan J, Shen C (2001) Statistical simulation of low-speed rarefied gas flows. J Computational Physics, 167: 393-412
32. Shen C, Jiang J Z, Fan J (2001) Information preservation method for the case of temperature variation. In: T J Bartel and M A Gallis (eds) Rarefied Gas Dynamics, edited by, AIP, 585:185-192
33. Fan J, Boyd ID, Cai CP, Hennighausen K, Candler GV (2001) Computation of rarefied gas flows around a NACA 0012 airfoil. AIAA J, 39: 618-626
34. Shen C (2004) The information preservation method in simulation of internal rarefied gas flows of MEMS, invited paper, Proceedings of the 24th International Symposium on Rarefied Gas Dynamics, 11-16 July 2004, Bari, Italy, to appear in Rarefied Gas Dynamics, AIP conference proceedings
35. Jiang J Z, Shen C, Fan J (2004) Information preservation method with temperature variation, to appear
36. Sone Y, Takata S, Ohwada T (1990) Numerical analysis of the plane Couette flow of a rarefied gas. Eur J Mech., B/Fluids, 9:273
37. Gross EP, Ziering S (1958) Kinetic theory of linear shear flow. Phys. of Fluids, 1:215
38. Dong W (1956) Univ. California Report, UCRL-3353
39. Ohwada T, Sone Y, Aoki A (1989) Numerical analysis of Poiseuille flow and thermal transpiration flows. Phys. of Fluids, A1:2042
40. Gross EP, Jackson EA (1958) Kinetic theory of impulsive motion of an infinite plane. Phys. of Fluids, 1:318
41. Xie C, Fan J, Shen C (2000) Direct statistical simulation of gas flow in microchannels. Proceedings of the mass and heat transfer conference of the Chinese engineering thermophysics, p. 388 (in Chinese)

42. Cai CP, Boyd ID, Fan J, Candlir GV (2000) Direct simulation method for low speed microchannel flows. J Thermophysics Heat Transfer,14:368
43. Xie C, Fan J, Shen C (2003) Rarefied gas flows in microchannels. In: AD Ketsdever, EP Muntz edited Rarefied Gas Dynamics, AIP conference proceedings 663: 800-807
44. Sun QH, Boyd ID, Fan J (2001) Development of an IP method for subsonic, microscale gas flows. In: TJ Bartel, MA Gallis edited Rarefied Gas Dynamics, AIP conference proceedings 585: 547-553
45. Sun QH, Boyd ID, Candler GV (2001) Numrical simulation of gas flow over micro-scale airfoils, AIAA paper 2001-3071
46. Sun QH, Boyd ID (2002) A direct simulation method for subsonic, microscale gas flows. J Computational Physics, 179:400-425
47. He HG, et al (2004) Private communication
48. Fan J et al (2001) Computation of rarefied gas flows around a NACA 0012 airfoil, AIAA J, 39:618
49. Jiang JZ, Fan J, Shen C (2003) Statistical simulation of micro squire cavity flows. In: AD Ketsdever, EP Muntz edited Rarefied Gas Dynamics, AIP conference proceedings 663: 784-791
50. Jiang J, Shen C, Fan J (2003). Statistical simulation of non-circular cross section Poiseuille flows. Proceedings of the 1st Intern. Conference on micro and mini-channels, ASME, 411-418
51. Liu HL, Xie C, Shen C, Fan J (2001) Flow in membrane filter simulated as microchannel flow with diaphragm. In: TJ Bartel and MA Gallis edited Rarefied Gas Dynamics, AIP conference proceedings 585: 524-530
52. Reynolds O (1886) On the theory of lubrication and its application to Mr. Beauchamp Tower's experiments including an experimental determination of the viscosity of the olive oil. Philos Trans R Soc London, A177:157-234
53. Burgdorfer A (1959) The influence of the molecular mean free path on the performance of hydrodynamic gas lubrication bearings. Trans ASME 81:94
54. Gross WA, Matsch LA, Castelli V, Eshel A, Vohr JH, Wildman M (1980) Fluid Film Lubrication. Wiley, New York
55. Cercignani C, Daneri A (1963) Flow of a rarefied gas between two parallel plates. J Applied Physics, 34:3509-3513
56. Fukui S, Kaneko R (1990) A database for interpolation of Poiseuille flow rates for high Knudsen number lubrication problems. J Tribology 112:78-83
57. Jiang JZ, Shen C, and Fan J (2004) Statistical simulation of thin-film bearings. Paper presented at 24th international symposium on Rarefied Gas Dynamics, 10-16 July 2004, Bari, Itaty
58. Shen C, Use of the degenerated Reynolds equation in solving the microchannel flow problem, submitted to Phys. of Fluids.
59. Veijola T, Kuisma H, Lahdenpera J, Equivalent-circuit model of the squeezed gas film in a silicon accelerometer, Sensors and Actuators, A48:239 (1995)
60. Pham-van-Diep G, Keely P, Muntz EP, Weaver DP (1995) A micromechanical Knudsen compressor. In J Harvey, G Lord edited Rarefied Gas Dynamics, Oxford Univ. Press, 1:715
61. Sun QH, Boyd ID, Candler GV (2003) A hybrid continuum/particle approach for microscale gas flows. In: AD Ketsdever and EP Muntz edited Rarefied Gas Dynamics, AIP 663:752-759

APPENDIX I GAS PROPERTIES

Table I.1 Characteristic temperatures of some diatomic molecules

Gas	Rotation Θ_r (K)	Vibration Θ_v (K)	Dissociation Θ_d (K)	Ionization Θ_i (K)
N_2	2.88	3371	113500	181000
NO	2.44	2719	75500	108000
O_2	2.07	2256	59500	142000

Table I.2 The properties and reference molecular diameters of some molecules at standard conditions

Gas	Molecular Mass m $\times 10^{27} kg$	Viscosity Coefficient μ $\times 10^5 Nsm^{-2}$	Viscosity Coefficient Index ω	α	VHS Model d_{ref} $\times 10^{10} m$	VSS Model d_{ref} $\times 10^{10} m$	HS Model d_{HS} $\times 10^{10} m$
N_2	46.5	1.656	0.74	1.36	4.17	4.11	3.784
NO	49.88	1.774	0.79		4.20		3.72
O_2	53.12	1.919	0.77	1.40	4.07	4.01	3.636
H_2	3.34	0.845	0.67	1.35	2.92	2.88	2.745
He	6.65	1.865	0.66	1.26	2.33	2.30	2.193
Ne	2.975	2.975	0.66	1.31	2.77	2.72	2.602
Ar	66.3	2.117	0.81	1.40	4.17	4.11	3.659
Kr	139.1	2.328	0.80	1.32	4.76	4.70	4.199
Xe	218.0	2.107	0.85	1.44	4.76	5.65	4.939

Table I.3 Collision diameters of some molecules for HS model and reference diameter for VHS model at $273K$ that should be adopted when applying the IP method [2]

	N_2	O_2	He	Ar	CO_2
$d_{HS} \times 10^{10} m$	4.128	3.948	2.365	3.963	5.013
$d_{VHSref} \times 10^{10} m$	4.312	4.295	2.463	4.283	5.620

Table I.4 Force constants in the Lennard-Jones 6-12 model, ρ and ω in Eq. (2.117) [3]

	N_2	NO	O_2	H_2	He	Ne	Ar	Kr	Xe
$\rho \times 10^{10} m$	3.681	3.47	3.433	2.968	2.576	2.789	3.418	3.61	4.055
ε/k (K)	915	119	113	33.3	10.22	35.7	124	91.5	229

REFERENCES

1. Chapman S and Cowling TG. (1970) The Mathematical Theory of Non-uniform Gases, 3rd Edition, Cambridge University Press
2. Fan J and Shen C. (2001) Statistical simulation of low-speed rarefied gas flows. J. of Comp. Physics, 167: 394
3. Hirschfelder JO, Curtiss CF and Bird RB. (1954) Molecular Theory of Gases and Liquids. John Wiley and Sons

APPENDIX II SOME INTEGRALS

II.1   GAMMA FUNCTION AND ERROR FUNCTION

1. *The gamma function* is defined as

$$\Gamma(j) = \int_0^\infty x^{j-1} \exp(-x)\, dx . \qquad (II.1)$$

The integration by parts yields the recursion formula

$$\Gamma(j+1) = j\Gamma(j) . \qquad (II.2)$$

Obviously, $\Gamma(1)=1, \Gamma(2)=1$, hence, when n is an integer, one has

$$\Gamma(n+1) = n! . \qquad (II.3)$$

The incomplete gamma function is defined as

$$\Gamma(j,a) = \int_a^\infty x^{j-1} \exp(-x)\, dx . \qquad (II.4)$$

Again, the integration by parts yields the reduction formula

$$\Gamma(j+1,\alpha) = j\Gamma(j,\alpha) + \alpha^j \exp(-\alpha) . \qquad (II.5)$$

The values of the Γ function with the values of the argument between 1.0 and 2.0 can be found in the mathematical hand books (p. 1312, [1]; p. 267, [2]; p. 75, [3]), the values out of this range can be obtained by using Eq. (II.2).

2. *The error function* is defined as

$$erfa = \frac{2}{\sqrt{\pi}} \int_0^a \exp(-x^2)\, dx . \qquad (II.6)$$

The *complementary error function* is defined as

$$erfca = 1 - erfa. \tag{II.7}$$

Some times the integral complementary error function is of use, which is defined as

$$ierfca = \frac{1}{\sqrt{\pi}} \exp(-a^2) - a\,erfca \tag{II.8}$$

The table of the error function can be found in [1, 2, 3] (the values given are $\phi(t) = erf(t/\sqrt{2})$).

II.2 SOME DEFINITE INGTEGTRALS

1. Here the values of *the following integrals* are given

$$\int_{\pm a}^{\infty} x^n \exp(-\beta^2 x^2) dx, \quad n = 1, 2, 3, \cdots. \tag{II.9}$$

The starting point of evaluation of the above integrals is the following integral

$$\int_0^{\infty} \exp(-x^2) dx = \frac{\sqrt{\pi}}{2}, \tag{II.10}$$

the easiest way of derivation is to find the squire of the above formula

$$\left(\int_0^{\infty} e^{-x^2} dx\right)^2 = \int_0^{\infty}\int_0^{\infty} e^{-(x^2+y^2)} dx\,dy = \int_0^{\infty}\int_0^{2\pi} e^{-r^2} r\,d\rho\,dr = \frac{\pi}{4}.$$

Making use of Eq. (II.10) and integration by parts, it is easy to prove successively (supposing $a = 0$ in Eq. (II.9), starting from $n = 0$)

$$\int_0^{\infty} \exp(-\beta^2 x^2) dx = \frac{\sqrt{\pi}}{2\beta}, \tag{II.11}$$

$$\int_0^{\infty} x \exp(-\beta^2 x^2) dx = \frac{1}{2\beta^2}, \tag{II.12}$$

$$\int_0^\infty x^2 \exp(-\beta^2 x^2)\,dx = \frac{\sqrt{\pi}}{4\beta^3}, \tag{II.13}$$

$$\int_0^\infty x^3 \exp(-\beta^2 x^2)\,dx = \frac{1}{2\beta^4}, \tag{II.14}$$

$$\int_0^\infty x^4 \exp(-\beta^2 x^2)\,dx = \frac{3}{8}\frac{\sqrt{\pi}}{\beta^5}, \tag{II.15}$$

$$\int_0^\infty x^5 \exp(-\beta^2 x^2)\,dx = \frac{1}{\beta^6}, \tag{II.16}$$

$$\int_0^\infty x^6 \exp(-\beta^2 x^2)\,dx = \frac{15}{16}\frac{\sqrt{\pi}}{\beta^7}, \tag{II.17}$$

$$\int_0^\infty x^7 \exp(-\beta^2 x^2)\,dx = \frac{3}{\beta^8}. \tag{II.18}$$

Making use of Eq. (II.10), the definition of the error function Eq. (II.6) and integration by parts, it is easy to prove successively

$$\int_{\pm a}^\infty \exp(-\beta^2 x^2)\,dx = \frac{\sqrt{\pi}}{2\beta}\left(1 \mp \mathrm{erf}(\beta a)\right), \tag{II.19}$$

$$\int_{\pm a}^\infty x \exp(-\beta^2 x^2)\,dx = \frac{1}{2\beta^2}\exp(-\beta^2 a^2), \tag{II.20}$$

$$\int_{\pm a}^\infty x^2 \exp(-\beta^2 x^2)\,dx = \frac{\sqrt{\pi}}{4\beta^3}\left[1 \mp \mathrm{erf}(\beta a)\right] \pm \frac{\beta a \exp(-\beta^2 a^2)}{2\beta^3}, \tag{II.21}$$

$$\int_{\pm a}^\infty x^3 \exp(-\beta^2 x^2)\,dx = \frac{\exp(-\beta^2 a^2)}{2\beta^4}\left(1 + \beta^2 a^2\right), \tag{II.22}$$

$$\int_{\pm a}^{\infty} x^4 \exp(-\beta^2 x^2) dx = \frac{3\sqrt{\pi}}{8 \beta^5} [1 \mp erf(\beta a)] \pm \frac{\beta a \exp(-\beta^2 a^2)}{2\beta^5} \left(\frac{3}{2} + \beta^2 a^2\right), \quad (II.23)$$

$$\int_{\pm a}^{\infty} x^5 \exp(-\beta^2 x^2) dx = \frac{\exp(-\beta^2 a^2)}{\beta^6} \left(1 + \beta^2 a^2 + \frac{1}{2}\beta^4 a^4\right). \quad (II.24)$$

2. According to the definition of the error function Eq. (II.6) and integration by parts, it is easy to obtain

$$\int_0^a x^2 e^{-x^2} dx = \frac{\sqrt{\pi}}{4} erfa - \frac{1}{2} ae^{-a^2}, \quad (II.25)$$

$$\int_0^a x\, erfx\, dx = \frac{1}{2}\left(a^2 - \frac{1}{2}\right) erfa + \frac{1}{2\sqrt{\pi}} ae^{-a^2}, \quad (II.26)$$

$$\int_0^a x^3 erfx\, dx = \frac{1}{4}\left(a^4 - \frac{3}{4}\right) erfa + \frac{1}{4\sqrt{\pi}} ae^{-a^2}\left(a^2 + \frac{3}{2}\right). \quad (II.27)$$

3. The following integrals can be expressed through *the modified Bessel functions* (definition of the latter see p.376, [2]; p.350, [4])

$$\int_0^\pi e^{z\cos\phi} \cos\phi\, d\phi = \pi I_1(z), \quad (|\arg z| \leq \frac{\pi}{2}), \quad (II.28)$$

$$\int_0^\pi e^{z\cos\phi} d\phi = \pi I_0(z). \quad (II.29)$$

Making use of Eqs. (II.28), (II.29), after some manipulation one obtains

$$\int_0^\pi \cos^2\theta \exp(-S^2 \cos^2\theta) d\theta = \frac{\pi}{2} \exp\left(-\frac{S^2}{2}\right) \left[I_0\left(\frac{S^2}{2}\right) - I_1\left(\frac{S^2}{2}\right)\right], \quad (II.30)$$

$$\int_0^\pi \cos\theta\, erf(S\cos\theta) d\theta = \sqrt{\pi} S \exp\left(-\frac{S^2}{2}\right) \left[I_0\left(\frac{S^2}{2}\right) + I_1\left(\frac{S^2}{2}\right)\right], \quad (II.31)$$

$$\int_0^\pi \cos^3\theta\, erf(S\cos\theta)\, d\theta =$$

$$\sqrt{\pi}\, S \exp\left(-\frac{S^2}{2}\right)\left[\frac{2}{3}I_0\left(\frac{S^2}{2}\right)+\frac{1}{3}I_1\left(\frac{S^2}{2}\right)+\frac{J_1(S^2/2)}{3S_2}\right]. \tag{II.32}$$

The tables of $I_0(x)$ and $I_1(x)$ can be found in p.416, [2]; p.76. [3].

II.3 BETA FUNCTION

The beta function is defined as

$$B(a,b) = \int_0^1 t^{a-1}(1-t)^{b-1}\, dt. \tag{II.33}$$

It is related with the gamma function

$$B(a,b) = \frac{\Gamma(a)\cdot\Gamma(b)}{\Gamma(a+b)} = B(b,a). \tag{II.34}$$

The incomplete beta function is defined as

$$B_x(a,b) = \int_0^x t^{a-1}(1-t)^{b-1}\, dt. \tag{II.35}$$

or, the normalized by $B(a,b)$ incomplete beta function is

$$I_x(a,b) = B_x(a,b)/B(a,b). \tag{II.36}$$

The equation

$$I_x(a,b) = \frac{1}{2}, \tag{II.37}$$

can be inverted (see p.945, [2])

$$x = \frac{a}{a+b^{2w}}, \tag{II.38}$$

where

$$w = \frac{1}{3}\left(\frac{1}{2a-1} + \frac{1}{2b-1} - 1\right)\left(\frac{1}{2b-1} - \frac{1}{2a-1}\right). \tag{II.39}$$

REFERENCES

1. Handbook of mathematics (1979) Peoples Education press, Beijing (in Chinese)
2. Abramowitz M and Stegun IA (1965) Handbook of Mathematical Functions, National Bureau of Standands, Appl. Math. Series 55
3. Bronshtein YN, Semendygev KA (1965) Handbook of Mathematics, 10th ed., 'Nauka' Press (in Russian)
4. Reizhik YM and Gradshtein YS (1951) Tables of Integrals, Sums, Series and Products, Moscow (in Russian)

APPENDIX III SAMPLING FROM A PRESCRIBED DISTRIBUTION

In the DSMC simulation and usual statistical simulations the sampling of variates with certain probabilistic distributions is required, that is, obtaining the representative values of the variates successively is required. The basis of such sampling is the set *ranf* of uniformly distributed between 0 to 1 variates (random fractions). If the integral of the probabilistic distribution (the cumulative distribution function) can be inverted relative to the argument, the sampling of such rariate can be accomplished by the *method of inversion of the cumulative distribution function*. If the cumulative distribution function can not be inverted, the sampling can be accomplished by the *acceptance-rejection method*. If the probabilistic distribution function has singularities, that is, it attains infinite values at some points, the sampling is accomplished by the *generalized acceptance–rejection method* or the *combined cumulative distribution and acceptance-rejection method*.

III.1 INVERSION OF CUMULATIVE DISTIRBUTION FUNCTION

The probability that a variate x is lying between x and dx is

$$f_x dx, \qquad (\text{III.1})$$

f_x is the normalized distribution function. If the values of x is confined between a and b, then obviously

$$\int_a^b f_x dx = 1. \qquad (\text{III.2})$$

The *cumulative distribution function* of the variate is defined as

$$F_x = \int_a^x f_x dx. \tag{III.3}$$

Generate a random fraction *ranf* and let it equal to F_x, the sampled value of x is found according to

$$F_x = ranf. \tag{III.4}$$

If (III.3) can be inverted to find a function for x, then from (III.4) one can obtain an explicit expression of x through *ranf*.

Example 1 Variate x is uniformly distributed between a and b, the condition Eq. (III.2) gives

$$f_x = \frac{1}{b-a}.$$

Equation (III.3) gives

$$F_x = \frac{x-a}{b-a}.$$

From (III.4) one obtains

$$\frac{x-a}{b-a} = ranf.$$

Solve it relative to x, it is found that x is sampled according to

$$x = a + (b-a) ranf. \tag{III.5}$$

If a number of represented values of the variate x uniformly distributed between a and b are required simultaneously, the sampling by repeatedly utilizing Eq. (III.5) would lead to large variance of scatter. The so called *variance reduction technique* is to be employed, i.e., the specificity of the uniform distribution is utilized and set a grosso modo uniform distribution and sample according to the random distribution only in small ranges. For example, m representative values of x uniformly distributed between a and b are to be sampled, one should sample m times starting from $n=1$

III.1 INVERSION OF CUMULATIVE DISTIRBUTION FUNCTION

$$x = a + (b - a)\frac{n - ranf}{m}, \quad n = 1, 2, 3, \cdots, m. \tag{III.5}'$$

Example 2 Variate x is distributed between a and b with probability proportional to x. Following the steps in the above example and utilizing Eqs. (III.2), (III.3) and (III.4), it is easy to obtain the sampling formula of x

$$x = \left[a^2 + (b^2 - a^2) ranf \right]^{1/2}. \tag{III.6}$$

If multiple sampling is needed, e.g., when one needs to m times distribute uniformly r in a ring with radius varying from a to b, r is to be sampled according to the following formula (starting from $n = 1$ for m times)

$$x = \left[a^2 + (b^2 - a^2)\frac{n - ranf}{m} \right]^{1/2}, \quad n = 1, 2, 3, \cdots, m \tag{III.6}'$$

The variance reduction method explained in examples 1 and 2 is to be used when allocate the initial positions of the incident simulated molecules (for the uniform linear and uniform axi-symmetric flow cases).

Example 3 The distribution of the value V of the resultant of the tangential velocity of the molecules after the diffuse reflection (see Eq. (3.17)) is

$$f_{\beta^2 V^2} = \exp(-\beta^2 V^2),$$

the cumulative distribution function is

$$F_{\beta^2 V^2} = 1 - \exp(-\beta^2 V^2).$$

As $ranf$ and $(1 - ranf)$ are equivalent, set $F_{\beta^2 V^2} = (1 - ranf)$ according to Eq. (III.4), from where the result of inversion for V is

$$V = \left(-\ln(ranf) \right)^{1/2} / \beta \tag{III.7}$$

The distribution function of the component u of the normal velocity of the molecules after the diffuse reflection is identical with the distribution function of

V (see the discussion in section 3.2 leading to Eq. (3.20)), so the sampling of u is the same as that of V.

III.2 ACCEPTANCE-REJECTION METHOD

The method of inversion of the cumulative distribution described in the above section can be employed only when Eq. (III.4) can be solved relative to x, but in many cases the inversion of Eq. (III.4) relative to x is impossible. In such case a usually adopted method of sampling is the so called acceptance-rejection method. The distribution function is normalized by its maximum

$$f'_x = f_x / f_{x\max} . \tag{III.8}$$

First by using Eq. (III.5) a value of x is sampled, a and b are the upper ad lower limits of x, the value of f_x' under this x is calculated and is compared with a newly generated random fraction $ranf$. Examine

$$f'_x > ranf ? \tag{III.9}$$

If Eq. (III.9) is satisfied, then this value of x is accepted (selected), if Eq. (III.9) is not satisfied, then this value of x is rejected (not selected), this process is repeated until the appropriate value of x is selected.

III.3 GENERALIZED ACCEPTANCE-REJECTION METHOD

When the distribution function $f(x)$ has singularities, i.e., when the values of the function are infinite somewhere between the upper and lower limits, the acceptance-rejection method can not be applied, we developed a generalized acceptance-rejection method, or the combined cumulative distribution and acceptance-rejection method [1, 2] and resolved the problem of sampling random values with singular distributions. In the following the method is explained separately for the distribution with single singularity ad with double singularities.

1. The case of distribution with single singularity

III.3 GENERALIZED ACCEPTANCE-REJECTION METHOD

Suppose the distribution has a singularity and can be expressed as

$$f(x) = f_1(x) f_2(x), \quad 0 \le x \le 1, \tag{III.10}$$

where $f_2(x)$ has a singularity, but the cumulative distribution of it can be found that can be inverted, $f_1(x)$ has no singularity. The random sampling can be accomplished as follows: Find the normalized cumulative distribution function $F_2(x)/F_2(1)$ of $f_2(x)$, generate a uniform random fraction, set it equal to $F_2(x)/F_2(1)$. Inverse $f_2(x)$ to determine x_1, then apply to $f_1(x)$ the acceptance-rejection method to see if this x_1 is selected.

Example 4 Sample a value from the variate x with the distribution

$$f(x) = x^{1-\omega}(1-x)^{\zeta-1}, \quad 0 \le x \le 1, \quad \omega < 1, \quad \zeta < 1. \tag{III.11}$$

Let

$$f_1 = x^{1-\omega}, \quad f_2 = (1-x)^{\zeta-1},$$

one has

$$F_2(x) = \int_0^x f_2(t)dt = \frac{1-(1-x)^\zeta}{\zeta},$$

$$F_2(x)/F_2(1) = 1 - (1-x)^\zeta.$$

Generate an uniform random fraction $ranf_1$, set it equal to the above formula to find x_1

$$x_1 = 1 - (1 - ranf_1)^{1/\zeta}.$$

Whether this x_1 is selected depends on the result of applying the acceptance-rejection method to $f_1(x)$, i.e., generate a $ranf_2$, and consider

$$x_1^{1-\omega} > ranf_2? \tag{III.12}$$

The acceptance or rejection of x_1 depends on the satisfaction or non-satisfaction of Eq. (III.12).

2. *The case of distribution with double singularities*

Suppose the variate x has a distribution with two singularities which can be expressed in the form

$$f(x) = f_1(x) f_2(x), \quad 0 \le x \le 1, \tag{III.13}$$

where $f_1(x), f_2(x)$ have singularities in intervals $[0, x_D], [x_D, 0]$, respectively, but both yield cumulative distribution functions which can be inverted. Firstly, the probability that x is less than x_D is determined

$$P_x(x < x_D) = \frac{\int_0^{x_D} f_1(t) f_2(t) dt}{\int_0^1 f_1(t) f_2(t) dt} \equiv \alpha. \tag{III.14}$$

Generate a random fraction $ranf_1$. If $ranf_1 > \alpha$, x lies in the interval $[x_D, 0]$, the sampling is accomplished by the method in the previous section with $f_2(x)$ having single singular distribution. F $ranf_1 < \alpha$, x lies in the interval $[0, x_D]$, the sampling is accomplished by the method in the previous section with $f_1(x)$ having single singular distribution.

Example 5 Sample a value of x according to the distribution

$$x^{a-1}(1-x)^{b-1}, \quad 0 \le x \le 1, \quad a < 1, \quad b < 1. \tag{III.15}$$

Let $f_1 = x^{a-1}, f_2 = (1-x)^{b-1}$, the singularities are at $x = 0$ and $x = 1$, respectively. First determine x_D, in the assumption $\alpha = 1/2$. According to

$$I_{x_D}(a,b) = \frac{1}{B(a,b)} \int_0^{x_D} t^{a-1}(1-t)^{b-1} dt = \frac{1}{2},$$

one can determine (see Eq. (II.38) and Eq. (II.39))

$$x_D = \frac{a}{a + be^{2w}},$$

$$w = \frac{1}{3}\left(\frac{1}{2a-1} + \frac{1}{2b-1} - 1\right)\left(\frac{1}{2b-1} - \frac{1}{2a-1}\right).$$

When $a = b$, one has $x_D = 1/2$. Generate a random fraction $ranf_1$, when $ranf_1 > 0.5$, the following manipulation in the assumption that $f_2(x)$ has singularity is accomplished (when $ranf_1 < 0.5$, the manipulation is similar and is omitted here):

Generate $ranf_2$, set it equal to

$$F_2(x)/F_2(1) = 1 - \left(\frac{1-x}{1-x_D}\right)^b,$$

invert the above expression to obtain

$$x_2 = 1 - (1 - x_D) ranf_2^{1/b}.$$

Apply the acceptance-rejection method to $f_1(x)$, to see if this x_2 is selected. Generate $ranf_3$, and consider

$$\left(\frac{x_2}{x_D}\right)^{a-1} > ranf_3 ?. \tag{III.16}$$

The acceptance or rejection of this x_2 depends on the satisfaction or non-satisfaction of (III.16).

REFERENCES

1. Shen C, Wu WQ, Hu ZH, Xu XY (1991) The direct statistical simulation of excitation, relaxation of internal energy and chemical reaction. Acta Aerodynamica Sinica 9: 1-7 (in Chinese)
2. Shen C, Hu ZH, Xu XY and Fan J (1995) Monte Carol simulation of vibrational energy relaxation in rarefied gas flow. In: J Harvey and G Lord edited Rarefied Gas Dynamics 19, 1: 467

APPENDIX IV PROGRAM OF THE COUETTE FLOW

```
c       APPENDIX-IV
c
c       The program of Couette problem (DSMC,*IP method)
c
c       The present DSMC program will be converted to the program for the IP
c       method, all one has to do is to use the sentence marked with * to replace
c       the sentence before it and add the sentences marked with ** anew.
c
        implicit double precision (a-h,o-z)

        parameter(no_cell=50)
c         no_cell: the number of the cells
        parameter(no_molecule_each_cell=30)
c         no_molecule_each_cell: the number of the molecules in each cell
        parameter(no_molecule=1500)
c         no_molecule : the total number of the molecules
        parameter(jnis=1000,nloop=11000)
c         jnis: the number of loops, after which the sampling starts
c         nloop: the total number of loops
        common /cons/ pi,boltz,eta51,y_length,dtm,
     1      area,drag_ns,cell_height,
     2      rmass,tkom,amda,rkn
c       For the explanation of these symbols see subc1.
        common /init/ t_ini,pressure,fnd,t_wall,u_wall,vm_ini,
     1      vm_wall,vrm_ini,vrmax_eta51
c       For the explanation of these symbols see subc1.
        dimension p(4,no_molecule),vmean(3,no_molecule),
     &              lcr(no_molecule)
c       p(i,n), i=1,2,3:  the velocity components of the molecule n in x, y, z
c           directions, respectively
c       p(4,n) is the y coordinate of the molecule n
c       vmean(i,n), i=1,2,3:  the IP velocity components of the molecule n
c           in x, y, z directions, respectively
c       lcr(n):  to which the original identifying number of the molecule is allo
c           cated, with n as its new position number
```

384 APPENDIX IV PROGRAM OF THE COUETTE FLOW

```
      dimension yc(no_cell),ic(2,no_cell),coll_remain(no_cell)
c     yc(i): the y coordinate of the center of the i-th cell
c     ic(1,i): the number of the molecules in the i-th cell
c     ic(2,i):  the order number of the first molecule in i-th cell -1
c     coll_remain(i): the remainder collision number in i-th cell

      dimension wall(7),up_wall(7),field(7,no_cell)
c     wall(i): the quantities added up on the lower wall:
c      (1): the number of the incident molecules;
c      (2),(5): incident and reflected tangential momentum;
c      (3),(6): incident and reflected normal momentum;
c      (4),(7): incident and reflected kinetic energy
c     up_wall(i): the above quantities added up on the upper wall
c
c     field(i,no_cell): the sum up of flow quantities:
c      (1,n): the sum of the molecules in n-th cell;
c      (2,n),(3,n): the sum of the velocity (* IP velocity) components of the
c        molecules in n-th cell in x, y directions, respectively;
c      (4,n),(5,n),(6,n): the accumulative totals of the square of the velocity
c        components of the molecules in n-th cell in x, y, zdirections,
c        respectively
c
      dimension op(8)
c     op(8):  output data units
c
c     molecular type: Ar

      rkn=10.0d0
c     rkn: the reciprocal of the KNUDSEN number

      open(9,file='qcou1.dat')

c     Set constants and initial variables
      call subc1(no_cell,no_molecule,yc,coll_remain)

      iloop=0
c     iloop: the number of loops at present

c     Clear store of flow quantities
      do ijk=1,7
         up_wall(ijk)=0.0d0
         wall(ijk)=0.0d0
         do icell=1,no_cell
            field(ijk,icell)=0.0d0
         end do
      end do
```

APPENDIX IV PROGRAM OF THE COUETTE FLOW

```
c       Set the initial velocities and positions of the molecules
        call subc2(no_molecule,no_cell,no_molecule_each_cell,vm_ini,
     1           cell_height,yc,p,vmean,y_length,u_wall)
c
100     continue
        iloop=iloop+1
        if(mod(iloop,100).eq.0) write(*,*)'iloop=',iloop
c
c       Calculate the molecular movements and their reflections on the walls
        call subc3(iloop,jnis,no_molecule,p,vmean,up_wall,wall)
c
c       Arrange the molecules in order in the cells, and number them
        call subc4(no_molecule,no_cell,cell_height,p,vmean,ic,lcr)


c       Calculate collisions
        call subc5 (no_cell,no_molecule,coll_remain,ic,lcr,p,vmean)
c
        if(iloop.gt.jnis) then
c
c       Sample the molecules on the walls and in the flow, if the number of the
c       current loops is larger than the number of loops, at which begins the
c       sampling
        call subc6(no_cell,no_molecule,ic,lcr,p,vmean,field)
        end if
c
        if(iloop.le.nloop) go to 100
c       Restart a new loop, if the number of current loops is less than the
c       total number of loops

c       Sample and output after attaining the total number of loops
        call subc7(nloop,jnis,no_cell,yc,no_molecule_each_cell,
     1           wall,up_wall,field,op)
        close(3)
        close(26)
24      format(1x,i4,f10.8,f15.5,7f10.5)
        stop
        end



        subroutine subc1(no_cell,no_molecule,yc,coll_remain)
c
c       Set constants and initial variables, prepare some parameters used in
c       the subroutines
c
        implicit double precision (a-h,o-z)
        common /cons/ pi,boltz,eta51,y_length,dtm,
```

```
     1      area,drag_ns,cell_height,
     2      rmass,tkom,amda,rkn
      common /init/ t_ini,pressure,fnd,t_wall,u_wall,vm_ini,
     1      vm_wall,vrm_ini,vrmax,eta51
      dimension yc(no_cell),coll_remain(no_cell)

      pi=3.1415926d0
c     the ratio of the perimeter of a circle to its diameter
      boltz=1.3805d-23
c     Boltzmann constant, k, (J/K)
      viscosity=2.117d-5
c     the coefficient of viscosity of Ar, (Ns/m**2), see APPENDIX-I, Table 2
      tref=273.0d0
c     reference temperature
      rmass=6.63d-26
c     atomic mass of Ar, (kg), see APPENDIX-I, Table 2
      diaref=3.659d-10
c     the diameter of Ar for the HS molecular model, (m), see APPENDIX-I,
c         Table 2
c         if using VHS model, set diaref= 4.17d-10
*     diaref=3.963d-10
*c    the diameter of Ar for the HS molecular model in IP method, (m), see
*c        APPENDIX-I, Table 3
*c        if using VHS model, set diaref= 4.283d-10
      power=999999.9d0
c     power, namely ETA, the power in the inverse power law, Eq. (2-78),
c        the value given here corresponds to the HS model
c        if using VHS model, power=7.452d0, see Eq. (2-99) and APPENDIX-I,
c        Table 2

      eta51=(power-5.0d0)/(power-1.0d0)
c     eta51:  the power of S(CR) in the inverse power law model (see (7-1)),
c        cf. Eq. (2-84)
      KSAI=2.0d0/(power-1.0d0)
c     KSAI:  see Eq. (2-99)
      amr=rmass/2.0d0
c     reduced mass
      cxsref=pi*diaref**2

c
c     The collision cross-section for VHS and VSS models can be calculated
c        according to Eq. (2-234).
c     Generally, if the power before this is taken as the value in the VHS
c        model, and in the following set VHS_COE=(GAMMA(2-KSAI))**2,
c        the collision cross-section of the molecules is obtained by
c        multiplying TKOM by CR**(-2.d0*KSAI), see Eq. (2-234), Eq.(2-99),
c        and TKOM multiplied by CR**ETA51 is S(CR), see Eq.(7-1).
c     For the HS model, VHS_COE=1.0. But VHS_COE is reserved here
c        in order to adapt to more common cases.
```

```
c
        VHS_coe=1.0
        tkom=cxsref*(2.0*boltz*tref/amr)**KSAI/VHS_coe
c
        t_ini=273d0
        pressure=101325d0
        fnd=pressure/(boltz*t_ini)
c       The three quantities above are initial temperature, initial pressure and ini-
c          tial number density.
        t_wall=273.0d0
        u_wall=100.0d0
*       u_wall=1.0d0
c       The two quantities above are the temperature and the velocity of the wall.

        vm_ini=dsqrt(2.0e0*boltz*t_ini/rmass)
        vm_wall=dsqrt(2.0e0*boltz*t_wall/rmass)
        vrm_ini=2.0*dsqrt(2.0/pi)*vm_ini
c         The three quantities above are initial most probable molecular thermal
c            speed, the most probable molecular thermal speed of the wall and the
c            initial relative velocity.
        vrmax_eta51=2.0d0*(vrm_ini)**eta51
c       SMAX:   the initial maximum value of S(CR) (see (7-1))

        amda=viscosity*16.0d0/(5.0d0*dsqrt(pi)*rmass*fnd*vm_ini)
c          the mean free path for the HS model, LAMBDA, see Eq. (2-221)
        write(*,*)'amda=',amda
        y_length=rkn*amda
        dtm=0.23*amda/vm_ini
c       Above are the scope of the flow and the time step.

        area=no_molecule/(fnd*y_length)
c       AREA is the representative sectional area of the one-dimensional
c          flow we are studying, corresponding to the number of the simulated
c          molecules
c
        cell_height=y_length/dble(no_cell)
c          cell height
        do n=1,no_cell
           yc(n)=(n-0.5d0)*cell_height
c       the coordinate of the center of the n-th cell
           coll_remain(n)=rf(0)
c       the remainder collision number in each cell, its initial value is a random
c          fraction
        end do
        write(9,'(2x,4e16.6)') rmass,eta51,tkom,dtm,
     1      amda,y_length,area,cell_height
     2
        write(9,'(2x,4e16.6)')   t_ini,t_wall,pressure,fnd,
```

```
          1    u_wall,vm_ini,vm_wall
          return
          end

          subroutine subc2(no_molecule,no_cell,no_molecule_each_cell,
         1     vm_ini, cell_height,yc,p,vmean,y_length,u_wall)
c
c         Set the initial velocities and positions of the molecules
c
          implicit double precision (a-h,o-z)
          dimension p(4,no_molecule),vmean(3,no_molecule),yc(no_cell)

          pi=3.1415926d0
c
          do n=1,no_cell
            do m=1,no_molecule_each_cell
              iall=(n-1)*no_molecule_each_cell+m
              do ijk=1,3
                abc1=dsqrt(-dlog(rf(0)))
                abc2=2.0d0*pi*rf(0)
                p(ijk,iall)=abc1*dsin(abc2)*vm_ini
**              vmean(ijk,iall)=0.0d0
              end do
c         The initial molecular velocities of the simulated molecules are generated
c         in this circle, i.e., the initial macro-velocities (0) are added by the ther-
c         mal velocity components in equilibrium gas, cf. Eq. (3-18), Eq. (3-19)
c         and the discussion followed. The initial IP velocity is 0.

              If (rf(0).gt.0.5d0) then
                p(4,iall)=yc(n)+0.5*rf(0)*cell_height
              else
                p(4,iall)=yc(n)-0.5*rf(0)*cell_height
c         The initial random positions of the simulated molecules in cells are
c         given, here the principle of variance reduction has been adopted, cf.
c         APPENDIX III, Eq. (III-5) and Eq.(III-5)'.
              end if
            end do
          end do
          return
          end


          subroutine subc3(iloop,jnis,no_molecule,p,vmean,up_wall,wall)
c
c         Calculate the molecular movements and their reflections on the upper
c         and lower walls, and sample
c
          implicit double precision (a-h,o-z)
```

APPENDIX IV PROGRAM OF THE COUETTE FLOW

```
      common /cons/ pi,boltz,eta51,y_length,dtm,
    1    area,drag_ns,cell_height,
    2    rmass,tkom,amda,rkn
      common /init/ t_ini,pressure,fnd,t_wall,u_wall,vm_ini,
    1    vm_wall,vrm_ini,vrmax_eta51
      dimension p(4,no_molecule),vmean(3,no_molecule)
      dimension wall(7),up_wall(7)

      do 180 m=1,no_molecule
        y=p(4,m)+p(2,m)*dtm
c       Y, the position arrived at by the molecule with a velocity P(2,M) after a
c       time step DTM

        if(y.ge.y_length) then
c       If Y is larger than Y_LENGTH, then the molecules reflect from the
c       upper wall.
        dtr=(y-y_length)/p(2,m)
c       the remainder movement time of the molecules after reflection
        if(iloop.ge.jnis) then
          p(1,m)=p(1,m)+u_wall/2.0d0
**        vmean(1,m)=vmean(1,m)+u_wall/2.0d0
c       Here, the reference framework is transferred to that associated with the
c       upper wall (moving with a velocity of -U_WALL/2.0D).
          up_wall(1)=up_wall(1)+1.0d0
          up_wall(2)=up_wall(2)+p(1,m)
*         up_wall(2)=up_wall(2)+vmean(1,m)
c       In the IP method, UP_WALL(2) is the sum of the IP velocities of the
c       incident molecules, see Eq. (8.7)
          up_wall(3)=up_wall(3)-p(2,m)
          up_wall(4)=up_wall(4)+0.5*(p(1,m)**2+p(2,m)**2+p(3,m)**2)
        end if
c       If the number (ILOOP) of the current loop is larger than the number
c       (JNIS) of loops, at which sampling starts, sample the molecules on the
c       upper wall. UP_WALL(I) sums up the contributions of incident mole-
c       cules: (1) molecular number, (2) tangential momentum,(3) normal mo-
c       mentum, (4) kinetic energy
        abc1=dsqrt(-dlog(rf(0)))*vm_wall
        abc2=2.0*pi*rf(0)
        p(1,m)=abc1*dsin(abc2)
        p(3,m)=abc1*dcos(abc2)
        p(2,m)=-dsqrt(-dlog(rf(0)))*vm_wall
**      vmean(1,m)=0.0d0
**      vmean(2,m)=0.0d0
**      vmean(3,m)=0.0d0
c       The molecular velocities after diffuse reflection from the upper
c       wall are calculated, cf. Eqs. (3-18), (3-19), (3-15), (3-20). The IP
c       velocity after diffuse reflection is 0 in the framework associated with
c       the upper wall.
```

APPENDIX IV PROGRAM OF THE COUETTE FLOW

```
              If (iloop.ge.jnis) then
                up_wall(5)=up_wall(5)+p(1,m)
*               up_wall(5)=up_wall(5)+vmean(1,m)
c         In the IP method, UP_WALL(5) is the sum of the IP velocities of the
c         reflected molecules, see Eq. (8.7)
                up_wall(6)=up_wall(6)+p(2,m)
                up_wall(7)=up_wall(7)+0.5*(p(1,m)**2+p(2,m)**2+p(3,m)**2)
              end if
c         In the above, the contributions of reflected molecules are summed up: (5)
c           tangential momentum (is 0), (6) normal momentum, (7) kinetic energy
              p(1,m)=p(1,m)-u_wall/2.0d0
**            vmean(1,m)=-u_wall/2.0d0
**            vmean(2,m)=0.0d0
              p(4,m)=y_length+p(2,m)*dtr
c         In the above, the tangential velocity of molecule reflected from the upper
c         wall is added by the velocity of the upper wall. The tangential and nor
c         mal IP velocities are assigned. The molecules arrive at new positions.
c
              goto 180
            end if

            if (y.le.0.0d0) then
c         If Y is smaller than 0.0, the molecules reflect from the lower wall.
c         The reflection and sampling of the molecules are analogous with those at
c         the upper wall.
              dtr=y/p(2,m)
              if (dtr.lt.1.0d-3*dtm) dtr=1.0d-3*dtm
c
              if (iloop.ge.jnis) then
                p(1,m)=p(1,m)-u_wall/2.0d0
**              vmean(1,m)=vmean(1,m)-u_wall/2.0d0
                wall(1)=wall(1)+1.0d0
                wall(2)=wall(2)+p(1,m)
*               wall(2)=wall(2)+vmean(1,m)
                wall(3)=wall(3)-p(2,m)
                wall(4)=wall(4)+0.5*(p(1,m)**2+p(2,m)**2+p(3,m)**2)
              end if
c
              abc1=dsqrt(-dlog(rf(0)))*vm_wall
              abc2=2.0d0*pi*rf(0)
              p(1,m)=abc1*dsin(abc2)
              p(3,m)=abc1*dcos(abc2)
              p(2,m)=dsqrt(-dlog(rf(0)))*vm_wall
**            vmean(1,m)=0.0d0
**            vmean(2,m)=0.0d0
**            vmean(3,m)=0.0d0
c
              if (iloop.ge.jnis) then
```

APPENDIX IV PROGRAM OF THE COUETTE FLOW

```
                wall(5)=wall(5)+p(1,m)
 *              wall(5)=wall(5)+vmean(1,m)
                wall(6)=wall(6)+p(2,m)
                wall(7)=wall(7)+0.5*(p(1,m)**2+p(2,m)**2+p(3,m)**2)
              end if
 c
                p(1,m)=p(1,m)+u_wall/2.0d0
 **             vmean(1,m)=u_wall/2.0d0
 **             vmean(2,m)=0.0d0

                p(4,m)=p(2,m)*dtr
                goto 180
              end if
              p(4,m)=y
 180        continue
            return
            end


            subroutine subc4(no_molecule,no_cell,cell_height,p,vmean,ic,lcr)

 c
 c      The molecule with the original identifying number M is re-indexed
 c         according to the cell it arrived at after movement. Its new order num
 c         ber is K. M is stored in LCR(K).
 c
            implicit double precision (a-h,o-z)
            dimension p(4,no_molecule),vmean(3,no_molecule),lcr(no_molecule),
           &          ic(2,no_cell)

 150        do icell=1,no_cell
              ic(1,icell)=0
            end do
            do m=1,no_molecule
              ncell=p(4,m)/cell_height+0.999999d0
              if (ncell.eq.0) ncell=1
              if (ncell.ge.no_cell) ncell=no_cell
 c          NCELL is the number of the new cell which molecule M arrived at
              ic(1,ncell)=ic(1,ncell)+1
            end do
 c      Here, the number IC(1,I) of the molecules in each cell of all NO_CELL
 c         cells has been counted. And this is done for obtaining IC(2,I).
            ic2=0
            do icell=1,no_cell
              ic(2,icell)=ic2
 c      Now, IC(2,I) is the order number of the first molecule in i-th cell -1.
              ic2=ic2+ic(1,icell)
              ic(1,icell)=0
```

```
          end do

          do m=1,no_molecule
            ncell=p(4,m)/cell_height+0.999999d0
            if (ncell.eq.0) ncell=1
            if (ncell.ge.no_cell) ncell=no_cell
            ic(1,ncell)=ic(1,ncell)+1
            k=ic(2,ncell)+ic(1,ncell)
            lcr(k)=m
          end do
c         In this circle, IC(1,I) is counted again, and the new order number K of the
c         molecule with original number M is obtained corresponding to its new
c         position. M is stored in LCR(K).
          return
          end

          subroutine subc5(no_cell,no_molecule,coll_remain,
         1     ic,lcr,p,vmean)
c
c         Calculate collisions
c
          implicit double precision (a-h,o-z)
          common /cons/ pi,boltz,eta51,y_length,dtm,
         1     area,drag_ns,cell_height,
         2     rmass,tkom,amda,rkn
          common /init/ t_ini,pressure,fnd,t_wall,u_wall,vm_ini,
         1     vm_wall,vrm_ini,vrmax_eta51
          dimension p(4,no_molecule),vmean(3,no_molecule),lcr(no_molecule)
          dimension ic(2,no_cell),coll_remain(no_cell)
          dimension vrc(3),vccm(3)
c         dimension vrc(3),vccm(3),vrcip(3),vccmip(3)

190       do 140 m=1,no_cell
          if (IC(1,m).lt.2) goto 140
c         If there is no molecule or only one molecule in cell M, skip collisions.

          VAVER=0.0
c         VAVER is the sum of CR**ETA51 in the cell
          icontr=IC(1,m)
          if (IC(1,m).eq.2) icontr=1
          if (IC(1,m).eq.3) icontr=2
c         ICONTR is the number of molecule pairs that need to be selected in
c         random in order to calculate the number of collisions in the cell
c         during a time step DTM. See section 7.2, RSF method
          do icoll=1,icontr
            k=int(rf(0)*IC(1,m)+IC(2,m)+0.9999999d0)
            if (k.eq.IC(2,m)) k=k+1
            l1=LCR(k)
```

APPENDIX IV PROGRAM OF THE COUETTE FLOW   393

```
c           A molecule L1 in cell M is selected randomly.
210         j=int(rf(0)*IC(1,m)+IC(2,m)+0.9999999d0)
            if (j.eq.IC(2,m)) j=j+1
            if (j.eq.k) goto 210
            l2=LCR(j)
c           Another molecule L2 in cell M is selected randomly
            do k=1,3
               vrc(k)=P(k,l1)-P(k,l2)
            end do

            VR=dsqrt(vrc(1)**2+vrc(2)**2+vrc(3)**2)
            VAVER=VAVER+VR**eta51
            end do
c           VR is the relative speed between molecules L1 and L2, the VR**ETA51
c           thus obtained multiplied by TKOM yields S(CR). See Eq. (7.1).
            VAVER=VAVER/icontr
c
            CNOIC=IC(1,m)*(IC(1,m)/(AREA*cell_height))*tkom*VAVER
     1              *DTM/2.0d0

c           The number density n of the flow is in the double brackets.
c           TKOM*VAVER is the mean value of S(CR).
c           CNOIC is the number NRSF of collisions that should occur in the cell
c           during DTM according to RSF method, see Eq. (7.10).
            cnoic=cnoic+coll_remain(m)
            NCOLL=anint(CNOIC)
c           NCOLL is the number of collisions which should happen factually in the
c           cell obtained after rounding off and with the remainder number of colli
c           sion during the last DTM taken into account.

            coll_remain(m)=CNOIC-NCOLL
            if (NCOLL.lt.1) goto 140
            NACOLL=0
C           NACOLL is used to count the number of collisions occurred factually.

300         k=int(rf(0)*IC(1,m)+IC(2,m)+0.9999999d0)
            if (k.eq.IC(2,m)) k=k+1
            l1=LCR(k)

310         j=int(rf(0)*IC(1,m)+IC(2,m)+0.9999999d0)
            if (j.eq.IC(2,m)) j=j+1
            if (j.eq.k) goto 310
            l2=LCR(j)
c           Two molecules L1,L2 are selected randomly in cell M.
c
            do k=1,3
               vrc(k)=P(k,l1)-P(k,l2)
            end do
```

```
            vr=dsqrt(vrc(1)**2+vrc(2)**2+vrc(3)**2)
c           VR is the relative speed of the collision pair.
            if (vr**eta51.gt.vrmax_eta51)vrmax_eta51=vr**eta51
            a=vr**eta51/vrmax_eta51
            b=rf(0)
            if (a.lt.b) goto 300
c           Judge whether the molecular pair is selected or not, using the acceptance-
c           rejection method, see Eq. (7.1). When ETA 51 is assigned appropriately,
c           VR**ETA 51 is suitable to include the VHS,VSS model.
            b=1.0-2.0*rf(0)
            a=dsqrt(1.0-b*b)
            VRC(1)=b*VR
            b=2.0*PI*rf(0)
            VRC(2)=a*dcos(b)*VR
            VRC(3)=a*dsin(b)*VR
c           VRC are the three components of the relative velocity after collision, see
c           Eq. (2-111). They can be used in the HS and VHS models.
            do k=1,3
              VCCM(k)=0.5*(P(k,l1)+P(k,l2))
              P(k,l1)=VCCM(k)+VRC(k)*0.5
              p(k,l2)=VCCM(k)-VRC(k)*0.5
c           VCCM are the velocity components of the centre-of-mass of the molecu-
c           le pair.
c           Here, the velocities of the molecules L1,L2 after collision are obtained,
c           see Eq. (2-41)
**            ve_macro=0.5*(vmean(k,l1)+vmean(k,l2))
**            vmean(k,l1)=ve_macro
**            vmean(k,l2)=ve_macro
c           IP collision rule: momentum is conserved and is distributed evenly be-
c           tween the two simulated particles, thus the post-collision IP velocities
c           are obtained, see Eq. (8.1).
            end do
            NACOLL=NACOLL+1
            if (NACOLL.lt.NCOLL) GOTO 300
c           Repeat the calculation of collisions until the number of collisions NRSF,
c           Eq. (7-10), is attained, then calculate the collisions in the next cell.
140         continue
            return
            end


            subroutine subc6(no_cell,no_molecule,ic,lcr,p,vmean,field)
c
c           Sum up of flow parameters. In FIELD(N,NO-CELL) stored are:
c           (1):   the number of the molecules;
c           (2),(3):  the sum of the $1^{st}$ and $2^{nd}$ component of the macro velocity,
c                  respectively;
c           (4),(5),(6):  the sum of the square of the three molecular velocity com-
```

```
c          ponents, respectively

           implicit double precision (a-h,o-z)
           dimension p(4,no_molecule),vmean(3,no_molecule),lcr(no_molecule),
          1    ic(2,no_cell),field(7,no_cell)

           do icell=1,no_cell
             do molecule=1,ic(1,icell)
               k=ic(2,icell)+molecule
               l=lcr(k)
               field(1,icell)=field(1,icell)+1.0d0
               field(2,icell)=field(2,icell)+p(1,l)
*              field(2,icell)=field(2,icell)+vmean(1,l)
c       In IP method, field(2,icell) is the sum of the IP velocity of the molecules,
c       see Eq. (8.7)
               field(3,icell)=field(3,icell)+p(2,l)
*              field(3,icell)=field(3,icell)+vmean(2,l)
               do ijk=1,3
                 field(3+ijk,icell)=field(3+ijk,icell)+p(ijk,l)**2
               end do
             end do
           end do
           return
           end


           subroutine subc7(nloop,jnis,no_cell,yc,no_molecule_each_cell,
          1         wall,up_wall,field,op)
c
c       Output surface parameters and write in surf_coue.dat
c       Output flow parameters and write in u_coue.dat
c
           implicit double precision (a-h,o-z)
           common /cons/ pi,boltz,eta51,y_length,dtm,
          1    area,drag_ns,cell_height,
          2    rmass,tkom,amda,rkn
           common /init/ t_ini,pressure,fnd,t_wall,u_wall,vm_ini,
          1    vm_wall,vrm_ini,vrmax_eta51
           dimension wall(7),up_wall(7),field(7,no_cell),op(8),
          1    yc(no_cell)

           open(26,file='surf_coue.dat')
c
c       In the following output the data on the upper surface
           skn=1.0d0/rkn
           sma=dsqrt(2.0d0/1.6670d0)*u_wall/vm_ini
           write(26,10)skn,sma
10         format(1x,'Knudsen number=',F6.3,'   Mach number=',F6.3/)
```

396 APPENDIX IV PROGRAM OF THE COUETTE FLOW

```
          write(26,18)
18        format(1x,'properties on upper surface'/)
          write(26,19)
19        format(1x,'   sample size   number flux      pressure(in,re)
         &       shear stress(in,re)       heat flux(in,re)')
c
          OP(1)=wall(1)
c     (1): the number of molecules being counted, sample size, the number
c          of molecules striking on the upper plate during the time of:
c          (nloop-jnis)*dtm=t
          a=(nloop-jnis)*dtm*(FND*AREA*vm_ini)
c     t*n*A*vm
          OP(2)=wall(1)/a
c     (2): the number flux normalized by n*vm, wall(1)/t*A
          OP(3)=wall(3)/(vm_ini*a)
          OP(4)=wall(6)/(vm_ini*a)
c     (3),(4): the incident and reflected normal momentum flux normalized
by
c     n*vm**2
          OP(5)=2.0*dsqrt(pi)*wall(2)/(u_wall*a)
          OP(6)=2.0*dsqrt(pi)*wall(5)/(u_wall*a)
c     (5),(6): the incident and reflected tangential momentum fluxes normal-
c          ized by TAUfm=m*n*U*vm/2*pi**0.5 (see Eq. (4-55)). Notice that
c          wall(2) and wall(5) have been the sums of the incident and reflected IP
c          velocities in the IP method, see subc3, cf. Eq. (8.7)
          OP(7)=wall(4)/(vm_ini**2*a)
          OP(8)=wall(7)/(vm_ini**2*a)
c     (7),(8): the incident and reflected energy fluxes normalized by n*vm**3

          write(26,21)OP
21        format(1x,e13.5,7e13.5)

          drag=rmass*wall(2)/((nloop-jnis)*dtm*AREA)
          write(26,*)
          write(26,9875)drag
9875      format(1x,'   drag=',e10.4)

c     In the following output the data of the lower surface
          write(26,*)
          write(26,81)
81        format(1x,'properties on lower suface'/)
          write(26,19)
c*
          OP(1)=up_wall(1)
          a=(nloop-jnis)*dtm*(FND*AREA*vm_ini)
          OP(2)=up_wall(1)/a
          OP(3)=up_wall(3)/(vm_ini*a)
          OP(4)=up_wall(6)/(vm_ini*a)
```

APPENDIX IV PROGRAM OF THE COUETTE FLOW

```
      OP(5)=2.0*dsqrt(pi)*up_wall(2)/(u_wall*a)
      OP(6)=2.0*dsqrt(pi)*up_wall(5)/(u_wall*a)
      OP(7)=up_wall(4)/(vm_ini**2*a)
      OP(8)=up_wall(7)/(vm_ini**2*a)

      write(26,21)OP

      drag=rmass*up_wall(2)/((nloop-jnis)*dtm*AREA)
      write(26,*)
      write(26,9875)drag


c
      open(3,file='u_coue.dat')
c     In the following output the data of the flow field
      write(3,123)
123   format(1x,'   N         y       sample size   rho          u         v
     1    Tx        Ty        Tz        T')

      do 270 n=1,no_cell

        OP(1)=field(1,n)
c     (1): the number of the molecules being counted in the cell, sample size
c
        OP(2)=OP(1)/((nloop-jnis)*no_molecule_each_cell)
c     (2) : the normalized number density
        OP(3)=field(2,n)/OP(1)
        OP(4)=field(3,n)/OP(1)
c     (3),(4): the average tangential, normal velocity, * the average of
c       the tangential, normal IP velocity,
        OP(5)=RMASS*(field(4,n)/OP(1)-OP(3)**2)/BOLTZ/t_ini
        OP(6)=RMASS*(field(5,n)/OP(1)-OP(4)**2)/BOLTZ/t_ini
        OP(7)=RMASS*field(6,n)/OP(1)/BOLTZ/t_ini
c     (5),(6),(7): the temperature in X,Y,Z directions, respectively
        OP(8)=(OP(5)+OP(6)+OP(7))/3

        write(3,1234)n,yc(n)/y_length,op
1234    format(1x,I4,F9.4,F12.1,2F9.4,F10.4,4F9.4)

270   continue
      return
      end

      FUNCTION RF(IDUM)
c
c     This section is to generate a uniformly distributed random fraction
c        between 0 and 1. Commonly IDUM is 0, however, negative values can
c        be used to restart.
```

```
      c
            implicit double precision (a-h,o-z)
            SAVE MA,INEXT,INEXTP
            PARAMETER (MBIG=1000000000,MSEED=161803398,MZ=0,
           1      FAC=1.E-9)
            DIMENSION MA(55)
            DATA IFF/0/
            IF (IDUM.LT.0.OR.IFF.EQ.0) THEN
               IFF=1
               MJ=MSEED-IABS(IDUM)
               MJ=MOD(MJ,MBIG)
               MA(55)=MJ
               MK=1
               DO 50 I=1,54
                  II=MOD(21*I,55)
                  MA(II)=MK
                  MK=MJ-MK
                  IF (MK.LT.MZ) MK=MK+MBIG
                  MJ=MA(II)
  50           CONTINUE
               DO 100 K=1,4
                  DO 60 I=1,55
                     MA(I)=MA(I)-MA(1+MOD(I+30,55))
                     IF (MA(I).LT.MZ) MA(I)=MA(I)+MBIG
  60              CONTINUE
 100           CONTINUE
               INEXT=0
               INEXTP=31
            END IF
 200        INEXT=INEXT+1
            IF (INEXT.EQ.56) INEXT=1
            INEXTP=INEXTP+1
            IF (INEXTP.EQ.56) INEXTP=1
            MJ=MA(INEXT)-MA(INEXTP)
            IF (MJ.LT.MZ) MJ=MJ+MBIG
            MA(INEXT)=MJ
            RF=MJ*FAC
            IF (RF.GT.1.E-8.AND.RF.LT.0.99999999) RETURN
            GO TO 200
            END
```

SUBJECT INDEX

acceptance-rejection method, 378
   generalized, 294, 378
accommodation coefficient, 138
activation energy, 300
adiabatic surface temperature, 173
aerodynamic force in free
  molecular flow
  cylinder, 166
  in hypersonic flow, 168
  plane plate, 165
  sphere, 167
angular momentum, 26
apse line, 66
Arrhenius formula, 300
Arrhenius-Kooil formula, 300
asymptotic theory at small Knudsen numbers, 203
average collision time, 8
average molecular weight, 125
average thermal speed, 113
averaging
  ensemble, 18
  time, 18
Avogadro's number, 59

basic equations of slip regime, 192

bearing number, 350
beta function, 373
BGK equation, 247
  test of, 251
binary collision modeling law, 16
binary elastic collision, 61
binary scaling law, 16
binomial distribution function, 241
BKW equation, 248
Bohr's hypothesis, 22
Boltzmann constant, 7, 10
Boltzmann distribution, 41
Boltzmann equation, 92, 96
  asymptotic analysis, 203
  Chapman-Enskog expansion, 193
  collision term, 96
  collisionless, 182
  discrete, 91, 257, 261
  for gas mixture, 126
Boltzmann's H theorem, 108
Boltzmann's relation, 32
Boley-Yip model equatiion, 251
Bose-Einstein statistics, 35
boson, 35
Burnett equations, 196
  augmented, 199
  experimental verification, 201

calculation of complicated flow field, 310
Cercignani-Lampis-Lord (CLL) model, 142
Chapman-Enskog expansion, 193
characteristic temperature, 367
   of dissociation, 12, 367
   of ionization, 12, 367
   of rotation, 10, 367
   of vibration, 10, 367
chemical reaction
   simulation of, 299
chemical reaction rate coefficient, 300
CLL model, 142
cold wall paradox of free molecular flow, 8
collision cross section
   differential, 70
   diffusion, 71
   in the IP method, 332
   reaction, 301
   total, 71
   viscosity, 71
collision frequency, 2, 4
   for VHS, VSS models, 118
   in gas mixture, 127
   of a molecule A with molecules B, 128
collision integral, 98
   evaluation of, 100
combined cumulative distribution
   acceptance-rejection method, 294, 378

complementary error function, 370
computation of complicated flow fields, 310
concentration jump, 212
conductance, 346
conductivity coefficient, 195
   in the BGK model, 249
conservation equations of mass, momentun and enenrgy, 192
conservative form of the continuity equation, 338
continuum model, 191
Couette flow
   by discrete velocity method, 257
   by DSMC method, 281
   by IP method, 333
   in free molecular flow, 177
   with slip, 213
cosine formula for the sides in the spherical trigonometry, 67
cumulative distribution function, 376

deflection angle, 64, 66
degeneracy, 27
degenerated Reynolds equation, 355
degrees of freedom, 47
   relative translational energy in collision, 295
   rotational, 47
   translational, 47
   vibrational, 47
density, 53
depletion of molecules, 89, 95

SUBJECT INDEX 401

diagram of the CLL model, 149
diatomic molecule, 21
differential collision cross-section, 70
differential effusion, 177
diffuse elastic reflection, 170
diffuse reflection, 132
   implementation in DSMC, 135
   with incomplete energy accommodation, 153
diffusion cross-section, 75
dilute gas, 62
direct simulation method, 264
direct simulation Monte Carlo method, 267, 275
discrete Boltzmann equation, 91, 257, 261
discrete energy Larsen-Borgnakke method, 295
discrete energy exchange, 151
discrete ordinate method, 257
discrete velocity method, 257
distance of closest approach, 62
distribution function
   equilibrium, 106
   for u component, 114
   internal energy, 49
   N particle, 51
   of discrete vibrational energy, 296
   of post collision energy, 292
   of the value of velocity, 113
   relative translational energy, 291
   rotational energy, 48
   single particle, 51
   translational energy, 48
   vibrational energy, 49
   velocity, 51
division of flow regimes, 5
DSMCmethod, 293

effusion, 175
effective temperature in the collision, 294
eigenvalues, 22
   of harmonic oscillator, 25
   of rigid rotator, 26
eight velocity gas model, 88
electron energy, 28
electro-kinetics, 318
entropy, 32
ergodic principle, 42
error function, 369
equilibrium distribution function, 106
Euler equations, 194
exchange of internal energy with surface, 150
exchange probability, 297
exchange reactions, 309
excitation and relaxation of the internal energy, 288
exponential factor of reaction rate constant, 308

FCT method, 326
feasibility of LBM, 324
Fermi-Dirac statistics, 35
fermion, 35
finite difference method, 255

flow conductance, 346
flux vector, 56
flux-corrected transport (FCT) method, 326
free expansion, 184
free molecular flow, 159
free molecular effusion, 175

gamma function, 369
gas kinetic unified algorithm, 255
gas kinetic scheme, 252
gas surface interaction, 131
generalized acceptance-rejection method, 294, 378
generalized hard sphere (GHS) model, 85
generalized Reynolds equation, 352
generalized soft sphere (GSS) model, 87
generalized theorem of mean value, 307
GHS model, 85
Grad 13 moment equation, 201
GSS model, 87

H theorem, 108
hard sphere (HS) model, 72
harmonic oscillator, 24
heat flux vector, 58, 125
heat conductivity, 195
   in the BGK model, 249
heat transfer
   between plane plates, 178
   to surface, 172
Heisenberg uncertainty principle, 11
HS model, 72

hybrid continuum/particle approach, 361

impact parameters, 62
incomplete beta function, 373
incomplete gamma function, 369
information preservation (IP) method, 326
information (IP) velocity, 326
integral method, 263
internal degrees of freedom, 171
internal energy flux, 172
inverse collisions, 64
inverse power law model, 74
inversion of cumulative distribution function, 375
IP collision rule, 327
   validation of, 330
IP method, 327
IP reflection rule, 326
   validation of, 329
IP simulation of
   case of temperature variation, 360
   microchannel flow, 338
   thin film air bearing, 353
   unidirectional flows, 333
IP velocity, 326

Kac method, 279
kinetic scheme, 252
Knudsen layer, 206
Knudsen minimum, 322, 335
Knudsen number, 2, 14
Knudsen paradox, 335
Kooij formula, 300

SUBJECT INDEX   403

Larsen-Borgnakke model, 289
   for discrete energy levels, 295
Lattice Boltzmann method (LBM), 323
lattice gas method, 323
LBM, 323
Leibnitz formula, 25
linearized Boltzmann equation, 234, 238
Loschmidt number, 1
Low speed non isothermal flows, 199

Mach number, 14
mass velocity, 54, 125
Maxwell distribution, 106
Maxwell molecule, 76
Maxwell slip velocity, 208
Maxwell-Smoluchovski expression of
   the temperature jump, 211
Maxwell transport equation, 104
Maxwell type boundary condition, 137
MD method, 265
mean collision frequency, 116
   in gas mixture, 127
mean collision time, 8
mean free path, 2, 4
   for VHS, VSS models, 118
   in gas mixture, 127
mean molecular spacing, 11, 62
mean molecular velocity, 54
mean squire root thermal speed, 114
mean thermal speed, 113
mean value of collision quantities, 120
mean value of relative velocity, 118

mean translational kinetic energy in
   collision, 120
MEMS, 318
microchannel flow problem, 338
micro electro mechanical systems
   (MEMS), 318
micromachined channel, 319
micromachining, 320
microscale slow gas flows, 317
microscopic state, 31
miss distance, 62
model equation, 247
   for multicomponent gas, 251
modeling of chemical reactions, 299
modified Bessel functions, 372
molecular chaos, 92, 96
molecular diameters
   hard sphere, 72
   reference values of, 123
   VHS, 80
   VSS, 82
molecular dynamics (MD) method, 265
molecular effusion, 175
molecular model of gases, 3
molecular speed ratio, 161
molecule surface interaction, 131
moment equation, 104
moment method, 239
moment of inertia, 25
momentum flux, 162
most probable molecular thermal speed,
   113
Mott-Smith solution, 241

Navier-Stokes equations, 194
normal momentum accommodation coefficient, 138
normalization condition of the scatter kiernel, 133
no time counter (NTC) method, 279
NTC method, 279
number density, 52
    component, 124
number flux, 161
number of microstates, 31, 34

partition function, 41
Pauli exclusion principle, 35
phenomenological chemical reaction model, , 300
Planck constant, 10, 21
Poiseuille flow
    by DSMC, 287
    by IP, 336
    in slip flow regime, 215
positioin element methpod, 311
post-collision velocity, 69
potential curve of the molecule, 23
Prandtl number, 250
pressure, 58, 125
pressure tensor, 59, 125
program demonstrating the DSMC, IP method, 282, 332, 383
projected distance, 62

quantum mechanics, 21
quantum number
    rotation, 10, 26
    second, 26
    vibration, 10, 25

randomly sampled frequency (RSF) method, 281
Rankine-Hugonio relations, 242
Rayleigh problem, 8
    by DSMC, 287
    by IP, 336
    in free molecular flow, 186
    in slip flow, 218
reaction cross section, 300
reaction probability, 300
reciprocity principle, 140
reduced mass, 24
reference diameters
    of VHS (VSS) models, 123, 367
relative velocity after collision, 292
relaxation collision number, 297
relaxation time, 12, 298
Reynolds equation, 348
    derivation of, 349
    degenerated, 355
    generalized, 352
    slip corrected, 351
Reynolds number, 14
rigid rotator, 25
root mean thermal speed, 114
rotational relaxation collision number, 295
rotational energy, 26, 28, 47
RSF method, 281

SUBJECT INDEX   405

sampling of inversion of the cumulative distribution, 136, 375
sampling from the singular distributions, 294
sampling of collisions, 278
sampling of components of thermal velocity, 136
scatter kernel, 132
   diffuse reflection, 134
   normal component in CLL, 145
   specular reflection, 133
   tangential component in CLL, 143
scale length of the gradient, 1
Schrodinger equation, 21
   for harmonic oscillator, 24
   for rigid rotator, 26
shear stress, 163
slip boundary conditions, 204
   concentration jump, 212
   in liquid, 318
   multicomponent nonequilibrium flow, 212
   second order, 227
   with injection, 212
slip coefficient, 214
slip corrected Reynolds equation, 351
slip velocity, 204
   derivation of, 207
   in the spherical coordinates, 224
S- model equation, 255
smoothing technique, 342
space quantization, 27
specific average kinetic energy, 47
specific rotational energy, 47
specific vibratioinal energy, 47
specular reflection, 132
speed ratio, 161
stagnation temperature, 9, 173
static pressure, 58
statistical mechanics, 30
steepest descend method, 305
steric factor, 300
sterically ndependent chemical reaction mpodel, 302
Stirling formula, 38
Stokes equation, 223
stream velocity, 54
stress tensor, 58
summational invariants, 101, 109
super relaxation factor, 340
super relaxation technique, 340
superthermal flow, 169

tangential momemtum accommodatioin coefficient, 138
TC method, see time counter method, 278
temperature
   adiabatic surface, 173
   characteristic
      of dissociation, 12, 367
      of ionization, 12, 367
      of rotation, 10, 367
      of vibration, 10, 367
   effective in collision, 294
   internal, 60
   kinetic, 59

stagnation, 9, 173
translational, 59
temperature jump, 209
temperature jump coefficient, 211
temperature stress convection, 200
ternary collision, 61
test molecule Monte-Carlo method, 268
thermal accommodation coefficient, 138
thermal creep, 220
thermal creep coefficient, 209
thermal stress slip flow, 201
thermal transpiration, 176
thermal velocity, 55
thermophoresis, 221
thermophoretic deposition, 227
thermophoretic force, 226
thin film air bearing problem, 319, 348
time counter (TC) method, 278
total collision cross section, 71
transitiona probability, 261
transitioanl regime, 231
translational kinetic temperature, 59, 126
translational enenrgy, 47
transpiration, 176

uncertainty principle (Heisenberg principle), 11
unidirectional flows, 333
unified algorithm, 255
unsteady free molecular flow, 182
unstructured body fitted grid, 310

validation of the IP method, 329
variable hard sphere (VHS) model, 78
variable soft sphere (VSS) model, 80
variance reduction technique, 376
velocity slip coefficient, 209, 228
VHS model, 77
vibrational energy, 25, 28, 46
vibrational exchange probability, 298
viscosity coefficient, 195
   in the BGK model, 249
viscosity cross-section, 71
viscous stress tensor, 58, 125
VSS model, 80

wave function, 21
Winchester hard disc drive, 319, 348
write/read head, 319

Printing: Strauss GmbH, Mörlenbach
Binding: Schäffer, Grünstadt